Principles of organic mass spectrometry

Other titles co-authored by Dudley H Williams

Interpretation of Mass Spectra of Organic Compounds, Holden-Day, 1964

Structure Elucidation of Natural Products by Mass Spectrometry, Holden-Day, 1964

Mass Spectrometry of Organic Compounds, Holden-Day, 1967

Applications of NMR in Organic Chemistry: illustrations from the steroid field, Holden-Day, 1964

Spectroscopic Methods in Organic Chemistry, McGraw-Hill, 1966

Spectroscopic Problems in Organic Chemistry, McGraw-Hill, 1967

Principles of Organic Mass Spectrometry

Dudley H Williams

Fellow of Churchill College, Cambridge

Ian Howe

Fellow of Churchill College, Cambridge

London · New York · St Louis · San Francisco
Düsseldorf · Johannesburg · Kuala Lumpur · Mexico
Montreal · New Delhi · Panama · Paris · Rio de Janeiro
Singapore · Sydney · Toronto

Published by McGRAW-HILL Book Company (UK) Limited
MAIDENHEAD · BERKSHIRE · ENGLAND

07 084003 2

PRINTED AND BOUND IN GREAT BRITAIN

Preface

Organic mass spectrometry underwent its remarkable growth between 1958 and 1966 almost entirely as an empirical subject. It was between 1966 and the present that a considerable increase occurred in our understanding of the chemistry and physics of the processes involved. None of the many texts dealing with organic mass spectrometry (published up to 1967) could therefore deal with this aspect of the subject. The present book aims to fill an obvious gap by dealing with the physics and chemistry of unimolecular (and occasionally bimolecular) reactions of positive ions in the gas phase. However, we have endeavoured also to produce a reasonably complete text for organic mass spectrometry by citing briefly earlier work on structure elucidation where general principles and/or methods are apparent. In addition, general methods which have increased in importance in the subject within the last few years (e.g., the use of computers, link-up with glc) are also covered. Throughout, we have concentrated on the principles involved and avoided citing references in any particular area *in extenso*.

In summary, it has been our aim to provide a textbook which should furnish the necessary background at the present time for all those interested in organic mass spectrometry, whether from the viewpoint of an analytical method, or the ion chemistry involved, or both.

We wish to thank numerous authors mentioned in the text for permission to cite their work. We are also indebted to the following journals and publishers for permission to reproduce figures: *Zeitschrift für Naturforschung*; *Journal of the American Chemical Society, Journal of Physical Chemistry, Accounts of Chemical Research, Analytical Chemistry and Inorganic Chemistry* (American Chemical Society); *Journal of Chemical Physics* (American Institute of Physics); *Organic Mass Spectrometry* (Heyden and Son Ltd.); *International Journal of Mass Spectrometry and Ion Physics* and *Journal of Chromatography* (Elsevier Publishing Company); *Angewandte Chemie and Chemische Berichte* (Verlag Chemie GmbH); *Experientia* (Birkhäuser Verlag); *Journal of the Chemical Society* and *Chemical Communications* (The Chemical Society, London); Interscience Publishers, John Wiley and Sons Inc.; The Institute of Petroleum, London; Holden-Day Inc.; *Canadian Journal of Biochemistry* (The National Research Council of Canada); *Analytical Biochemistry* (Academic Press).

Dudley H Williams
Ian Howe

Contents

1. The Mass Spectrometer

In its simplest form, the modern commercial mass spectrometer is designed to perform three basic functions. These are (a) to vaporize compounds of widely varying volatility, (b) to produce ions from the neutral molecules in the vapour phase, and (c) to separate ions according to their mass-to-charge ratios (m/e) and to record them. The devices used to perform these functions are briefly described in the following sections, dealing first with the production of positive ions.

1-1. The ion source

By far the most common technique currently in use for the production of positive ions in the mass spectrometer is electron bombardment. Let us assume that a few microgrammes of the substance to be examined have successfully been introduced as a vapour into the source at its operating pressure (ca. 10^{-6} mm Hg). The vapour is allowed to pass through a slit A into the ionization chamber (Fig. 1-1), where it is bombarded with a beam of

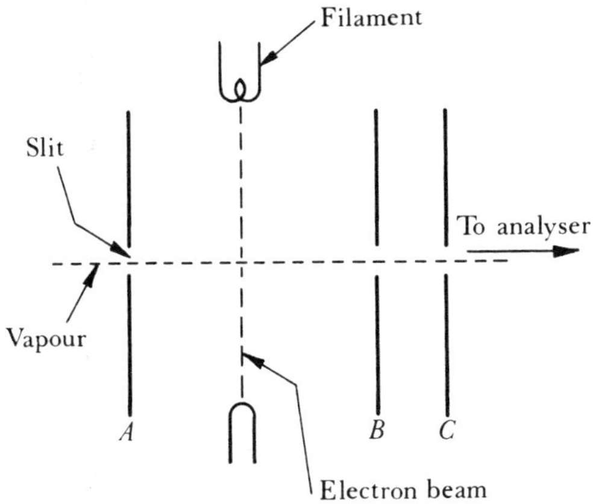

Figure 1-1

electrons accelerated from a hot filament. The energy of the electron beam can be varied from 0–100 eV. Since the ionization potentials of most organic molecules fall in the range 7–13 eV (1 eV $\simeq$ 23 kcal/mole $\simeq$ 96 kJ/mole), one can supply energy in excess of the ionization potential. The ionization potential corresponds to the energy required to remove an electron from the highest occupied molecular orbital and the reaction is represented by eq. (1-1).

$$M + e \longrightarrow M^{+\cdot} + 2e \qquad (1\text{-}1)$$

The $M^{+\cdot}$ symbolism is used to indicate that the molecular ion so formed is both a positive ion and a radical (since organic molecules are almost without exception even-electron species). The alternative process of capture of an electron by a molecule of vapour to give a negative ion radical (eq. 1-2) is less probable by a factor of about 10^{-2}.

$$M + e \longrightarrow M^{-\cdot} \qquad (1\text{-}2)$$

Negative-ion mass spectrometry has not been extensively developed as a method of structure elucidation of organic compounds, nor are energetic and kinetic studies of negative ions as well developed as the corresponding studies on positive ions. This book is concerned solely with positive-ion mass spectrometry.

Since the energy given by the bombarding electrons to the molecules of vapour is usually in excess of the ionization potential, excess energy remains in the molecular ion ($M^{+\cdot}$) and this excess energy may be used to break one (or possibly more) bonds. Depending on the energy content, molecular ions may remain as such or dissociate into fragment ions. In general, a molecular ion may dissociate by elimination of a radical.

$$M^{+\cdot} \longrightarrow A^{+} + B\cdot$$

or by loss of a molecule

$$M^{+\cdot} \longrightarrow C^{+\cdot} + D$$

In either event, the particles B· and D are thereafter ignored by the mass spectrometer since they do not possess a positive charge. In the former case the fragment ion is an even-electron species, while in the latter case the fragment is an odd-electron ion radical. Depending on their internal energies, the fragment ions A^{+} and $C^{+\cdot}$ do of course have the possibility of fragmenting further.

The next problem is to bring the various positive ions ($M^{+\cdot}$, A^{+}, $C^{+\cdot}$, etc.) out of the source into a region of the instrument where they can be separated according to their mass-to-charge ratios. First, the positive ions are directed through a slit in plate *B* by means of a small potential (*repeller potential*) applied between plates *A* and *B* (Fig. 1-1). Then they are accelerated through the slit in plate *C* by means of an accelerating voltage (usually a few kilovolts) applied between plates *B* and *C*. The ions are now *en route* to the collector, but before arriving there they must be separated according to their m/e-values.

1-2. Single- and double-focusing instruments

The term *mass spectrometer* is usually reserved for instruments in which the ion beams are measured electrically after mass analysis, while the term *mass spectrograph* refers to instruments in which the focused ion beams are recorded on a photographic plate. Mass spectrographs usually use double-focusing methods.

Dempster's first mass spectrometer[1] was of the single-focusing type in which the positive ions were deflected through 180° in a magnetic field H (see the schematic illustration in Fig. 1-2).

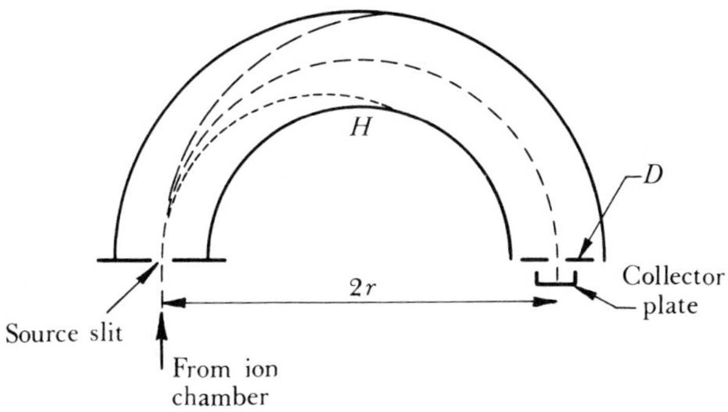

Figure 1-2

As one might expect intuitively, the heaviest ions are constrained into the circular paths of largest radius. Quantitatively, any ion having a charge e when accelerated through a voltage V will acquire a translational (kinetic) energy eV. Thus, the kinetic energy of the ions is independent of their masses, but since

$$\tfrac{1}{2}mv^2 = eV \tag{1-3}$$

(where m is the mass of the ion and v its velocity after acceleration), the more massive ions are travelling more slowly. In a magnetic field of strength H, any ion will experience a force of Hev, producing an acceleration of v^2/r in a circular path of radius r. Hence, from Newton's second law of motion

$$Hev = \frac{mv^2}{r} \tag{1-4}$$

From eqs. (1-3) and (1-4) one obtains

$$\frac{m}{e} = \frac{H^2r^2}{2V} \tag{1-5}$$

From eq. (1-5) it is evident that at a given magnetic field strength and accelerating voltage, ions of a given m/e-value will follow a particular path of radius r, and that we may sweep the ions of various m/e-values past the collector slit D (Fig. 1-2) either by varying H at constant V, or by varying V at constant H. The former possibility is called magnetic scanning and the latter electric (or voltage) scanning. Both modes may be used in practice.

The angle of deflection does not necessarily have to be 180° to retain the ion-beam focusing properties of magnetic fields. Indeed, sector fields are cheaper to produce and have the advantage that the ion source and collector are more accessible than in 180° instruments. A diagram of a 90° sector single-focusing mass spectrometer is given in Fig. 1-3.

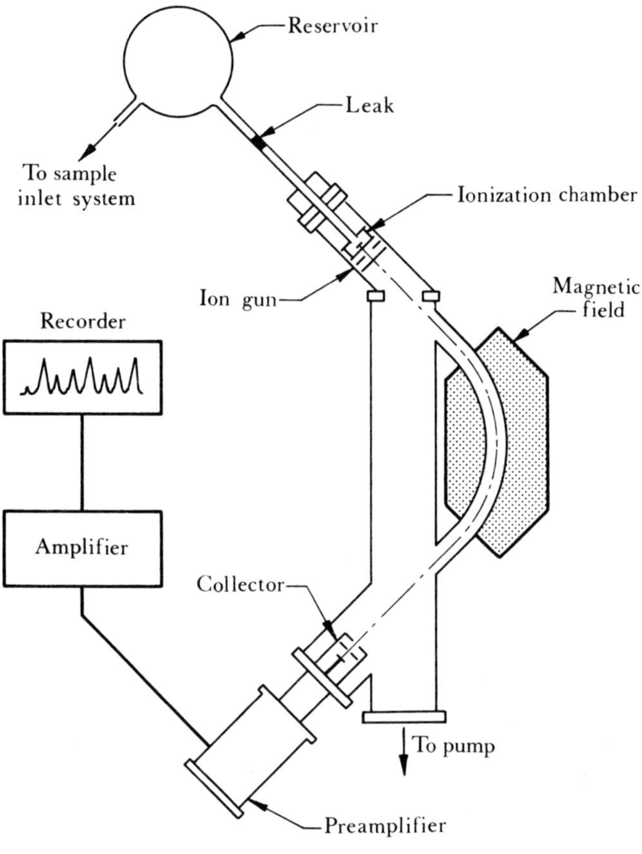

Figure 1-3

One of the problems of the sector instrument shown in Fig. 1-3 is that the resolving power is limited by the initial spread of translational energies of the ions on leaving the source (due to the Boltzmann distribution and field inhomogeneity in the source). This problem may be overcome by passing the ions through an electric field prior to the magnetic field. The electric field effects velocity focusing of the ions and by placing a slit between the electrostatic and magnetic analyser, ions of a closely defined kinetic energy may be selected prior to mass analysis. Instruments incorporating such a system are described as double-focusing mass spectrometers and are capable of attaining much higher resolving power than single-focusing instruments.

In the instrument of Mattauch and Herzog,[2] ions of all *m*/*e*-ratios are brought to a double focus in a plane and a photographic plate can then be used to record the ions (Fig. 1-4).

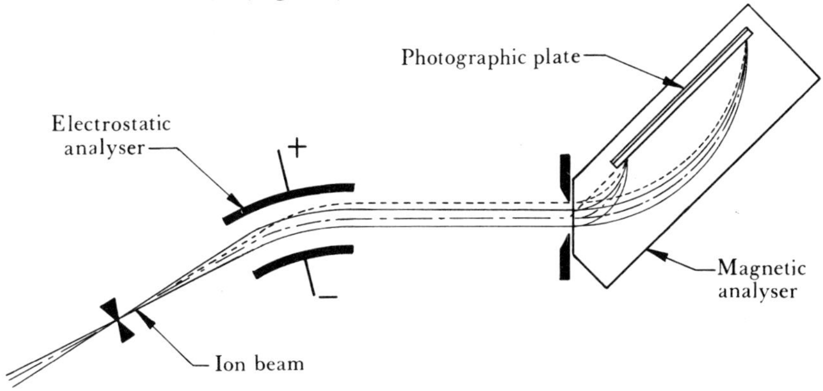

Figure 1-4

On the other hand, in the geometry used by Nier and Johnson,[3] shown in Fig. 1-5, only electrical detection is possible, since there is only one double-focusing point. An electron multiplier collector is usually employed.

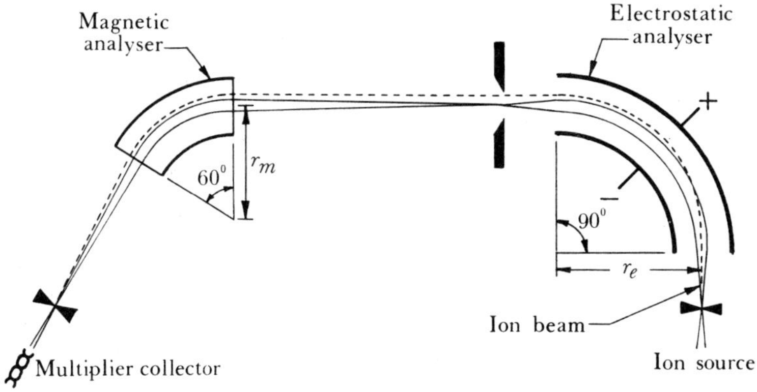

Figure 1-5

Instruments of both Mattauch–Herzog and Nier–Johnson geometries are commercially available and it is on such instruments that high-resolution studies on relatively high-molecular weight organic compounds have so far been carried out. Operating pressures for electrostatic and magnetic analysers are of the order of 10^{-7} mm of Hg.

1-3. Inlet systems

It will be obvious that no problems arise when a gas is to be introduced into a mass spectrometer. The introduction of samples of volatile liquids is also usually a simple matter, the inlet system illustrated in Fig. 1-6 being typical of the devices employed.

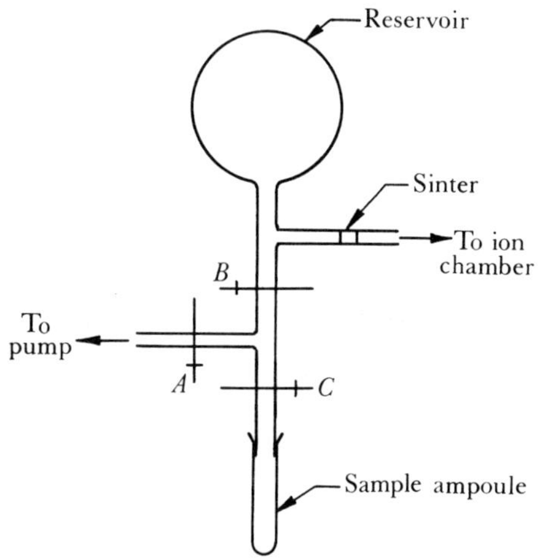

Figure 1-6

If the spectrometer is not being used to examine a sample, valve *C* may be kept closed; and when valves *A* and *B* are open, all parts of the inlet system above *C* are under vacuum. A volatile liquid sample in the ampoule may be retained in the ampoule by cooling with liquid nitrogen while air is removed with *A* and *C* open, and *B* closed. Once the whole inlet system is under vacuum, *A* is closed, *B* opened and the liquid nitrogen which is cooling the ampoule is removed; the sample then volatilizes into the reservoir from which it can diffuse at a suitable rate through a sinter to the ion chamber.

A large number of organic compounds are relatively thermally stable solids with molecular weights in the approximate range of 100 to 400. The majority of such samples may be introduced into the ion chamber by means of a system analogous to that illustrated in Fig. 1-6, but having the whole (except the sample ampoule) encased in an oven whose temperature can be varied in the range from room temperature to 300°C; the sample ampoule can, of course, be heated separately. In this way, adequate vapour pressures can be obtained from relatively 'involatile' materials. Sample size is of the order of 1 mg for such inlet systems.

Frequently we wish to deal with samples which have insufficient thermal stability to be heated to 200–300°C or have very low vapour pressures even at these temperatures. Many substances in the molecular weight range 400 to 1200 fall into the latter class. For these cases, advantage is taken of the extremely low pressures which can be achieved in the ion chamber. The sample is introduced directly into the ion chamber on the end of a probe. The end of the probe may be heated because of its proximity to the electron beam or, in addition, by means of a heater wire in the probe. A vapour pressure of about 10^{-6} mm Hg permits a spectrum to be obtained using this direct insertion technique on a sample of as little as a few μg.

1-4. Presentation and recording of mass spectra. Resolving power

The most usual presentation of a low-resolution mass spectrum (i.e., a mass spectrum in which only the integral *m*/*e*-values have been resolved) is in the form of a bar graph, in which the *m*/*e*-values are plotted horizontally and the ion abundances vertically. Such a bar graph for the 70 eV mass spectrum of *n*-decane (1) is reproduced in Fig. 1-7. The most abundant ion is arbitrarily given an intensity of 100 per cent and is called the base peak.

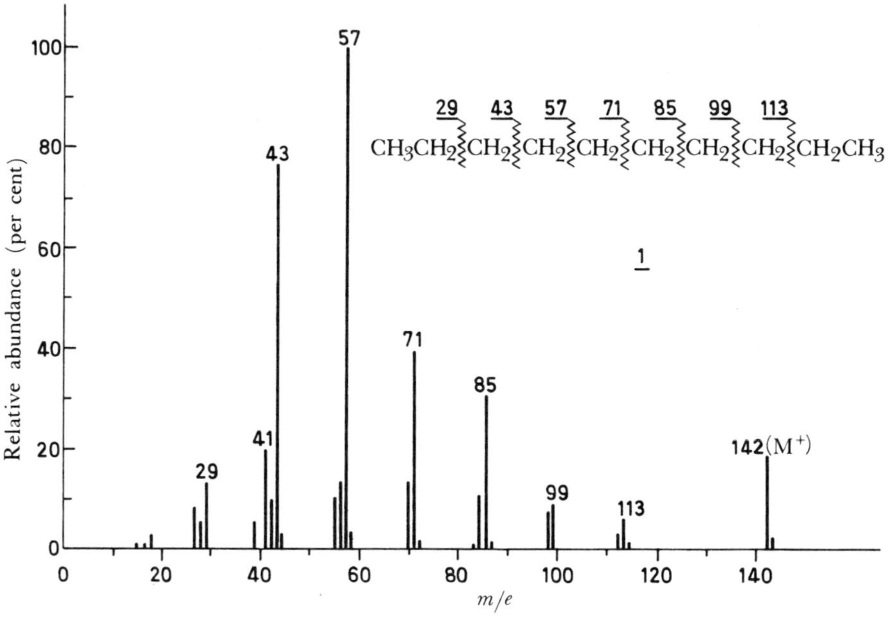

Figure 1-7

If electrical recording is used, then a common arrangement is to feed the amplifier output into a recorder containing several mirror galvanometers of varying sensitivities. A beam of ultraviolet light may be reflected onto ultraviolet-sensitive paper from the mirror galvanometers so that when the chart paper is driven by a motor at a constant rate during the scan of the mass spectrum, records of the mass spectrum corresponding to a wide range of sensitivities are traced out. Such a trace is given in Fig. 1-8;[4] the sensitivities of the galvanometers increase by a factor of approximately three on passing from any one trace to the one above.

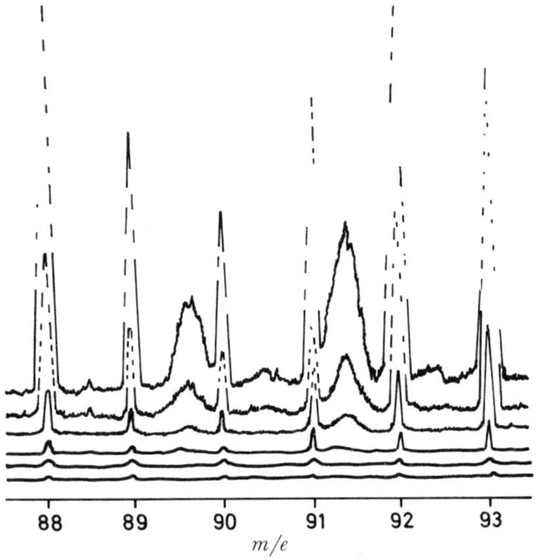

Figure 1-8

If a high-resolution spectrum is recorded on a photographic plate, then the exact position and density of the lines produced by the ions on the plate are measured by means of a microdensitometer.

A recording device sensitive to very different ion abundance ratios is essential in mass spectrometry, since ions of very low abundance will occasionally be of great importance in deducing a molecular formula or structure.

It is obvious from the large separation between the peaks in Fig. 1-8 (which is a typical *low-resolution* trace of the type which may be produced by commercially available single-focusing instruments) that the separation of m/e-values differing by one unit poses no problem at m/e-values of a few hundred or less. In general, a fundamental requirement of the mass spectrometer is that it resolve an ion beam of mass m from an ion beam of nearly the same mass, $m + \Delta m$. There is no single accepted definition for the case where m may be considered to be first resolved from $m + \Delta m$, but one method defines the resolving power as the value of $m/\Delta m$ at which the height of the valley between m and $m + \Delta m$ is 10 per cent of the peak height of m and $m + \Delta m$ (Fig. 1-9). This is the so-called 10 per cent valley definition. On the basis of this definition, single-focusing instruments can attain resolving powers in the region 300–3000, or in favourable cases as high as 9000. Double-focusing instruments give resolving powers from about 10 000 to 40 000, although the most recent technical advances have pushed the latter figure up to the region of 100 000. Since other definitions of resolving power (e.g., a 5 per cent valley definition) are also in use, it is important to be aware of the definition employed.

The most important additions to the collection methods employed in mass spectrometry have been the computerized data-collection systems, which most commonly print out the masses and relative abundances of the ions in the spectrum. The use of computers in mass spectrometry, both for data collection and for spectral interpretation will be discussed in chapter 12.

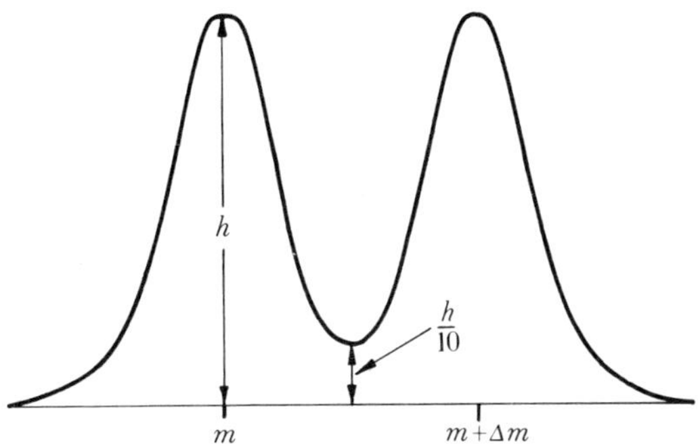

Figure 1-9

1-5. Spectrometers employing other mass-separation methods

The single- and double-focusing techniques described in section 1-2 are those most commonly used in commercial mass spectrometers for mass analysis. Mass spectrometers currently in use which employ other mass-separation methods include (a) the ion cyclotron resonance (icr) spectrometer, (b) the quadrupole mass spectrometer, and (c) the time-of-flight (tof) instrument. The importance and function of the icr instrument is covered in section 8-2, but here we briefly consider the instruments (b) and (c).

A schematic diagram of the quadrupole mass filter[5] is shown in Fig. 1-10. The gaseous sample is ionized by electron impact and the ions are extracted electrostatically from the ion source, then focused electrostatically into the quadrupole mass filter (along the z-axis). The precisely aligned metal rods are connected in pairs; one diametrically opposed pair to a radiofrequency (rf) supply and the other pair to a dc voltage supply.

To effect a mass separation, the rf and dc voltages are scanned while being kept in the same ratio. In the quadrupole field, most ions are deflected by the changing electric fields onto the rods. At any given time, ions of a given m/e-ratio are deflected as much towards one rod as another and are collected on an electron multiplier. Ions of other m/e-values are not collected and the entire mass range is scanned by sweeping the voltages from zero to their

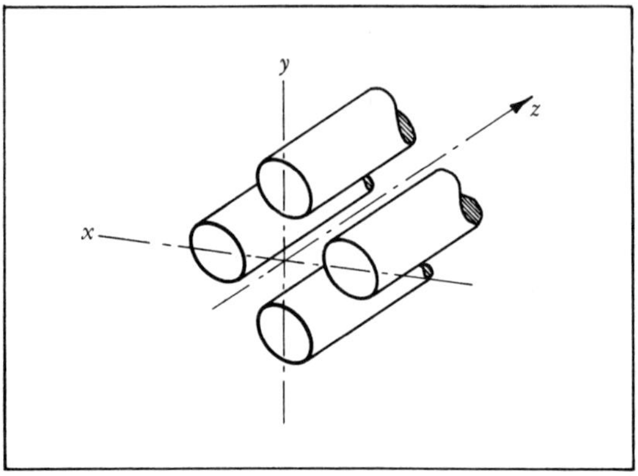

Figure 1-10

maximum values. Quadrupole mass spectrometers are capable of mass resolution (see section 1-4) up to about 1000. Because of their relative cheapness, these mass spectrometers are in common use, especially for gas-chromatographic work (see chapter 11).

The time-of-flight mass spectrometer[6] (see Fig. 1-11 for schematic diagram) employs as its method of mass separation the principle that ions of different mass accelerated to a uniform kinetic energy have different velocities, and hence different times of flight over a given distance. The commonest type of time-of-flight spectrometer uses a pulsed ion source and a voltage pulse on grid *A* draws the ions from the ion source. A potential gradient between *A* and *B* accelerates the ions to uniform kinetic energy and into the field-free flight tube. The ions are separated in time according to their *m*/*e*-values on arrival at the repeller grid *C* and are collected at *D*. It is common to have differences in arrival times between successive mass peaks of 10^{-7} s or lower; hence fast electronics are required to distinguish successive peaks.

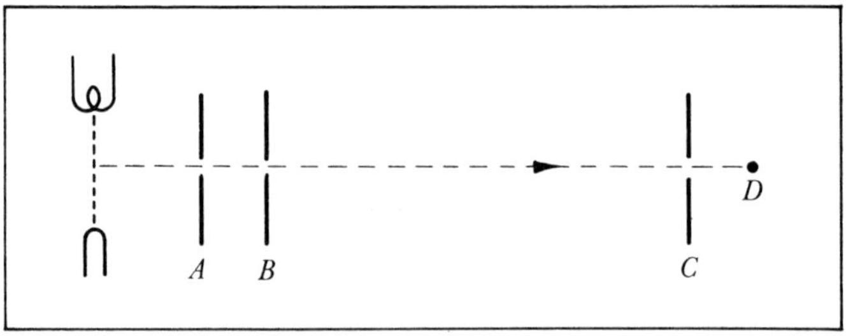

Figure 1-11

References

[1] A. J. Dempster, *Phys. Rev.* **11**, 316 (1918).
[2] J. Mattauch and R. F. K. Herzog, *Z. Physik* **89**, 786 (1934).
[3] E. G. Johnson and A. O. Nier, *Phys. Rev.* **91**, 10 (1953).
[4] J. H. Beynon, B. E. Job and A. E. Williams, *Zeitschrift für Naturforschg.* **20a**, 883 (1965).
[5] R. W. Kiser, *Introduction to Mass Spectrometry and Its Applications*, Prentice-Hall Inc., N.J., 1965, p. 86.
[6] *Time-of-Flight Mass Spectrometry*, D. Price and J. E. Williams, eds., Pergamon Press, 1969.

2. Ionization and Energy Transfer

2-1. Ionization and ionization potentials

It has been pointed out[1] that a 50 eV electron has a velocity of $4{\cdot}2 \times 10^8$ cm/s and will traverse molecular diameters of a few Å in around 10^{-16} s. The fastest vibrations found in organic molecules (C—H stretching vibrations) are almost 10^2 times slower and therefore the position of the nuclei in the molecule being ionized will not change during the time that the bombarding electron is in the vicinity of the molecule. Thus, when the electron-bombardment process results in the removal of a valence electron to give a molecular ion, the process will occur without changes in internuclear distances. That is, the transition will follow the Franck–Condon rule, which requires that the configuration and momenta of the nuclei do not alter during the transition. If, for convenience, we use the potential-energy diagrams applicable to diatomic molecules to illustrate the principle of the Franck–Condon rule with reference to polyatomic molecules (Fig. 2-1), then the ionization process

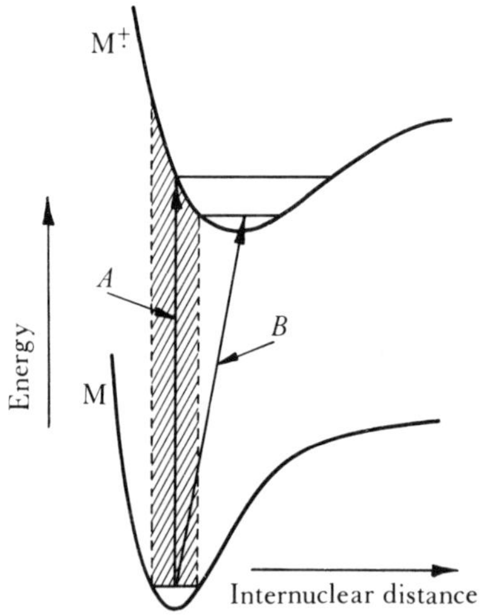

Figure 2-1

would be represented by the vertical transition *A*. The lowest-energy process that can occur corresponds to removal of an electron from the highest occupied molecular orbital and in that case *A* corresponds to the vertical ionization potential, which in turn is the energy lost by the bombarding electron.

In contrast, the adiabatic ionization potential is the energy required to remove an electron from a molecule in its lowest vibrational level, leaving the resultant ion in its lowest vibrational level (transition *B* in Fig. 2-1). The probability for such an adiabatic transition is usually very small. It has been commonly held that electron-impact ionization potentials are frequently slightly greater than spectroscopic ionization potentials because the former method may produce ions with some vibrational excitation even at the threshold (Fig. 2-1). In contrast, spectroscopic ionization potentials are obtained by extrapolation of vibrational bands and correspond to transitions to the ground state of the ion. Certainly, it should always be borne in mind that electron-impact ionization potentials may include some small contribution from excess vibrational energy.

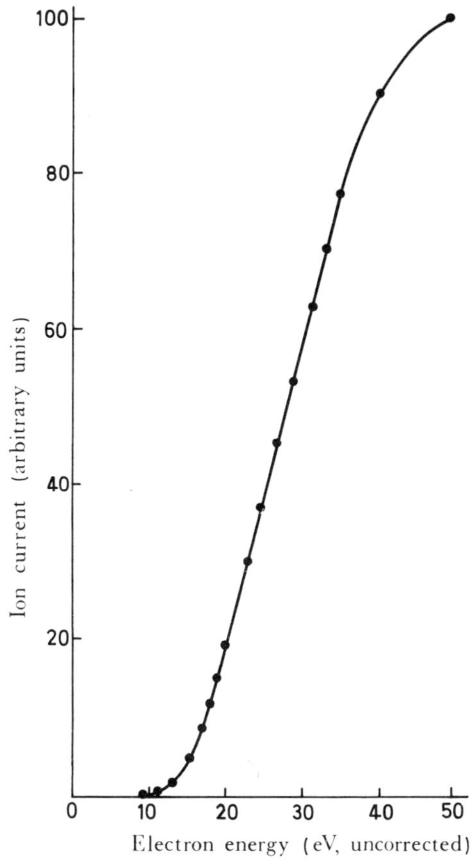

Figure 2-2

In practice, ionization potentials of organic molecules may conveniently be determined using commercial mass spectrometers in the following manner. The substance to be examined is introduced into the mass spectrometer concurrently with one or preferably two reference substances of accurately known ionization potential. Inert gases such as argon and krypton are commonly used. For each ion at, say, a nominal 50 eV the initial ion current is measured and is then recorded as the beam energy is decreased in regular steps until the ion current can no longer be detected. If the ion current for the substance under investigation is plotted against the electron beam energy, then the *ionization efficiency curve* for the production of $M^{+\cdot}$ from M is obtained and typically appears as shown in Fig. 2-2.

The region where the ion current falls off gradually to zero (usually within about 4 eV of the lowest energy at which measurements are possible) is

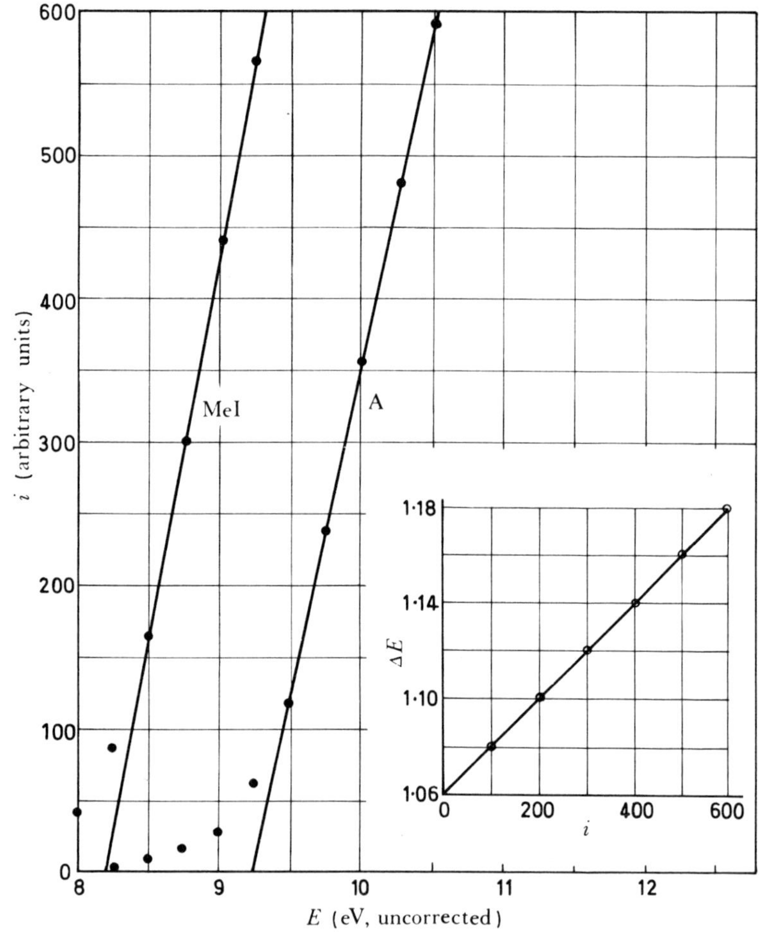

Figure 2-3

described as the *foot* of the ionization efficiency curve. This is followed by an approximately linear rise after which the ion current increases more slowly with increasing beam energy. The ion current is relatively insensitive to the beam energy in the range 70–150 eV.

A portion of the ion efficiency curves for methyl iodide and a substance A is reproduced in Fig. 2-3. The ion-current scale of either the methyl iodide or A must be arbitrarily adjusted to make the linear portions of the ionization efficiency curves roughly parallel.

The box in Fig. 2-3 contains a plot of difference along the energy axis for a given ion current against the appropriate values of the ion current. Extrapolation of the resulting plot to $i = 0$ gives $\Delta E = 1{\cdot}06$ eV. Since the ionization potential of methyl iodide (here used as an internal calibrant instead of an inert gas or gases) is 9·54 eV, *the extrapolated potential difference method*[2] of determining ionization potentials gives the ionization potential (I.P.) of substance A as 10·60 eV.

Alternatively, the log of the ion current for each species is plotted against the beam energy (eV) to give semi-log plots.[3] In practice, the ion currents for $M^{+\cdot}$, $^{132}Xe^{+\cdot}$ and $^{84}Kr^{+\cdot}$ may be normalized to the same value at, say, 50 eV and then I/I_{50} plotted against the nominal beam energy for each species on semi-log graph paper (I_{50} signifies the ion current at 50 eV). The semi-log plots obtained in this way for 1,1-dibromoethane are reproduced in Fig. 2-4. It is usually the case that the curves for the substance under investigation and the inert gases are parallel to a good approximation in the region where the ion current is 10^{-2} to 10^{-3} of its initial value (Fig. 2-4). Knowing the ionization potentials of xenon and krypton to be 12·13 and 14·00 eV, the ionization potential of 1,1-dibromoethane can be determined as 9·6 eV from these experiments. This technique is known as the semi-logarithmic plot method and, with the extrapolated potential difference method, is commonly employed to determine ionization potentials of organic molecules using commercial mass spectrometers.

Ionization potentials for a number of simple monofunctional aliphatic and aromatic compounds are given in Tables 2-1 and 2-2. The values in Table 2-1

Table 2-1
Ionization potentials (eV) of some monofunctional aliphatic compounds $CH_3CH_2CH_2X$

X	Ionization potential	X	Ionization potential
H	11·07	OH	10·17
ONO_2	11·07	CHO	9·86
Cl	10·82	$CH{=}CH_2$	9·5
CH_3	10·63	$COCH_3$	9·34
OAc	10·54	I	9·26
C_2H_5	10·34	NH_2	8·78
Br	10·18		

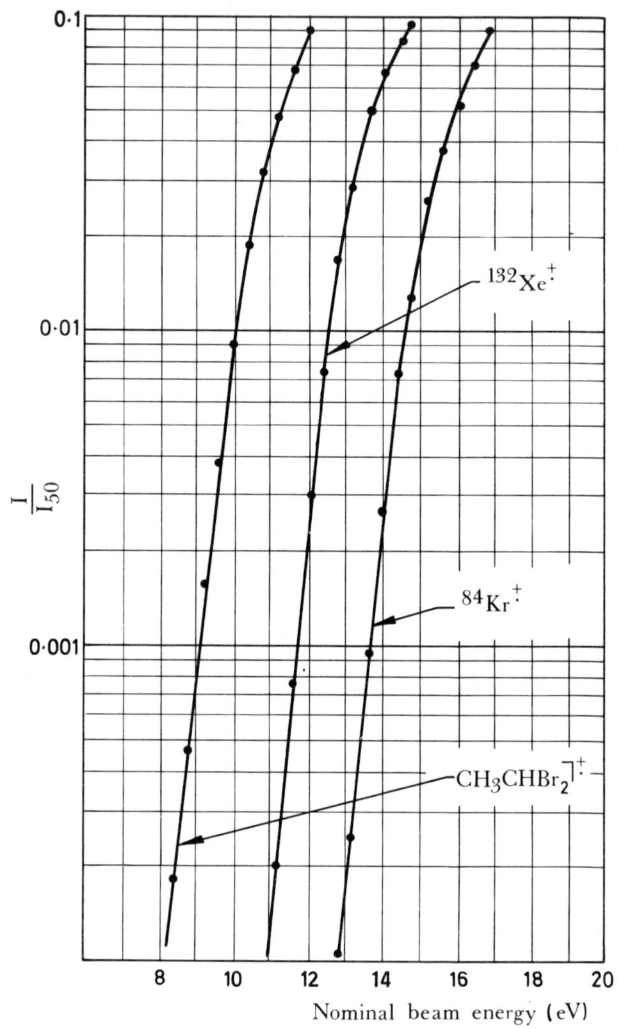

Figure 2-4

were obtained by photoionization, while values obtained by three methods (photoionization, electron impact, and photoelectron spectroscopy) are compared in Table 2-2.

Before the relative merits of the values compared in Table 2-2 are discussed, it is worthwhile to note the simple relationships which are found between ionization potentials and molecular structure. The simplest correlation is found in the case of the monosubstituted aromatic compounds C_6H_5X (Table 2-2), where it is observed that substituents which have strong electron-withdrawing mesomeric effects (NO_2, CN, CHO, $COCH_3$) increase the ionization potential relative to benzene. In contrast, those substituents with

Table 2-2
Ionization potentials of some monofunctional aromatic compounds C_6H_5X

X	Photoionization	Electron impact	Photoelectron spectroscopy
NO_2	9·92	10·18	10·26
CN	9·71	10·09	10·02
CHO	9·53	9·70	9·80
$COCH_3$	9·27	9·57	—
H	9·24	9·56	9·40
F	9·20	9·73	9·50
Cl	9·07	9·60	9·31
Br	8·98	9·52	9·25
CH_3	8·82	9·18	8·9
I	8·72	9·27	8·78
OH	8·50	9·16	8·75
SH	8·33	—	—
OCH_3	8·22	8·83	8·54
NH_2	7·70	8·32	8·04
$NHCH_3$	7·35	—	7·73
$N(CH_3)_2$	7·14	7·95	7·51

the largest electron-donating effects (OH, SH, OCH_3, NH_2, $NHCH_3$, $N(CH_3)_2$), cause a reduction in the ionization potential relative to benzene. A good correlation is found between the ionization potential and the σ^+-value for the substituent (Fig. 2-5)[4] in the cases of both monosubstituted benzenes and substituted toluenes.

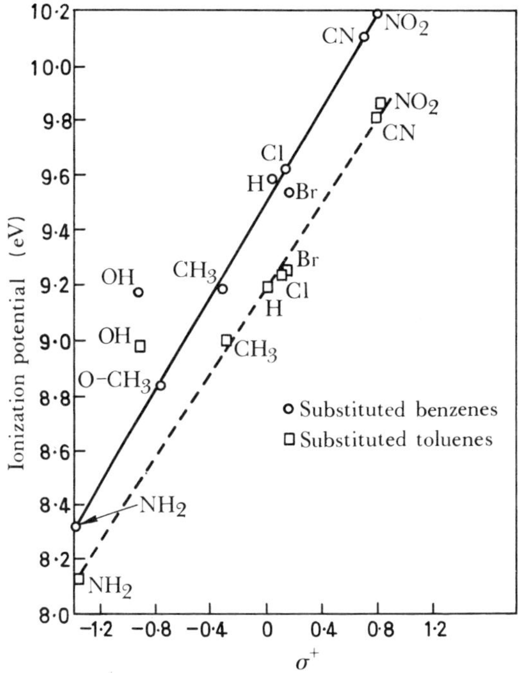

Figure 2-5

A relationship also exists between the number of π-electrons in a polynuclear aromatic compound and its ionization potential.[5] There is in general a smooth decrease in the ionization potential of a series of polynuclear aromatic hydrocarbons of the same symmetry as the number of π-electrons is increased. For example, Fig. 2-6 illustrates that the ionization potentials of the D_{2h} point group, starting with benzene and ending with pentacene, range smoothly from 9·38 to 6·55 eV.

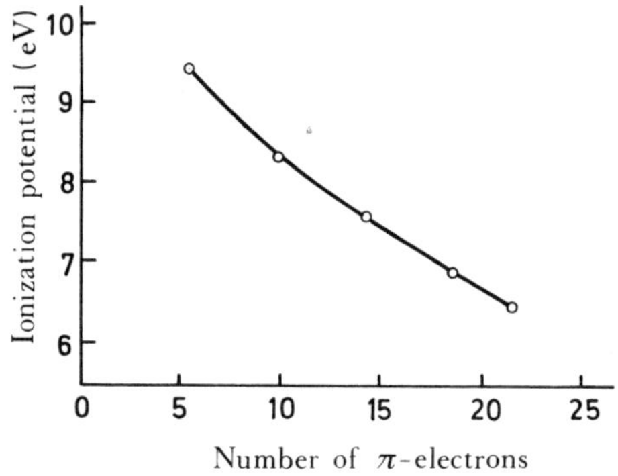

Figure 2-6

In the series of aliphatic compounds $CH_3CH_2CH_2X$ (Table 2-1) it may be seen that all substituents (with the exception of nitrate) lower the ionization potential with respect to the case where X = H. When the so-called non-bonding electrons are least tightly bound (e.g. in amines), then the effect of the substituent on lowering the ionization potential is greatest. Thus in amines the lowest-energy ionization process occurs by removal of a non-bonding electron from nitrogen and at these energies it is physically significant to regard the charge as being localized on the nitrogen atom (1). At the other extreme, in hydrocarbons there is no single site at which the charge resides with great preference. Here ionization occurs by removal of an electron from a molecular orbital distributed over the entire molecule, so that in an *n*-alkane all the bonds are weakened simultaneously. However, the calculations of Lennard-Jones and Hall[6] suggest the positive charge is not distributed completely randomly, but that the charge density in some bonds is greater than in others. Thus the best notation which we can use for the pentane molecular ion is 2. It will be seen subsequently that notations such as 1 are very useful in understanding the reactions of lowest activation energy in the molecular ion.

$$CH_3CH_2CH_2\overset{+\bullet}{N}H_2 \qquad CH_3CH_2CH_2CH_2CH_3\,]^{+\bullet}$$

$$\underline{1} \qquad\qquad \underline{2}$$

It is instructive to consider the spread of values obtained for the monofunctional benzenes (Table 2-2) when values obtained by three methods are compared. Photoionization values may be obtained by ionizing the molecule with a continuously variable and nearly monoenergetic beam of photons.[7] In the photoelectron spectroscopy experiments,[8] ionization is effected by irradiation with a beam of 21·21 eV photons (from a helium resonance lamp). Since the photon producing ionization must give up all its energy (in contrast to an electron) the kinetic energy of the electrons knocked out of the highest occupied molecular orbital will be 21·21 − I.P. eV. The kinetic energies of the photoelectrons may be measured with a magnetic-field deflection analyser, the kinetic energy scale being derived from the ionization potentials of the calibrating gases argon, xenon and krypton. All three methods compared (Table 2-2) are considered to give vertical ionization potentials, although in some cases the $0 \rightarrow 0$ transition may occur with sufficient probability to be detected. While there is usually a good degree of self-consistency within the values obtained by any one method, it is apparent (Table 2-2) that the spread of values for any one compound is about 0·3–0·6 eV (with the exception of dimethylaniline) if all three sets of values are considered. Electron-impact values lose some precision because the bombarding electrons are not energetically homogeneous and, in addition, the neutral molecules usually acquire a significant amount of thermal energy in the source prior to ionization. In careful work, electron-impact determinations may be reproducible within $\pm 0{\cdot}1$ eV, but unless favourable comparison with values obtained by other methods can be made it would seem unwise to rely on the absolute value to within less than $\pm 0{\cdot}3$ eV. Of the values listed in Table 2-2, those obtained by photoelectron spectroscopy are probably the most reliable vertical ionization potentials.

It should also be mentioned that if sufficient energy is transferred to the neutral molecule by electron (or photon) impact a doubly charged molecular ion (M^{++}) may be generated and recorded at an *m/e*-value one-half of that for M^{+}. The minimum energy required to generate M^{++} is known as the

Table 2-3
Some second ionization potentials

Molecule	Second I.P. (eV)
Ne	62·5
NH_3	36·8
C_6H_5D (benzene)	26·0
$C_{10}H_8$ (naphthalene)	22·8
$C_{22}H_{14}$ (pentacene)	19·6

second ionization potential and examples are shown in Table 2-3. It is obviously undesirable that two centres of repulsive positive charge should be confined in close proximity and it is not surprising to find that doubly charged molecular ions are only really important in large aromatic compounds.

2-2. Energy distributions in molecular ions

Within a few eV of the threshold for production of positive ions, ionization efficiency curves for polyatomic molecules rise rapidly (see Fig. 2-2). Hence the rate of production of ions is increasing rapidly due to the presence, in the electron beam, of electrons with energies in excess of the ionization potential. Such electrons can cause ionization by giving up most or all of their energy, in which case an electron may be ejected from a lower-lying molecular orbital of the species being ionized. Thus, if the three highest occupied molecular orbitals of the neutral molecule are represented as in A (see Fig. 2-7), then an energy equal to the ionization potential is needed to produce the molecular ion in its ground state (B), while a more energetic bombarding electron may generate the electronically excited molecular ion C.

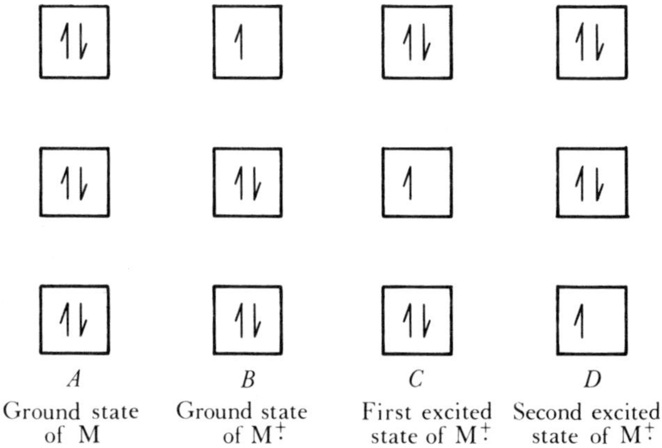

Figure 2-7

As the energy of the bombarding electrons is increased further, it is obvious that higher excited states of M$^{+\cdot}$ (e.g., D) can potentially be populated. Since a bombarding electron does not have to give up all its energy, a high-energy electron which interacts relatively weakly with the molecule to be ionized may transfer only as much energy as a low-energy electron which interacts strongly. It is not difficult to see that if we use, say, 20 eV electrons, a distribution of ion energies will be produced extending from those which are merely ionized and contain no internal electronic excitation energy and (with very low probability) almost no vibrational energy, to those (again with low probability) which were initially highly thermally excited and received 20 eV from the electron beam. Thus, in terms of internal energy of the ions produced, the formal limits of the energy distribution are 0 to $E_{el} + E_{th} -$ I.P., where E_{el} is the electron beam energy and E_{th} the maximum initial thermal energy of the molecules.

Obviously, it would be of great interest to know the energy distribution of the molecular ions (i.e., the fractions of ions having specific energies). Unfortunately, such information is not readily available for electron impact, although it has become available for photoionization.[8] If the kinetic energies of all the electrons expelled from a molecule upon irradiation with 21·21 eV photons are measured, then the difference between 21·21 and the measured kinetic energies gives the ionization potentials for removal of electrons from the various molecular orbitals. The photoelectron spectrum of benzene[8] is given in Fig. 2-8(a). The upper-energy scale gives the ionization potentials for ionization from the various molecular orbitals, whereas

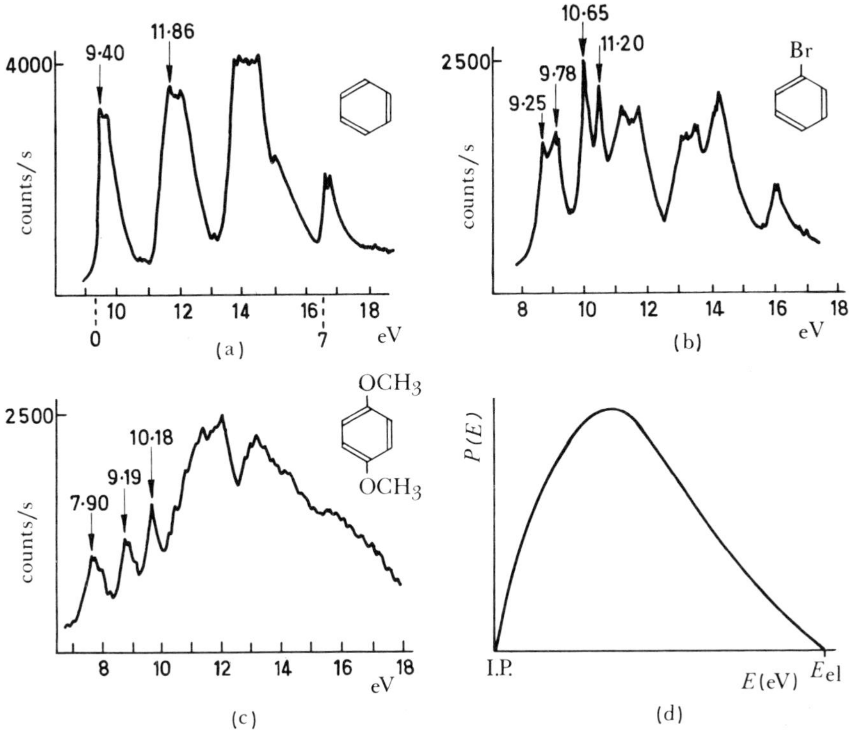

Figure 2-8

the lower-energy scale is referred to the ground state of the benzene molecular ion as zero. Thus, the band at 11·86 eV corresponds to benzene molecular ions possessing, at the instant of production, 2·46 eV of internal energy (largely as electronic energy) in excess of the ground-state energy of the benzene molecular ion. The broadness of the various bands is due to vibrational fine structure which is incompletely resolved. The intensities of the various bands are proportional to the numbers of photoelectrons with a given energy, and therefore also to the numbers of molecular ions with a given internal energy. Hence Fig. 2-8(a) gives the energy distribution of benzene molecular ions which is produced by 21·21 eV photons.

As the molecules studied become less symmetrical and more complex, the numbers of molecular orbitals of different energy levels become larger and the energy distributions lose much of the detail and fine structure. In the case of bromobenzene (Fig. 2-8(b)) many additional bands are already evident. Of particular interest are those at 10·65 and 11·20 eV which it is suggested[8] are due to removal of the 4_{p_x} and 4_{p_y} 'non-bonding' electrons on the bromine. The p_y-orbital, perpendicular to the C—Br bond and to the plane of the benzene ring, will overlap more effectively with the benzene *p*-orbitals than will the p_x-orbital and hence would have the higher ionization potential. In the case of *p*-dimethoxybenzene (Fig. 2-8(c)) little fine structure is resolved.

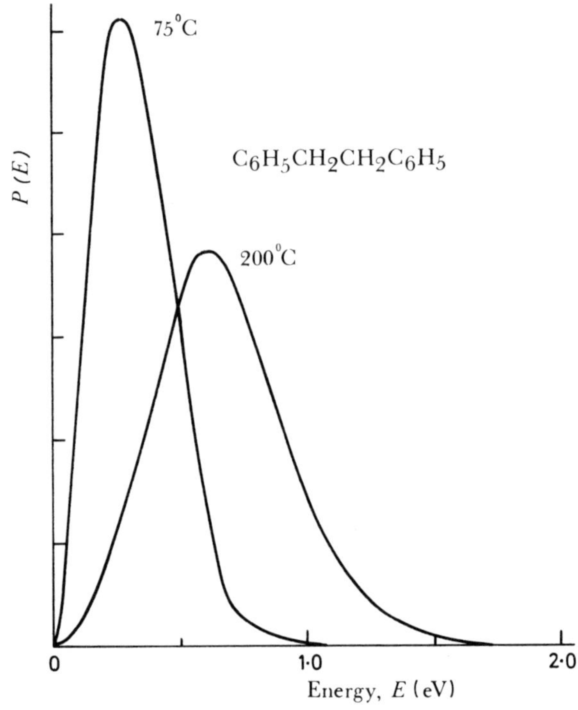

Figure 2-9

Such data are not readily available to give electron-impact energy distributions, although the distribution could in principle be obtained from the second derivative of the ionization efficiency curves if a linear threshold law is valid. In practice, the law holds for only a few eV above the threshold for producing ions, and after this region, the second derivative curve becomes negative.[9] The levels which are available to populate will be the same for both photon and electron impact, but the probabilities of the excitation processes will be different. Nevertheless, it remains a fact that where 21·21 eV photoionization and 20 eV electron impact mass spectra of complex molecules have been compared, they are in general not grossly different. Photoelectron energy distributions for some monosubstituted 1,2-diphenylethane derivatives have been approximated by summing the photoelectron spectra[8] of toluene and the corresponding substituted toluene, and convoluting the resultant distribution with the calculated thermal-energy distribution.[10] For example, the thermal-energy distribution for 1,2-diphenylethane has been calculated for 75° and 200° (Fig. 2-9); in Fig. 2-10,

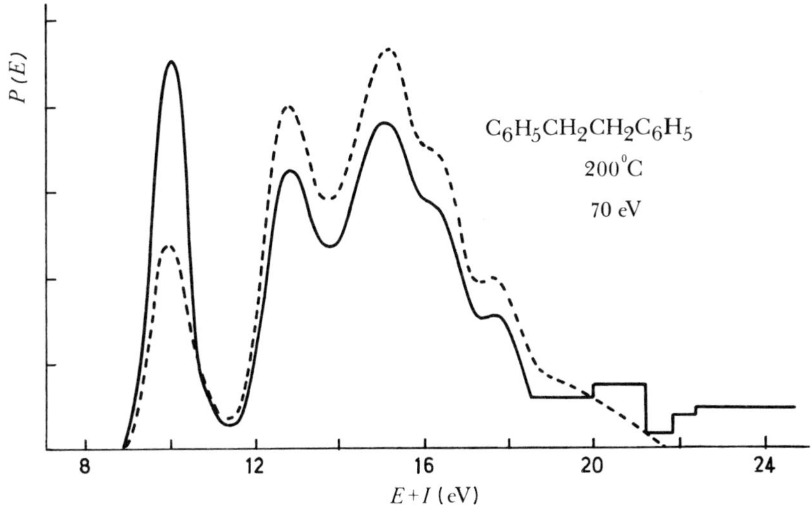

Figure 2-10

the dashed lines represent the photoelectron spectrum of toluene convoluted over the 200° thermal-energy distribution, while the solid lines show the distribution which is consistent with the observed ion abundances in the 70 eV mass spectrum of 1,2-diphenylethane. Note that the energy scale of Fig. 2-10 gives the internal energy of the ions only after subtraction of the ionization potential of 1,2-diphenylethane (8·7 eV). It is evident that the photoelectron data, when adjusted to allow for the thermal excitation of molecules which is important in electron-impact sources, do seem to weight less those energies within 2 eV of the I.P. relative to the electron-impact situation; similar results were obtained for substituted diphenylethanes.[10] For semi-empirical calculations of 20 eV electron-impact mass spectra, an energy distribution such as that shown in Fig. 2-8(d) may be used for complex organic molecules (the graph is actually a plot of $y = (1 - x)(x - x^2)$); $P(E)\,dE$ is the fraction of ions having energies between E and $E + dE$.

Figure 2-8(c) establishes that for photo-impact the probability of transferring 10 eV in excess of the I.P. of *p*-dimethoxybenzene is small. In Fig. 2-8(d) it is assumed that in the general case, the probability of transferring, say, 10 eV in excess of the I.P. with 20 eV electrons is also small (in the case of a large molecule). Indeed, there is experimental evidence to show that the most probable energy transfer is between 1–8 eV in excess of the ionization potential.[11] Even in 70 eV spectra only a small fraction of the molecular ions will be produced with energies in excess of 10 eV of the ionization potential.

2-3. Internal conversions and radiative transitions

At the instant of production of molecular ions with high internal energies, much of the excess energy will be present as energy of electronic excitation, although as a consequence of the Franck–Condon principle some ions will also be produced as vibrationally-excited species within a given electronic state. Since ions in the source are non-interacting, intermolecular energy transfer cannot occur and in the absence of radiative transitions, the internal energy (E) of each ion would be fixed at the value produced in the initial excitation. For ions that do not undergo radiative transitions (loss of energy by photon release), we know that the total internal energy (electronic, vibration and rotational) must be constant and hence $E = E_{el} + E_{vib} + E_{rot}$. In the absence of collisions, E_{vib} and E_{rot} are not interconvertible (angular momentum restriction) and hence $E_{el} + E_{vib}$ is constant. The significance of this conclusion may be seen by considering two electronic states whose potential surfaces cross and therefore are in equilibrium for ions of internal energy E (Fig. 2-11; a schematic illustration employing the analogy to diatomic molecules).

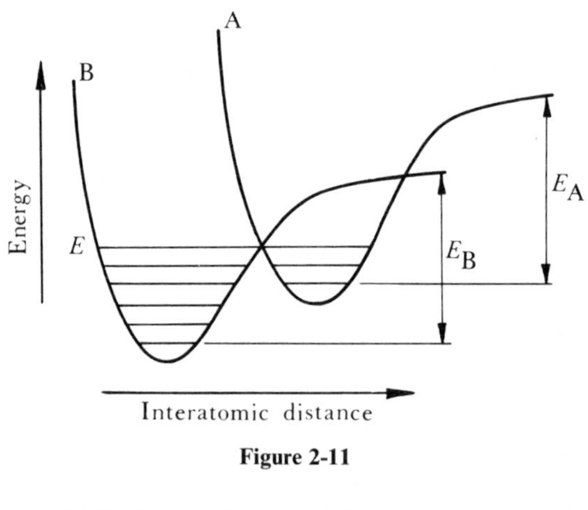

Figure 2-11

The process by which electronic state A is converted into electronic state B is known as internal conversion. Since state B is the lower-lying electronic state of the two, the crossing of potential surfaces here causes a conversion of electronic to vibrational energy and can occur, if the oscillators are anharmonic and efficiently coupled, in times corresponding to vibrational periods (10^{-14} s). Hence, if the density of electronic states is large, it is possible to populate low-lying electronic states (e.g., perhaps the ground electronic state of $M^{\dagger}$) from higher electronic states by the crossing of potential surfaces. The low-lying electronic states so produced will however be highly vibrationally excited. Since decomposition will occur when sufficient vibrational quanta have been gathered in the reaction coordinate to provide the activation energy for reaction, decomposition may occur from such highly vibrationally-excited states relatively rapidly (e.g., in times less than 10^{-9} s or even comparable to one vibration). However, if decomposition does not occur, then it must be recognized that state B will be able to re-convert to state A since there is no mechanism to lose the vibrational energy. This situation is in marked contrast to normal thermal and photochemical reactions in the presence of collisions, where collisional deactivation of the vibrationally excited species can occur. Thus, in the absence of the radiative transitions, the amount of reaction occurring from the electronic states A and B will depend on the relative activation energies E_A and E_B for decomposition from A and B and the relative populations of the levels. Given comparable activation energies, then decomposition from the lower-lying electronic state will be more probable since (a) more vibrational quanta are available in state B than A (alternatively, one may simply consider that more internal energy is needed to produce dissociation from state A than state B) and (b) it is much more probable for a molecule of internal energy E to find itself in state B rather than A since the degeneracy of states is greater in B. The latter situation follows from the case of a set of s similar harmonic oscillators containing n quanta, for which the number of states is $(n + s - 1)!/n!(s - 1)!$. This arises simply because there are more ways of accommodating the larger number of vibrational quanta in a given number of bonds.

If there is a low activation energy for decomposition from singly and doubly excited molecular ions, it is quite possible that these will decompose in times within the range 10^{-14}–10^{-9} s. If, on the other hand, the activation energies for decomposition from such states are larger, and they survive $\sim 10^{-8}$ s, then it is probable that they will lose energy by ejection of a photon and fall to the ground ionized state. This follows since the molecular ions are ion-radicals usually with a spin (s) of $\frac{1}{2}$ and hence a multiplicity $(2s + 1)$ of 2 (i.e., they are doublet states) and the radiative lifetimes of doublet states are of the order of 10^{-8} s.[12] It should be noted that quartet states cannot be entirely ruled out, since double excitation with spin inversion is possible in the ionization process.

To summarize, the ground electronic state can in principle become densely populated by internal conversion and the resulting ground states will be (often highly) vibrationally excited. Electronically excited doublet states may react as such within 10^{-8} s or less, or alternatively lose energy radiatively and convert to the ground state of the radical ion. Hence, reactions occurring at relatively slow rates ($k \simeq 10^6\ s^{-1}$) are likely to occur from the ground state of the ion or from an associated low-lying state. The initial energy distribution in the molecular ions is fixed except insomuch as the highly excited species may lose energy radiatively. Some possible transitions are summarized schematically in Fig. 2-12.

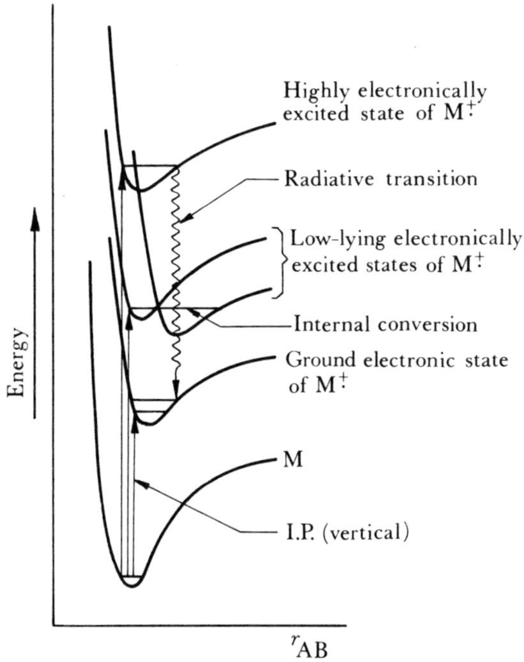

Figure 2-12

For complex molecules, no information is yet available about the numbers of electronic states and associated vibrational states to which excitation may occur. The number of such states per unit of energy is referred to as the density of states and it is generally held that this density of states increases rapidly with increasing internal energy. Hence, the number of crossings of potential surfaces will increase with increasing energy. In addition, vibrations become more anharmonic with increasing energy and thus the vibrational quanta which may be initially located in certain bonds can be randomized among other vibrational modes. In this way, the initial excitation energy, which may be localized in certain parts of the molecular ion, can rapidly be randomized throughout the ion by intramolecular energy transfer. This randomization of a given internal energy E throughout the ion, at a rate assumed to be rapid relative to the rates of subsequent fragmentation, is one of the basic assumptions of the original quasi-equilibrium theory (QET) of mass spectra.[13]

Although our knowledge of the relative importance of the various phenomena described in determining the nature of mass spectra is currently limited, it is worthwhile to consider their possible relevance by a specific example. The amino ketone 3 when ionized by 20 eV electrons will produce molecular ions in various electronic states, some of which may be represented as in 4, 5, 6 and 7.

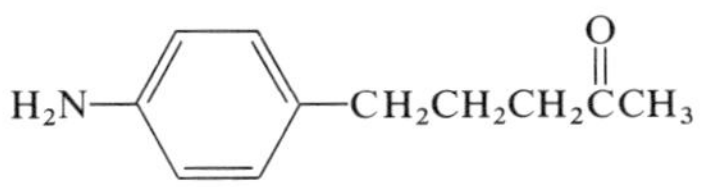

3

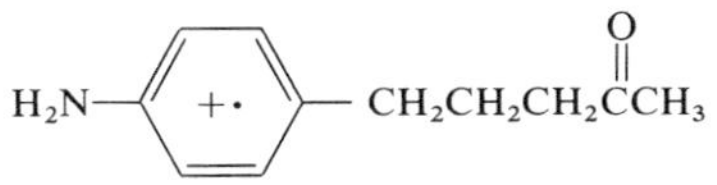

4

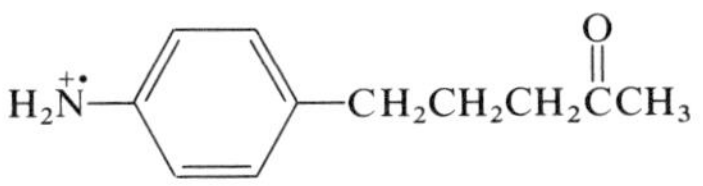

5

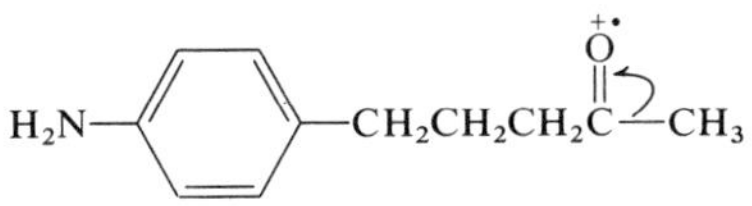

6

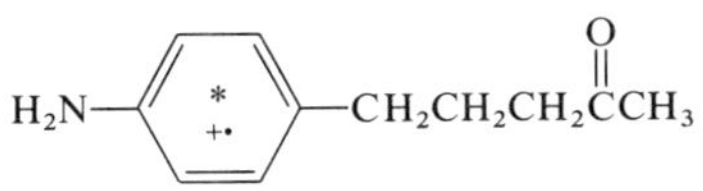

7

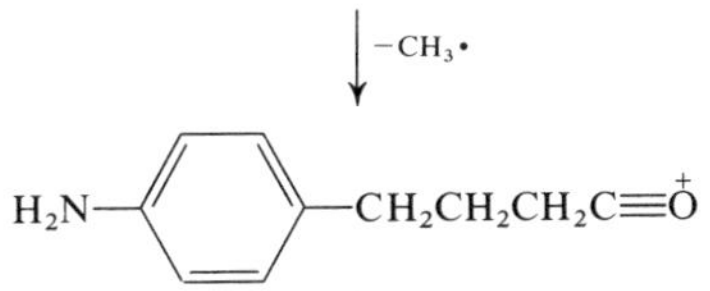

8

The representations are meant to imply that 4 is generated by removal of an electron from the highest occupied molecular orbital of the benzene ring (to give the molecular ion in its ground electronic state), while 5 and 6 are generated by removal of lone-pair electrons on nitrogen and oxygen respectively. Structure 7 represents a highly excited electronic state obtained by removal of, say, an electron from a σ_{CC} orbital of the benzene ring. It is true that conversions between some of these electronic states may be able to occur at rates which are rapid relative to dissociation, and rapid relative to the mass spectrometer time-scale (section 3-3), thus meaning that for all effective observational purposes we may regard the charge as being 'smeared out' over the molecule. Nevertheless, it remains a fact that once the requisite number of vibrational quanta (required to supply the activation energy for decomposition) have been gathered in the reaction coordinate, dissociation occurs in one vibration and does so from a specific electronic state (e.g., 4 or 6). In addition, the photoelectron spectra of substances such as aniline[8] clearly establish that it needs more energy to cause ionization of electrons largely localized in lone-pair-type orbitals (cf. 5) than it does to remove an electron from the highest occupied molecular orbital of the benzene ring (cf. 4). Hence, the bands which appear in photoelectron spectra support the physical significance of such considerations.

In their respective ground vibrational states, 6 will be about 1·3 eV higher in energy than 4 and it is from a vibrationally-excited form of 6 that a methyl radical would be most likely to be lost to give the acylium ion 8. Since 6 represents the ground ionized state of the *carbonyl group*, we can assume radiative transitions from such *low-lying electronic states* (see Fig. 2-12) to be less probable than from highly electronically-excited states such as 7. Thus we may for convenience define a low-lying electronic state as one from which a radiative transition is sufficiently improbable that it may in principle survive up to several μs before undergoing decomposition. It will be seen later (section 3-3) that this means we may expect so-called metastable transitions from such low-lying electronic states (as well as from ground states).

References

[1] J. H. Beynon, R. A. Saunders and A. E. Williams, *The Mass Spectra of Organic Molecules*, Elsevier, Amsterdam, 1968.
[2] J. W. Warren, *Nature* **165**, 810 (1950).
[3] F. P. Lossing, A. W. Tickner and W. A. Bryce, *J. Chem. Phys.* **19**, 1254 (1951).
[4] G. F. Crable and G. L. Kearns, *J. Phys. Chem.* **66**, 436 (1962).
[5] E. J. Gallegos, *J. Phys. Chem.* **72**, 3452 (1968).
[6] J. Lennard-Jones and G. G. Hall, *Proc. Roy. Soc.* **213A**, 102 (1952).
[7] K. Watanabe, *J. Chem. Phys.* **26**, 542 (1957).
[8] A. D. Baker, D. P. May and D. W. Turner, *J. Chem. Soc.* (**B**), 22 (1968).
[9] H. Ehrhardt, F. Linder and T. Tekaat in *Advances in Mass Spectrometry*, Vol. 4, E. Kendrick, ed., The Institute of Petroleum, London, 1968, pp. 705–712.
[10] F. W. McLafferty, T. Wachs, C. Lifshitz, G. Innorta and P. Irving, *J. Am. Chem. Soc.* **92**, 6867 (1970).
[11] See, for example, W. A. Chupka and M. Kaminsky, *J. Chem. Phys.* **35**, 1991 (1961).
[12] R. C. Dougherty, *J. Am. Chem. Soc.* **90**, 5780 (1968).
[13] H. M. Rosenstock, M. B. Wallenstein, A. L. Wahrhaftig and H. Eyring, *Proc. Natl. Acad. Sci.* **38**, 667 (1952).

3. Metastable ions

3-1. The origins of metastable peaks

If a singly charged molecular ion of mass m_1 does not dissociate before arriving at the collector, it is of course recorded as a molecular ion (sometimes called *parent ion*). Alternatively, if the reaction $m_1^+ \rightarrow m_2^+$ occurs in the source, then m_2^+ may travel the whole of the analyser region as a mass m_2 and be recorded as an m_2^+ daughter ion. However, it is possible that the transition $m_1^+ \rightarrow m_2^+$ will occur after the source exit slit but before arrival at the collector. Ions undergoing such transitions are termed metastable ions. Under normal operating conditions, the pressure in the analyser (or analysers) is less than the source and around 10^{-7} mm Hg. Under such conditions, the vast majority of metastable transitions are due to true unimolecular reactions, although if the pressure in the analyser is increased ($>10^{-6}$ mm Hg), collision-induced reactions may occur (section 8-3).

In double-focusing mass spectrometers, there are two field-free regions (also called drift regions), through which ions pass after acceleration, that are in common use for the detection of metastable transitions. The so-called first field-free region refers to the portion of the ion path immediately prior to the electrostatic analyser, and the second field-free region lies between the electrostatic and magnetic analysers (Fig. 3-1, which refers to the Nier–Johnson geometry). Daughter ions formed in these first and second field-free regions will be identified as $^1m^*$ and $^2m^*$, respectively.

A. The second field-free region. Since an ion with a charge e will possess a kinetic energy eV† after being accelerated through a potential difference of V volts, all ions arriving at point A (Fig. 3-1) after acceleration will possess a translational energy eV, independent of their mass. While it is true that the initial Boltzmann distribution of translational energies and field inhomogeneities in the source will cause a slight velocity spread, which is significant when trying to achieve very high resolving powers (section 1-2), in the present context this spread is insignificant with respect to the kinetic energies of several thousand eV. The voltage E applied between the radial plates of the electrostatic analyser is fixed to pass the ions of kinetic energy eV. If one

† It is important to remember that this kinetic energy is completely independent of the internal energy of the ion.

is determining spectra with an accelerating voltage of 8 kV, then determinations at say 6, 4 or 2 kV require alteration of V and E in exactly the same ratio.

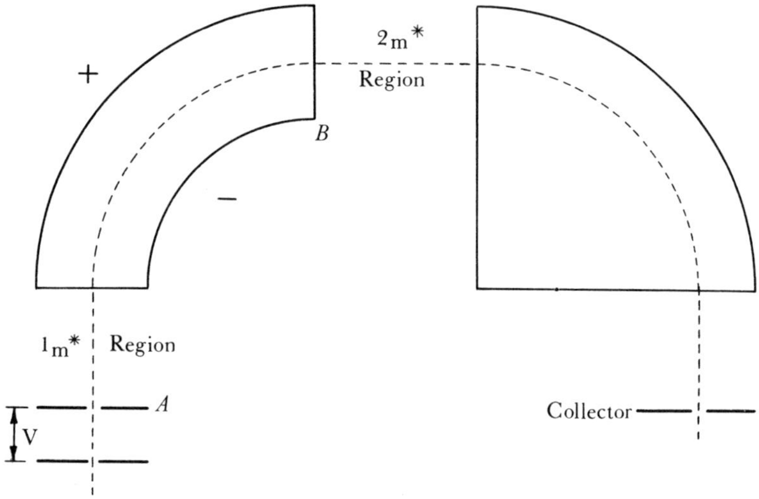

Figure 3-1

Hence, ions of energy eV arrive in the second field-free region (B). However, if the reaction $m_1^+ \rightarrow m_2^+$ occurs in the second field-free region, then since the precursor ion m_1^+ possessed energy eV, the daughter ion m_2^+ will possess kinetic energy $(m_2/m_1)eV$. This follows since the kinetic energy is partitioned in the ratio of the masses of the two product particles m_2^+ and $m_1 - m_2$, the neutral particle carrying off $[(m_1 - m_2)/m_1]eV$. Therefore, m_2^+ ions formed in the second field-free region possess less kinetic energy than 'normal' m_2^+ ions formed in the source and are accordingly deflected more readily when traversing the magnetic analyser. They do not appear at the normal position for m_2 on the mass scale, but at $m_2 \times m_2/m_1$, i.e., m_2^2/m_1.[1] Thus, the position, $^2m^*$, of *metastable peaks* due to the reaction $m_1^+ \rightarrow m_2^+$ occurring in the second field-free region is such that

$$^2m^* = m_2^2/m_1. \qquad (3\text{-}1)$$

In a single-focusing instrument (i.e., in the absence of the electrostatic analyser), the metastable peaks appearing in the normal spectrum are formed after acceleration of m_1^+, when m_1^+ decomposes to m_2^+ in the field-free region in front of the magnetic analyser. In general, remarks applicable to metastable transitions occurring in front of the magnetic analyser in a double-focusing mass spectrometer are applicable to the corresponding transitions in a single-focusing instrument.

Although the vast majority of metastables result from decompositions of singly charged species, examples of metastable decompositions of doubly charged ions are known—to form either two singly charged ions or a doubly charged ion plus a neutral. An example of the former is found in the mass spectrum of benzene,[2] where the metastable transition $C_6H_6^{++} \rightarrow CH_3^{+} + C_5H_3^{+}$ occurs, as evidenced by broad metastable peaks centred at $m/e = 5{\cdot}77$ and 101·8. Since doubly charged ions possess twice the kinetic energy of singly charged ions after acceleration, it may readily be shown that for the reaction $m_1^{++} \rightarrow m_2^{+} + m_3^{+}$, *two* metastable peaks will occur at *m/e*-values of $2m_2^2/m_1$ and $2m_3^2/m_1$. It is noteworthy in the above example that one of the metastables is found at higher mass than the molecular ion.

Metastable peaks are usually readily identifiable in the normal mass spectrum because they (a) need not occur near integral *m/e*-values, (b) are nearly always much broader than normal peaks and (c) are of relatively low abundance. On a scale that is roughly linear in *m/e*-values, it is useful to remember that, for the reaction $m_1^{+} \rightarrow m_2^{+}$, m* lies a distance below m_2 on the mass scale which is similar to the distance that m_2 lies below m_1. The abundance of m* is frequently only of the order of 10^{-2} (or less) of the abundance of the parent and daughter ions in 70 eV spectra, because, as we shall see, it is usual for many more ions to react in the source than in the second field-free region. Two metastable peaks in part of the spectrum (m/e = 50–100) of *p*-aminoanisole are evident in Fig. 3-2. The assignments of the metastable peaks are as shown in Table 3-1.

Table 3-1

Assignments of metastable peaks in the partial spectrum of p-aminoanisole

^{2}m*	Transition	Inferred reaction
m/e 94·8	*m/e* 123 → 108	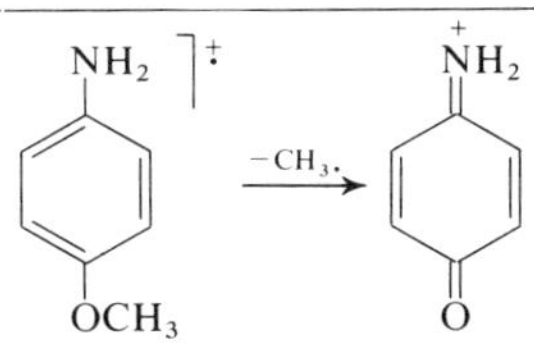
m/e 59·2	*m/e* 108 → 80	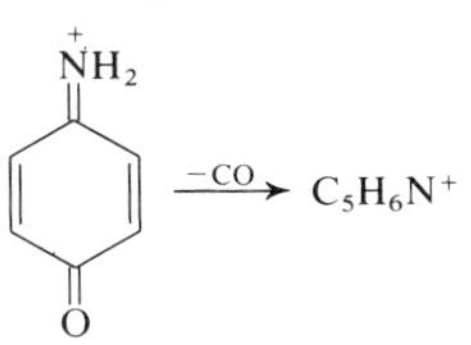

The peak at *m/e* 94·8 is roughly Gaussian in shape and partially overlaps with the *normal* peak at *m/e* 95, while that at *m/e* 59·2 is much broader and flat-topped.

For the vast majority of cases where metastable transitions are observed, there is no reasonable doubt that the reaction $m_1^{+} \rightarrow m_2^{+}$ occurs in one step, but the position of the metastable peak is independent of the number of steps.

A documented example of a two-step metastable transition is found in the spectrum of toluene, where $C_5H_5^+$ ions formed from $C_7H_7^+$ ions in the first drift region were specifically transmitted by the electrostatic analyser (see part B of this section) and found to decompose to $C_3H_3^+$ ions in the second drift region.[3] Many instances of consecutive collision-induced metastable transitions are also known (see section 8-3).

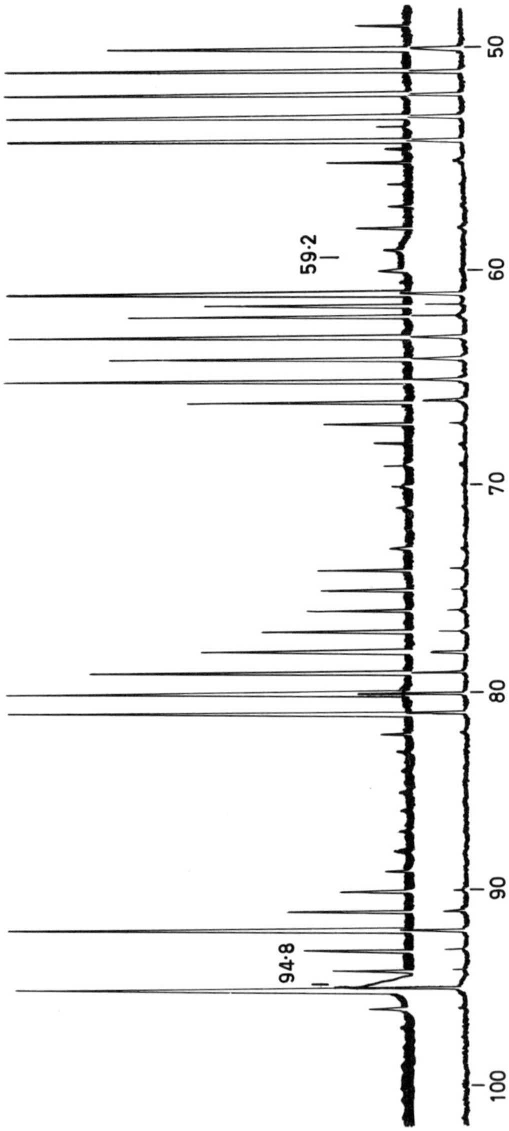

Figure 3-2

Factors which influence the shapes of metastable peaks have been discussed by Beynon and Fontaine.[4] Let us suppose that when the reaction $m_1^+ \rightarrow m_2^+$ occurs there is a release of T eV of kinetic energy, rather than the ions drifting apart with almost identical velocities. This kinetic energy is available for release in some cases because of the electronic nature of the reaction, i.e., the implication is that, in the reverse reaction, some collisional activation energy is necessary when m_2^+ and $(m_1 - m_2)$ react. Since the laws of conservation of energy and momentum must apply, the only form that the kinetic-energy release (originating from internal energy of m_1^+) can take is a mutual recoil between m_2^+ and the neutral particle $m_1 - m_2$. When this process is occurring in the second field-free region, it can be seen that the mutual recoil between the product particles can occur in any direction (Fig. 3-3).

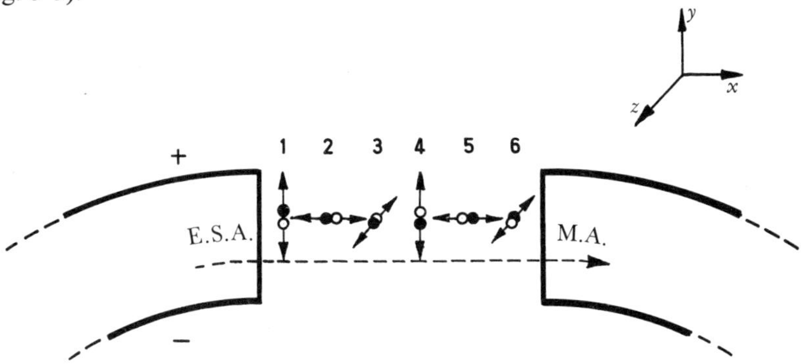

Figure 3-3

At any point in the field-free region where a decomposition occurs with release of kinetic energy T eV, the result will be an exploding sphere of ions. It might reasonably be expected from our knowledge of chemical reactions that the kinetic-energy release will be at the most only of the order of 1 eV or less, and therefore small compared with the kinetic energies along the line of flight of, say, 8000 eV prior to decomposition. If the incipient charged particle m_2^+ is represented by the open circles, then reaction in orientation 2 would give m_2^+ an additional velocity component along the line of flight, while reaction in orientation 5 will reduce the velocity of m_2^+ slightly relative to the situation where $T = 0$. The slightly accelerated m_2^+ ions will be collected at a slightly higher mass than the normal m* position, and the slightly decelerated m_2^+ ions at a correspondingly lower mass. Reaction in orientations 3 and 6 will give m_2^+ a small velocity component perpendicular to the pole face of the magnet, i.e., if an x,y-plane contains the pole face, then such m_2^+ ions have a velocity component in the z-direction which is perpendicular to the pole face. The lengths of the slits of the mass spectrometer lie in the z-direction and therefore it can be seen that such m_2^+ ions will not be collected if the distance travelled in the z-direction finally exceeds

$l/2$ from the centre of the beam, where l is the collector slit length. Therefore discrimination is possible against these ions (which have not changed their component of velocity along the original flight path), especially if the slit length is small. Reaction in orientations 1 and 4 will give the resulting m_2^+ ions a component of velocity in the y-direction (Fig. 3-3), but discrimination does not occur against these ions since they will still traverse the collector slit as the magnetic field is swept.

A quantitative treatment of the problem[4, 5] allows one to relate the expected broadening of the metastable peak $^2m^*$ to the kinetic energy T released. The resulting metastable peaks are usually more or less flat-topped and extend over a small range of m/e-values. In Fig. 3-4(a), the shape of the

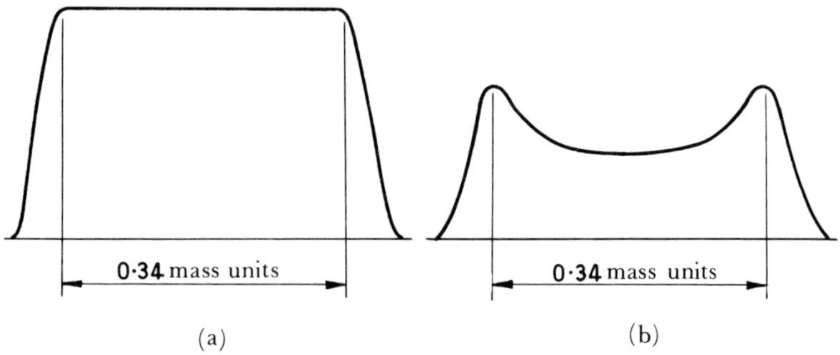

Figure 3-4

peak computed from the data, $T = 0{\cdot}1$ eV, $V = 8000$, $m_1 = 79$ and $m_2 = 52$ is given, assuming no discrimination by the collector slit.[4] In Fig. 3-4(b), the peak shape has been calculated including the effect of discrimination by a collector slit of length 0·080 inch (the collector slit length of the AEI MS 9 mass spectrometer). The positions on the mass scale at which the *ends* of the flat-topped region occur are related to the energy release T eV by the expression

$$^2m^* = \frac{m_2^2}{m_1}\left(1 \pm \sqrt{\frac{\mu T}{Ve}}\right)^2, \tag{3-2}$$

where V is the accelerating voltage, e is the electronic charge, and $\mu = (m_1 - m_2)/m_2$. These ends correspond to the cases in which the whole energy is released along an axis coincident with the ion beam (see 2 and 5 in Fig. 3-3). If the width of the flat-topped region (i.e., 0·34 mass units in Fig. 3-4) is put equal to $2\,\Delta m$, eq. (3-2) can be rearranged to give

$$T = \frac{Ve}{4\mu}\left(\frac{\Delta m}{m_2^2/m_1}\right)^2, \tag{3-3}$$

since $\mu T/Ve$ is very much less than unity. It is noteworthy that since the kinetic energy released will be characteristic of the reaction which is occurring and independent of the accelerating voltage V, the width of the flat-topped region (or the distance between the end peaks in Fig. 3-4(b)) will be inversely proportional to the square root of the accelerating voltage. As an example, we may cite the mass spectrum of vinyl fluoride (1) in which the elimination of HF occurs from the molecular ion with release of kinetic energy.

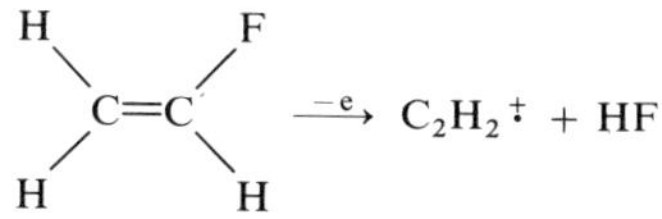

1

At different accelerating voltages, the ends of the flat-topped region of the associated metastable peak were found to be as follows (see also Fig. 5-7):

Accelerating voltage (kV)	8	4	3	2
m/e	14·47–14·95	14·36–15·04	14·31–15·11	14·20–15·19

Substitution of these values into eq. (3-3) gives the kinetic energies released as 0·69, 0·69(5), 0·72 and 0·74 eV respectively in 8, 4, 3 and 2 kV spectra. Thus, the average value of the kinetic energy released is 0·71 eV.

We shall defer discussion of the physical significance of such energy release and its mechanistic significance (see sections 5-3 and 5-4). Discussion as to why metastable peaks are broader (possessing less steeply sloping sides) than normal peaks even when not flat topped will also be deferred (section 5-5).

B. The first field-free region. If an ion m_1^+ decomposes to m_2^+ after acceleration but before entering the electrostatic analyser, then the electrostatic analyser will not transmit m_2^+ because it possesses a kinetic energy $(m_2/m_1)eV$ instead of the required kinetic energy eV. However, m_2^+ ions formed exclusively in the first field-free region can be transmitted at the pre-set electrostatic-analyser voltage of E if their kinetic energy is increased in the ratio m_1/m_2. To do this, the accelerating voltage must be increased from its normal value V_0 to V_1, such that

$$\frac{V_1}{V_0} = \frac{m_1}{m_2}, \tag{3-4}$$

while keeping the electrostatic-analyser voltage constant. Since the V/E-ratio is normally fixed and must be kept constant for the transmission of normal ions formed in the source, increase of the accelerating voltage from V_0 to V_1 at constant E prevents the transmission of all ions formed in the source.[6] If V_0 were set at 8 kV, then V_1 would be greater than 8 kV and since such large voltages will possibly cause breakdown in the source, it is usual to operate with V_0 initially set at 2 or 4 kV.[7] In practice, the mass spectrometer is set up for normal, say, 2 kV operation and the magnet current adjusted so that the m_2^+ ion of interest is arriving at the collector. The accelerating voltage V is then scanned slowly towards higher values and m_2^+ will only be transmitted to the collector again when $m_1^+ \rightarrow m_2^+$ occurs in the first field-free region and the accelerating voltage V_1 is such that eq. (3-4) is satisfied. Since m_2^+ is known and the ratio V_1/V_0 is measured experimentally, it is possible to determine all values of m_1^+ ions which are precursors of m_2^+ ions in the first field-free region. This method of observation of metastable transitions is known as the *metastable refocusing* technique. For example, all precursors of *m/e* 43 ($C_3H_7^+$ and/or $CH_3C{\equiv}O^+$) in the mass spectrum of methyl *n*-pentyl ketone (2) can be determined by this method and are summarized in Fig. 3-5(a), where the relative abundances of the

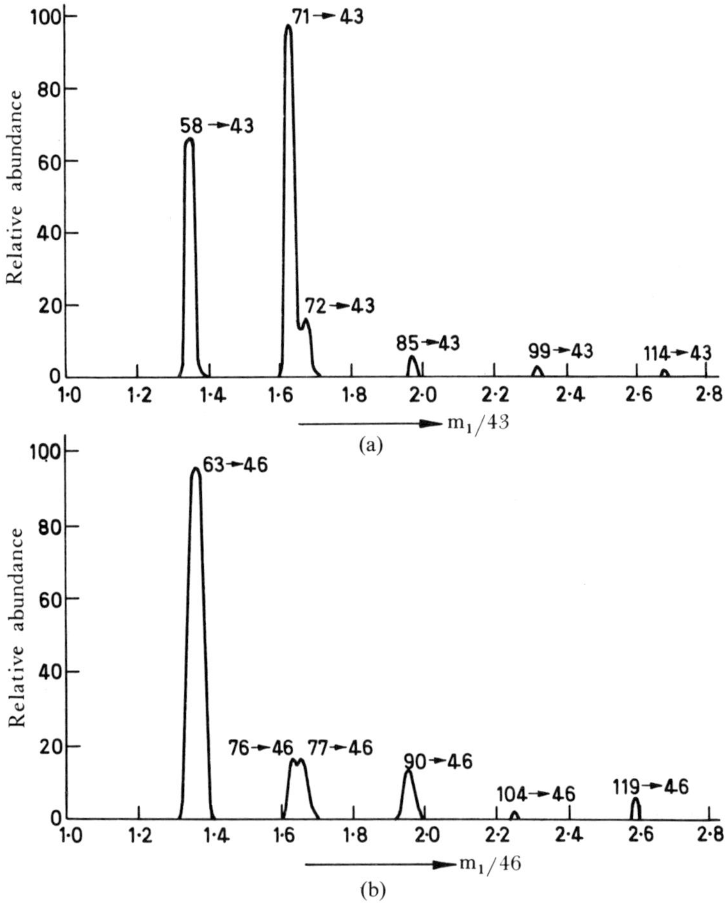

Figure 3-5

various transitions are also indicated. If a similar experiment is carried out on the d_5-ketone 3, then the precursors of $CD_3C{\equiv}O^+$ (*m/e* 46) alone can be determined (Fig. 3-5(b)).

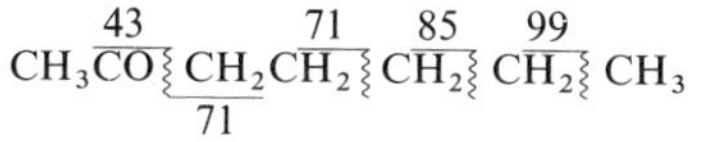

2

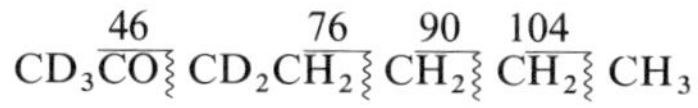

3

From Fig. 3-5(b) it is evident that 3 decomposes directly to CD_3CO^+ (*m/e* 46) or *via* $M^+{-}CH_3$, $M^+{-}C_2H_5$ and $M^+{-}C_3H_7^+$ ions (see 3). In addition, $M^+{-}C_3H_6$ (*m/e* 77) and $M^+{-}C_4H_8$ (*m/e* 63) ions act as precursors of CD_3CO^+. The main difference between Figs. 3-5(a) and (b) (apart from the expected mass shifts) is the large reduction in intensity on passing from the 71 → 43 transition in A to the analogous 76 → 46 transition in B. This arises because the major portion of the 71 → 43 reaction is associated with the step $C_5H_{11}^+ \rightarrow C_3H_7^+ + C_2H_4$ and in this case no shift of *m/e* 43 to *m/e* 46 is possible upon deuteriation.

The advantages of using the first field-free region for the study of metastable transitions are that (a) each transition is unequivocally identified in terms of the mass of parent and daughter ions, (b) all the precursor ions of m_2^+ may rapidly be found in one scan of the accelerator voltage, (c) normal ions do not interfere, and (d) in the absence of normal ions, the electron multiplier (collection sensitivity) can be increased and so weak transitions may be detected (down to abundances corresponding to 1 part in 10^5 of the base peak intensity).

In passing it should be mentioned that the alternative refocusing method of decreasing the potential of the electrostatic analyser at constant accelerating voltage is also in use.[8] This mode of operation has the disadvantage that the magnetic field must first be set as if to focus an ordinary metastable, not a normal peak, as in the previous method. An advantage is that the penetration of the source by the accelerating field is constant, and therefore the source tuning does not change during the scan.

This technique has been used to demonstrate that doubly charged ions may arise by other than the mechanism

$$m_1 + e \rightarrow m_1^{++} + 3e$$

and that charge separation (autoionization) may occur:

$$m_1^+ \rightarrow m_1^{++} + e$$

For example, ions which undergo the latter transition in the first drift region will be transmitted when the electrostatic analyser voltage is reduced to 0·5 of its normal value and will occur at a position $m_1/4$ on the mass scale. Such charge separation occurs in the 1,3,5-trichlorobenzene molecular ion, as evidenced by weak signals at *m/e* 45, 45·5 and 46·0 (in the correct intensity ratios for the presence of three chlorine atoms and a 3:1 ratio of $^{35}Cl:^{37}Cl$), when the above operation is performed.[9]

The main use of the variation of the electrostatic analyser voltage, however, lies in the measurement of ion kinetic energy (IKE) spectra (see section 3-2) by monitoring the total ion current leaving the electrostatic analyser.

As in the case of metastable transitions occurring in the second field-free region, those occurring in the first field-free region are broadened upon release of kinetic energy and the extent of the broadening may be used to measure the kinetic energy released.[10] However, here a slight complication arises because in the case of dish-shaped and flat-topped metastable peaks, the accelerating voltages V_L and V_H (needed to transmit ions receiving, respectively, small 'pushes' against and along the line of flight) are found to vary slightly but systematically with the value of the electrostatic-analyser voltage E. Thus in the mass spectrum of 2-fluoropropene, a flat-topped metastable peak in the region of $m/e = 25{\cdot}8$ arises from the decomposition

$$C_3H_4F^+ \longrightarrow C_3H_3^+ + HF$$

and the kinetic energy release, calculated *via* the procedure outlined in section 3-1A, is $0{\cdot}57 \pm 0{\cdot}04$ eV. However, the V/V_0-ratios required to focus $C_3H_3^+$ ions arising from $C_3H_4F^+$ ions in the region between the source and electrostatic analyser vary with E and give values of kinetic energy T apparently released[10] as shown in Table 3-2.

Table 3-2

Data for the reaction $C_3H_4F^+ \rightarrow C_3H_3^+ + HF$ *occurring in the first field-free region*

E (V)	V_L/V_0	V_H/V_0	T_L (eV)	T_H (eV)	Average T (eV)
67·5	1·471	1·542	0·53	0·53	0·53
135	1·484	1·537	0·48	0·43	0·455
202·5	1·492	1·530	0·35	0·35	0·35
270	1·496	1·527	0·29	0·33	0·31
337·5	1·498	1·523	0·28	0·23	0·255

All the values of T in Table 3-2 are less than the value calculated from the normal metastable peak, and the apparent value of T decreases with increase in E. If T is plotted against E (Fig. 3-6), then the value of T (0·59 eV) obtained by extrapolating to zero electrostatic-analyser voltage is in good agreement with the value obtained from the second field-free region.

The origin of the complication in determining kinetic-energy release from $^1m^*$ transitions appears to be as follows. The direction-focusing properties of the electrostatic analyser partially correct for any divergence in the paths of the daughter ions caused by energy release other than solely along the original line of flight. Therefore, ions having a *small* velocity component

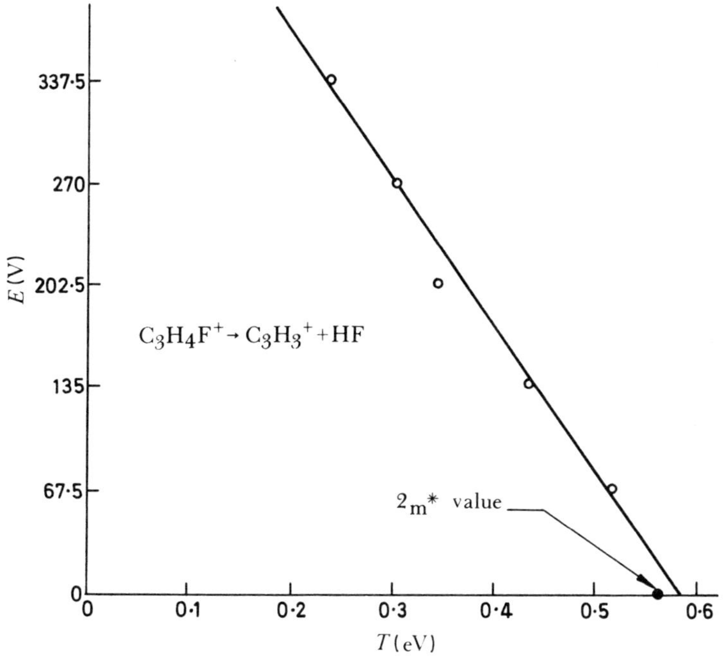

Figure 3-6

perpendicular to the line of flight may be collected almost as efficiently as ions not possessing such components, and more efficiently than for comparable decompositions occurring in ^{2}m* transitions. In ^{1}m* transitions at high voltages, the kinetic energy released decreases in significance relative to the original translational energy of the ion and hence very few ions with components of velocity perpendicular to the line of flight are lost. The result of collecting the latter type of ions is to superimpose the dotted profile

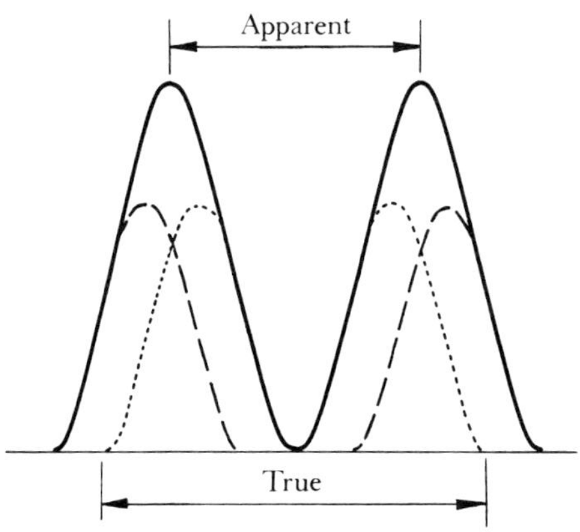

Figure 3-7

(Fig. 3-7) on the dashed profile (which gives the true energy release) and the observed peak (solid profile) thus gives an apparent value of T which is less than the true value.[10]

At low accelerating voltages (low E), the kinetic energy released becomes larger relative to the original translational energy and the direction-focusing properties of the electric sector are no longer sufficient to result in efficient collection of the ions giving rise to the dotted profile (Fig. 3-7). Hence, by extrapolating to $E = 0$, the true value of T is obtained.

3-2. Ion kinetic-energy spectroscopy

We have noted earlier (in section 3-1) that for the reaction $m_1^+ \rightarrow m_2^+$ in the metastable drift regions, the m_2^+ ions possess only a fraction m_2/m_1 of the kinetic energy of the main ion beam. Therefore, due to decompositions in the first drift region, the ions entering the electrostatic analyser will have a spectrum of kinetic energies, varying downwards from the energy of the normal ion beam (neglecting those decompositions of doubly charged ions which produce metastable ions of kinetic energy above that of the ion beam). If now an electron-multiplier detector is placed at the energy-resolving (β) slit at the exit from the electrostatic analyser, and the electrostatic voltage is continuously varied, a spectrum of ion kinetic energies can be obtained.[11] As an example, the ion kinetic energy (IKE) spectrum of naphthalene is shown in Fig. 3-8.[12]

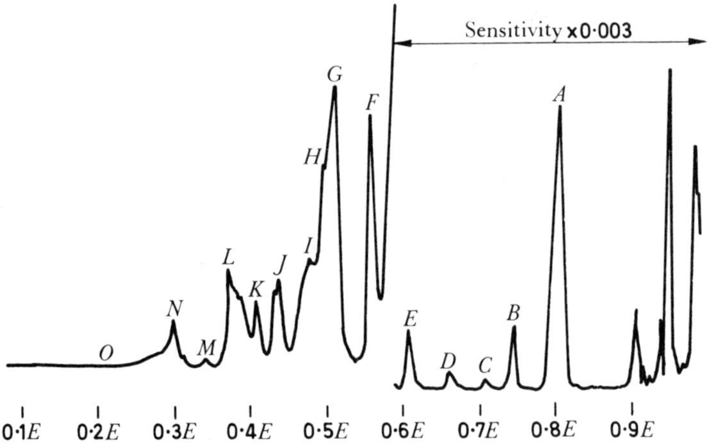

Figure 3-8

Each peak in the IKE spectrum is assignable to a transition $m_1^+ \rightarrow m_2^+$. For example, the sharp peak at 0·56E in Fig. 3-8 (E is the voltage necessary to focus the main ion beam at the β-slit) represents the reaction $m/e = 113 \rightarrow 63$ (i.e., $C_9H_5^+ \rightarrow C_5H_3^+$). Some other assignments of transitions in the IKE spectrum of naphthalene are shown in Table 3-3.

Table 3-3
Mass analysis of some peaks in the IKE spectrum of naphthalene

Peak	Transition
A	$128^{++} \rightarrow 102^{++} + 26$
	$128^{+} \rightarrow 102^{+} + 26$
B	$102^{+} \rightarrow 76^{+} + 26$
	$101^{+} \rightarrow 75^{+} + 26$
F	$113^{+} \rightarrow 63^{+} + 50$
G	$102^{+} \rightarrow 52^{+} + 50$
N	$126^{+} \rightarrow 39^{+} + 87$

Although IKE spectra give detailed 'fingerprints' of organic compounds and appear to be sensitive to small differences in structure, their diagnostic usefulness in structure elucidation appears to be strictly limited.

3-3. The mass-spectrometer time scale

We have seen that molecular ions are those species which survive intact to arrive at the collector, that normal daughter ions are formed in the source, and that metastable ions arise from decompositions between source and collector. It is next important to be aware of the rate constants appropriate to decompositions in the regions so far discussed.

In representative commercial mass spectrometers currently in use for organic analysis, ions typically spend somewhere in the region of 1–5 μs in the source. Unfortunately, this time is difficult to define precisely. Obviously, it will be reduced if the repeller potential is increased. In addition, the field due to the accelerating voltage will penetrate into the source, and hence the average source residence time of ions decreases with increasing accelerating voltage.

If instrument dimensions are known, then since an ion of given mass has a calculable velocity at a given accelerating voltage, the time of flight needed to reach the collector and the second field-free region ($^2m^*$) can also be estimated. Thus in the AEI MS 9 double-focusing mass spectrometer,[13] an ion of m/e 100 takes approximately 15 μs to reach the collector at an accelerating voltage of 8 kV,[14] and around half that time to reach the second field-free region. Since for an ion of mass m with velocity v after acceleration through a voltage difference V, we may write

$$eV = \tfrac{1}{2}mv^2,$$

it follows that $v \propto \sqrt{V}$, and hence the time t needed to reach any point on the flight path after acceleration is such that $t \propto 1/\sqrt{V}$. Thus, at an accelerating voltage of 2 kV, the time needed for an ion m/e 100 to reach the collector would be about 30 μs.

More sophisticated estimates of source residence times and times of flight of ions are however possible. For the loss of HCN occurring from the molecular ion of pyrazine (4) in the $^2m^*$ region, the source residence times t_1, and times t_2 taken at various accelerating voltages, to reach the second field-free region after leaving the source, have been calculated (Table 3-4).[15] The data refer to zero repeller voltage and various plausible assumptions are made about the extent of field penetration into the source etc.

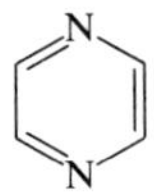
$+ e \longrightarrow C_3H_3N^{+\cdot} + HCN + 2e.$

4

Table 3-4

Source residence times and ion lifetimes for the metastable transition

$C_4H_4N_2^{+\cdot} \rightarrow C_3H_3N^{+\cdot} + HCN$ *in pyrazine*

Accelerating voltage (kV)	t_1 (μs)	t_2 (μs)	Lifetime of dissociating ion ($t_1 + t_2$) (μs)
8	1·4 ± 0·7	8·1 ± 2·2	9·5 ± 2·9
4	1·9 ± 1·1	11·5 ± 3·1	13·4 ± 4·2
2	2·5 ± 1·7	16·1 ± 4·4	18·6 ± 6·1
1	3·4 ± 2·6	22·9 ± 6·2	26·3 ± 8·8

The effect of penetration of the accelerating field into the ion source is clearly to decrease the source residence time. However, the important feature of these data from our point of view is that ion residence times in the source are of the order of 1 μs, and that around 10 μs is the time needed to reach the second field-free region for an ion of mass 100 (singly charged) at 8 kV. The latter time will depend on the mass of the ion and the accelerating voltage. Times needed to reach and traverse the first field-free region will be slightly greater than source residence times (i.e., say 2–5 μs at 8 kV).

We now face the problem of relating these approximate times to rate constants, which is slightly complicated by the fact that the rate constant is defined as being the reciprocal of the *mean* lifetime of the ion. An ion of a given internal energy will decompose with a certain rate constant, but that rate constant will be associated with a spread of lifetimes. Let us assume for convenience that the size of the vibrational quantum is a certain single value in a polyatomic molecule, the internal energy corresponding to 10 such quanta, and that 5 quanta must be gathered in the reaction coordinate to cause decomposition. Decomposition will occur when a fluctuation of 5 quanta into the reaction coordinate occurs; the problem is a statistical one and obviously there will be a most probable time in which this will occur, but in some molecules (of the same internal energy) it will occur in a somewhat shorter or longer time. Thus there is a finite possibility that an ion decomposing with a unimolecular rate constant (k) of 10^5 s^{-1} (i.e., a mean lifetime of 10 μs) will decompose in 1 μs (and therefore give rise to a daughter ion in the source), or even survive long enough to be recorded as an intact particle at the collector.

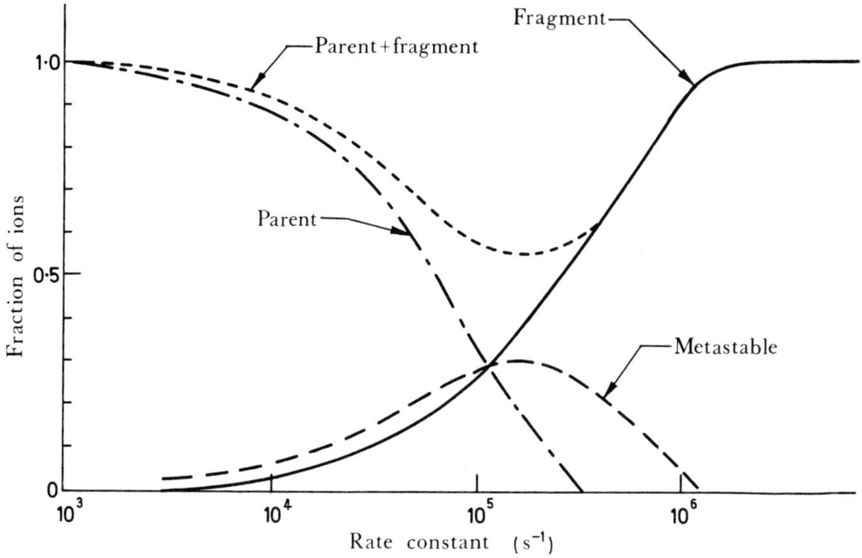

Figure 3-9

The situation may be clarified and put on a semi-quantitative basis by recognizing that parent ions with sufficient energy to decompose do so unimolecularly and that the decay is exponential. If a number of molecular ions are generated at a time $t = 0$ and then undergo unimolecular decay with a rate constant $k\ s^{-1}$, the fractions of ions (I_P, I_F and I_M) detected respectively as parent, fragment and metastable ions ($^2m^*$) will be given by the following equations:

$$I_P = e^{-kt_4}$$

$$I_F = 1 - e^{-kt_1}$$

$$I_M = e^{-kt_2} - e^{-kt_3},$$

where t_4 is the total time needed to reach the exit of the magnetic analyser, t_1 is the source residence time, and t_2 and t_3 are the total times needed to enter and to leave (respectively) the second field-free region.[16, 17] Further decomposition of the fragment ion is not considered. If the fraction of ions recorded as parent, metastable and fragment ions is plotted against the unimolecular rate constant k, the curves shown in Fig. 3-9 are obtained,[16] and may be regarded as representative for most commercial mass spectrometers.

Figure 3-9 shows quite clearly that ions decomposing with a rate constant of $10^5\ s^{-1}$ contribute significantly to the parent and fragment ions as well as to the metastable ions. The rate constants contributing to the metastable transitions lie, to a good approximation, in the range 10^4–10^6 s^{-1}. An ion decomposing with a rate constant greater than 10^6 s^{-1} will give rise virtually exclusively to normal daughter ions, while those ions with rate constants less than 10^4 s^{-1} will almost all be recorded as parent ions. Additionally, it is worthwhile to recall at this stage that all parent ions with insufficient energy to decompose (not considered above) must be recorded exclusively as parent ions.

From our discussion it is evident that if we are going to detect a fragment ion produced by the decomposition of a molecular ion in a conventional mass spectrometer, we cannot do so until the rate constant for its production is $\sim 10^4$ s^{-1} and the reaction is occurring in the second field-free region. Since the lowest reaction rates correspond to the lowest internal energies sufficient to bring about reaction, we come to the important conclusion that it is ions with internal energies least in excess of the activation energy that undergo metastable transitions. It is the ions of higher internal energies that decompose in the source.

In conclusion, it is worth stating at this stage that this discussion of ion lifetimes has been appropriate to a typical electron-impact mass spectrometer. Ions of various lifetimes may be sampled by complementary techniques. For example, ions of very short lifetimes ($\sim 10^{-12}$ s) may be sampled by a field-ionization source (see section 9-3) and 'fast' metastables in such an instrument have lifetimes as low as 10^{-8} s. At the other end of the scale, ion cyclotron resonance spectroscopy (see section 8-2) is suited for ions of very long lifetime (up to 10^{-3} s). The range of ion lifetimes encompassed by these various techniques is illustrated in Fig. 9-6.

References

[1] J. A. Hipple and E. U. Condon, *Phys. Rev.* **69**, 347 (1946).
[2] W. Higgins and K. R. Jennings, *Chem. Comm.* 99 (1965).
[3] K. R. Jennings, *Chem. Comm.* 283 (1966).
[4] J. H. Beynon and A. E. Fontaine, *Zeitschrift für Naturforschg.* **22a**, 334 (1967), and references cited therein.
[5] J. H. Beynon, R. A. Saunders and A. E. Williams, *Zeitschrift für Naturforschg.* **20a**, 180 (1965).
[6] M. Barber and R. M. Elliott, 12th Annual Conference on Mass Spectrometry and Allied Topics, Committee E.14 A.S.T.M., Montreal, June 1964.
[7] K. R. Jennings in *Some Newer Physical Methods in Structural Chemistry*, R. Bonnett and J. G. Davis, eds., United Trade Press, London, 1967, pp. 105–109.
[8] F. W. McLafferty, J. Okamoto, H. Tsuyama, Y. Nakajima, T. Noda and H. W. Major, *Org. Mass Spec.* **2**, 751 (1969).
[9] J. Seibl, *Org. Mass Spec.* **2**, 1033 (1969).
[10] M. Barber, K. R. Jennings and R. Rhodes, *Zeitschrift für Naturforschg.* **22a**, 15 (1967).

[11] J. H. Beynon, J. W. Amy and W. E. Baitinger, *Chem. Comm.* 723 (1969).
[12] J. H. Beynon, R. M. Caprioli, W. E. Baitinger and J. W. Amy, *Org. Mass Spec.* **3**, 455 (1970).
[13] R. D. Craig, B. N. Green and J. D. Waldron, *Chimia (Aarau)* **17**, 33 (1963).
[14] I. Howe and D. H. Williams, *Chem. Comm.* 220 (1968).
[15] J. H. Beynon, J. A. Hopkinson and G. R. Lester, *J. Mass Spectrometry and Ion Physics* **1**, 343 (1968).
[16] W. A. Chupka, *J. Chem. Phys.* **30**, 191 (1959).
[17] B. Steiner, C. F. Giese and M. G. Inghram, *J. Chem. Phys.* **34**, 189 (1961).

4. The quasi-equilibrium theory

4-1. Fundamental assumptions

It is not intended here to deal in depth with the quasi-equilibrium theory as proposed in its original form,[1] nor with more up-to-date aspects of the approach.[2, 3] Rather it is intended to emphasize that some of its fundamental assumptions are extremely helpful in understanding the mass spectra of complex organic molecules.

The most important of these fundamental assumptions from our point of view is that the molecular processes leading to the formation of a mass spectrum consist of a series of competing, consecutive unimolecular decomposition reactions of energetically excited parent ions. It is further assumed that the initial excitation energy randomizes throughout the molecule at a rate which is fast relative to the rate of dissociation. That is, although a certain amount of energy may be produced in a specific bond at the instant of ion production, this energy may flow into other bonds in times comparable to molecular vibrational periods; and decomposition results only when sufficient energy has concentrated in the necessary bond(s).

If we are to calculate the extent of competition between two or more reactions, then, since there is a distribution of energies in the parent ions (section 2-2), we must be able to calculate the relative values of the unimolecular rate constants (k) for any given internal energy (E) sufficient to bring about the various reactions. It is important to realize that a fundamental difference exists between the kinetic situation found in the mass spectrometer and that normally encountered in solution chemistry or most gas-phase reactions. In the mass spectrometer, the operating pressure ($\sim 10^{-6}$ mm Hg) effectively precludes collisions and the internal energies of the ions therefore remain in a non-equilibrium distribution from the instant of ionization. This contrasts with the common kinetic situation, where molecules are continually energized and de-energized by collisions and a Maxwell–Boltzmann distribution of energies results, defined by a temperature. The concept of temperature is therefore quite meaningless for ions in the mass spectrometer.

In the original and simplest form of the quasi-equilibrium theory (QET),[1] the relationship between the rate constant k and the internal energy E was expressed as follows:

$$k = \nu\left(\frac{E - E_0}{E}\right)^{s-1}, \qquad (4\text{-}1)$$

where E_0 is the activation energy for the reaction in question, ν may be regarded as a frequency factor and s is the effective number of oscillators. With a little consideration of the physical significance of ν and s, it is evident that eq. (4-1) is an intuitively reasonable relationship. If the internal energy of the ion is very much larger than the activation energy, then $(E - E_0)/E$ tends to unity and the rate constant approaches the frequency factor. So for a simple single-bond cleavage reaction in the hypothetical case where the internal energy is extremely large, reaction will occur in one vibration and $k \simeq \nu \simeq$ vibrational frequency of the bond in question (10^{13}–10^{14} s^{-1}). For the purposes of our approximations, ν is a constant for any given reaction and for a simple bond cleavage may be approximated by the bond vibrational frequency.

If E is greater than the activation energy E_0 but of the same order of magnitude, then reaction will not occur in one vibration, but only when the requisite number of vibrational quanta (equal to the activation energy) have gathered in the reaction coordinate. The time taken for this will depend on the total number of vibrational quanta available (equivalent to an energy E in the ground electronic state of the ion), the activation energy (E_0), and the molecular size. Obviously, the larger the molecule the fewer vibrational quanta per bond there will be for a given E; fluctuation of E_0 into the reaction coordinate will be less probable and hence k will be smaller. Since $(E - E_0)/E < 1$, eq. (4-1) allows for this situation because a large value of s will reduce k. In the derivation of eq. (4-1), s is the number of oscillators (i.e., vibrational modes) in the molecule and is given by $3n - 6$ for the large molecules we shall be considering, where n is the number of atoms in the molecule. However, the derivation of eq. (4-1) involves numerous approximations and it is found in practice that, in molecules so far investigated, s must be taken as the *effective* number of oscillators, usually one-half or one-third of the actual value, i.e., $(3n-6)/2$ or $(3n-6)/3$. Even taking the effective number of oscillators as one-fifth the total may be most appropriate for energies very near threshold, resulting in a rapid rise of k with E.

In rearrangement reactions, such as the loss of HCN from benzonitrile $\underline{1}$ or of formaldehyde from anisole $\underline{2}$, it is not a sufficient condition for reaction to take place merely that the energy of activation for reaction be collected in the appropriate coordinates. In addition, in the transition state, the hydrogen atom which is to rearrange, with formation of a new bond, must take up a specific orientation with respect to the atom which is to receive it.

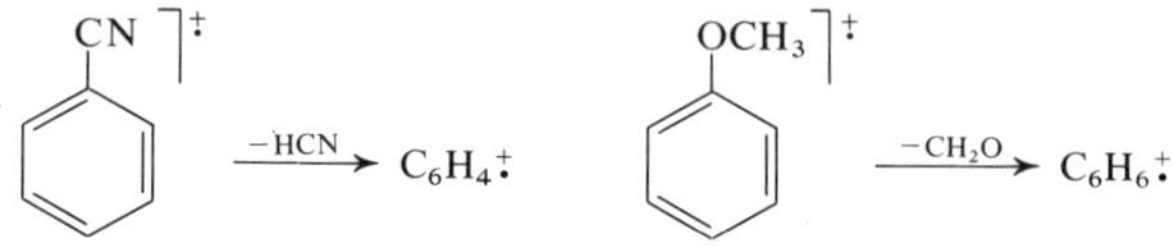

1   m/e 76   2 m/e 78

The relatively low probability of attaining such a specific orientation lowers the reaction rate, and therefore the rates of such rearrangement reactions do not approach bond vibrational frequencies even when the internal energies are extremely large. Rather, ν will be lowered by some characteristic amount and this amount will be larger the more precise the geometrical orientation which is necessary for reaction. The situation is analogous to the negative entropies of activation for the formation of highly order transition states in ordinary thermal kinetics.

Three k vs. E curves are reproduced in Fig. 4-1. The curves apply to a 12-atom molecule, so that $3n - 6 = 30$ and one-third of the total number of oscillators is taken to give the effective number of oscillators s, so that $s = 10$. The parameters used for curve A are $\nu = 10^{13}\,s^{-1}$, $E_0 = 2$ eV; for curve B, $\nu = 10^9 s^{-1}$, $E_0 = 1$ eV; and for curve C, $\nu = 10^9 s^{-1}$, $E_0 = 2$ eV.

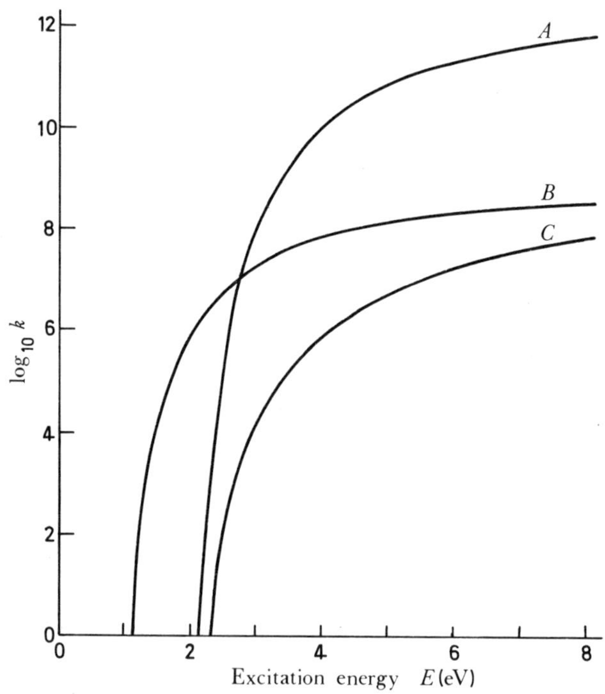

Figure 4-1

Suppose that the reactions corresponding to curves A, B and C are competing unimolecular reactions occurring from a given molecular ion. It is apparent that at any given value of the internal energy E sufficient to cause decomposition on the mass-spectrometer time scale, reaction C is several orders of magnitude slower than reaction A, and also much slower than reaction B except at the highest internal energies. Hence the abundance of the metastable ion ($\log_{10} k \simeq 4$–6) and of the daughter ion ($\log_{10} k \geqslant 6$) resulting from reaction C will be negligible. Indeed, the existence of reaction C is purely hypothetical since it is effectively unable to compete with reactions A and B.

Ions possessing internal energies of about 1·5–2·3 eV in excess of the ionization potential of the molecule undergo reaction B with rate constants in the range $\log_{10} k = 4$–6 and no other reaction at a significant rate (Fig. 4-1). Hence molecular ions of these internal energies give rise to a metastable peak for reaction B. It is interesting to note that the occurrence of metastable transitions indicates that an appreciable fraction of ions with sufficient energy to decompose undergo around 10^8 vibrations of the reaction coordinate before decomposing. The band of energies which causes reaction A to occur with the same range of rate constants ($\log_{10} k = 4$–6) is much narrower (~2·3–2·5 eV, see Fig. 4-1), because this reaction has a larger frequency factor and the rise of k with E is much sharper.[4] On these grounds alone we would anticipate a less abundant metastable peak for reaction A.[5] However, even more important in this context is the fact that at internal energies of 2·3–2·5 eV in excess of the ionization potential, reaction A is on the average going much more slowly than B. Therefore the intensity of the metastable peak due to reaction A will either be very low or negligible.

When internal energies in excess of 2·7 eV are considered then reaction A is always occurring faster than B. Accordingly the daughter ion due to the A process will be abundant compared with that due to the B process at these higher energies.

The foregoing considerations demonstrate quite clearly that when a daughter ion due to one reaction is more abundant than a daughter ion due to a competing second reaction, it cannot be unequivocally stated that the first reaction is proceeding at a faster rate. It may be proceeding more slowly at some internal energies and faster at others. This situation arises because the observed ion abundances arise by integration of the relative k-values for each reaction over the whole of the energy distribution. A useful generalization is that for competing reactions, the relative activation energies are most important in determining relative ion abundances near threshold, i.e., in low-eV spectra and metastable transitions. The most abundant metastable ion will normally occur for the lowest activation energy process. In contrast, in ions of high internal energies, relative reaction rates are largely influenced by the relative values of the frequency factors (Fig. 4-1). Very frequently in mass spectra it is noted that ions arising from simple single-bond cleavages (large ν) increase in relative abundance as the electron beam energy is increased. Conversely, rearrangement reactions are often dominant in low (e.g., 10–12 eV) spectra,[6] since bond formation, concurrent with bond breaking, in the transition state tends to lower activation energies relative to those found in simple cleavage processes. From another viewpoint, we can surmise that reactions with low frequency factors will only compete successfully (i.e., be observed to a significant extent) if they have lower activation energies than competing reactions of high frequency factors.

It is quite common for k vs. E curves for competing unimolecular reactions from a given molecular ion to cross. For example, the following four compounds undergo the reactions shown from their molecular ions to give abundant daughter ions. In each case, one reaction is a simple bond cleavage and the other a rearrangement process.

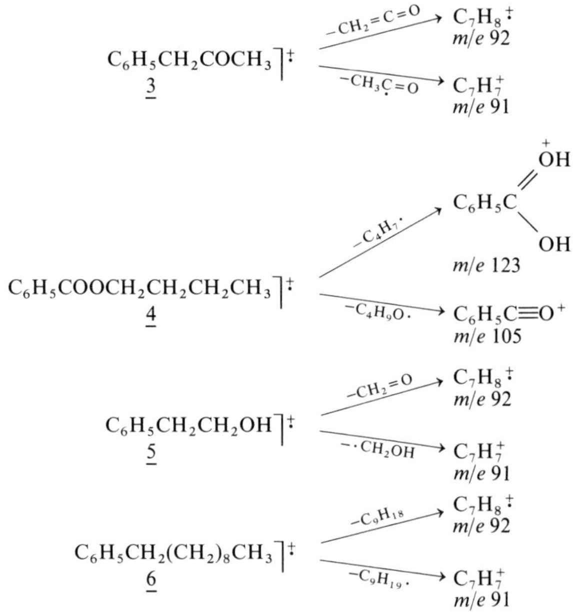

In Table 4-1, the abundance of each of the daughter ions (per cent, relative to the base peak) in 75 eV spectra is given, as is also the ratio [metastable]/[daughter] (per cent) for each reaction.[5]

In 3, 4 and 5, the simple bond cleavages give rise to the more abundant daughter ion, but the metastable-ion abundances for these reactions are very low (as also for simple cleavage in 6). In all cases, it appears that the rearrangement reaction has the lower activation energy and so gives rise to the more abundant metastable ion. The simple cleavage reactions, with their associated higher frequency factors, are weighted more heavily at higher

Table 4-1

Relative metastable and daughter-ion abundances in some simple cleavage and rearrangement reactions

Compound	Daughter ion	[Daughter ion] (per cent)	$\frac{[\text{Metastable}]}{[\text{Daughter ion}]}$ (per cent)
3	*m/e* 92 (rearr.)	25	3·2
	m/e 91	86	<0·01
4	*m/e* 123 (rearr.)	57	0·18
	m/e 105	100	0·003
5	*m/e* 92 (rearr.)	60	0·58
	m/e 91	100	<0·01
6	*m/e* 92 (rearr.)	100	0·45
	m/e 91	82	<0·001

internal energies. Thus we see that metastable-ion abundances are not simply proportional to daughter-ion abundances but, where competing reactions are involved, are predominantly determined by the relative activation energies.

A comparison of relative daughter-ion abundances due to some simple cleavage reactions at low voltage (10–16 eV) and at 70 eV is given in Table 4-2.[6] The lower activation energy processes dominate at low eV and the high frequency factor processes dominate at 70 eV.

Table 4-2

Relative abundances of daughter ions produced by simple cleavage and rearrangement reactions at low voltage (10–16 eV) and at 70 eV

Compound	Processes	Abundance ratio†	
		Low eV	70 eV
n-$C_4H_9COC_6H_5$	$M^+—C_3H_6:C_6H_5CO^+$	10:1·0	1·0:2.0
$CH_3CO(CH_2)_4C_6H_5$	$M^+—H_2O:C_7H_7^+$	2·0:1·0	1·0:13
$CH_3CH_2CH_2CH_2C_6H_5$	$M^+—C_3H_6:M^+—C_3H_7$	13:1·0	1·0:2·0
$CH_3COCH_2COOC_2H_5$	$M^+—C_2H_5OH:M^+—OC_2H_5$	22:1·0	1·0:2·0

† In the quoted abundance ratios, the value for the less abundant ion is arbitrarily taken as 1·0 unit.

The above kinetic properties of rearrangements may be also diagnostically very useful in identifying reactions which may be formulated as direct cleavages but which involve anchimeric assistance by another group. For example, in the spectra of *meta*- and *para*-substituted β-phenylethyl bromides (7), the abundance of $M^+ - Br$ increases relative to that of the $M^+ - CH_2Br$ ion as the electron-beam energy is lowered.[7] These results implicate a lower frequency factor for the elimination of a bromine atom and hence a tighter transition state, probably involving aryl participation (8).

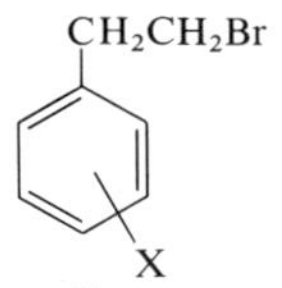

7

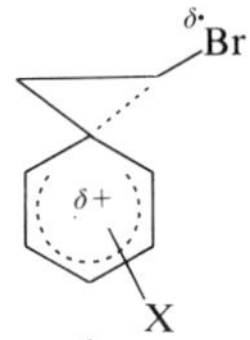

8

The classical equation relating k and E (eq. (4-1)) is thus capable of explaining in qualitative terms spectral features such as those discussed above. This equation (which is usually modified empirically) can be replaced by the more correct Rice–Ramsperger–Kassel–Marcus equation for unimolecular reactions, which employs an exact enumeration of states:[2, 3]

$$k(E) = \frac{1}{h}\frac{Z\dagger}{Z^*}\frac{\sum P\dagger(E-E_a)}{\rho^*(E)}. \tag{4-2}$$

In eq. (4-2), h is Planck's constant, Z is the partition function for the adiabatic degrees of freedom, † refers to the activated complex, * refers to the active molecule, $P(E-E_a)$ is the number of states in the energy range $E-E_a$, and $\rho(E)$ the density of states. Unfortunately, the use of such equations introduces increased mathematical complexity into the problem. However, it is interesting to note that in this equation, the procedure of estimating frequency factors for reaction is, in effect, replaced by estimating the changes in relative numbers of vibrational and rotational states in passing from the active molecule to the activated complex. In rearrangement reactions, rotational states are frozen out in the activated complex so that the required geometry for reaction may be attained, and this has the effect of decreasing the rate of rise of k with E. For example, in the decomposition of p-aminobibenzyl to $H_2NC_7H_6^+$, the reaction appears to have a slower rise of k with E than the corresponding benzylic cleavage in bibenzyl, suggesting a relatively tight activated complex for a single bond cleavage. In the transition state the orbital overlap which is necessary to minimize the energy required for reaction freezes out the N—C and one C—C rotation which are possible in the active molecule.[8] Although this reaction requires eq. (4-2) for an explanation, the example falls within a general scheme which requires a slower rise of k with E for reactions involving a restricted geometry in the transition state.

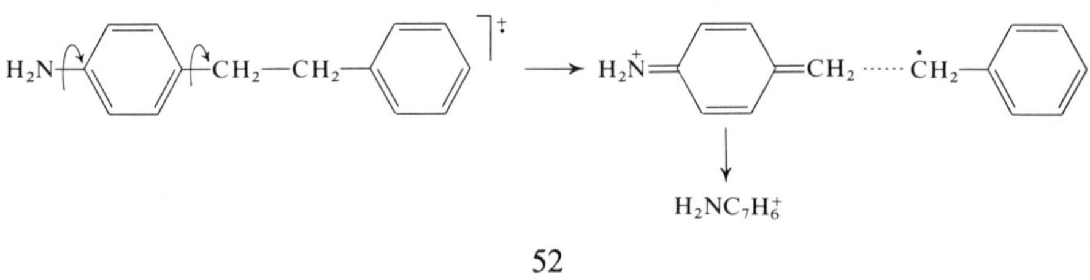

4-2. Factors influencing ion abundances

We have already seen (section 3-3) that for molecular ions decomposing with a rate constant k, the relative fractions of ions collected as parent, fragment and metastable ions are given by the following equations (for the reaction $M^+ \xrightarrow{m^*} A^+$):

$$[M^+] = e^{-kt_4} \tag{4-3}$$

$$[A^+] = 1 - e^{-kt_1} \tag{4-4}$$

$$[m^*] = e^{-kt_2} - e^{-kt_3}, \tag{4-5}$$

where t_4 is the time needed to reach the exit of the magnetic analyser, t_1 is the source residence time and t_2 and t_3 are the times at which the ions enter and leave the field-free region in question. These times are usually known with an accuracy of about 20 per cent. However, since there is a distribution of energies in the mass spectrometer, there is a distribution of rate constants. Therefore the relative fractions of ions with sufficient energy to decompose which are recorded as M^+, A^+ and m* are determined by integrating eqs. (4-3)–(4-5) over the whole energy range from the activation energy to the maximum internal energy, i.e., from E_0 to E_{max}.

$$[M_E^+] = \int_{E_0}^{E_{max}} P(E)e^{-kt_4}\,dE \tag{4-6}$$

$$[A^+] = \int_{E_0}^{E_{max}} P(E)(1 - e^{-kt_1})\,dE \tag{4-7}$$

$$[m^*] = \int_{E_0}^{E_{max}} P(E)(e^{-kt_2} - e^{-kt_3})\,dE, \tag{4-8}$$

where $P(E)\,dE$ is the fraction of ions having energies between E and $E + dE$. In eq. (4-6), $[M_E^+]$ is the fraction of molecular ions with sufficient energy to react given an appropriate time, but which do not do so on the mass spectrometer time-scale (*vide infra*). The fraction of molecular ions which have insufficient energy to decompose is given by

$$[M_0^+] = \int_0^{E_0} P(E)\,dE. \tag{4-9}$$

It is evident from eqs. (4-6)–(4-9) that if energy distributions and $k(E)$-curves are known or can be estimated, then the computation of ion abundances is a simple matter. However, the functions in the integrands are not measurable with sufficient reliability at present, although reasonable approximations may be made for qualitative work. Furthermore, the 'one-decomposition' situation defined by eqs. (4-6)–(4-9) is only attainable in a few selected compounds at low electron beam energies and in the vast majority of cases both competitive and consecutive reactions must be considered, thus complicating the kinetics. However, the complexity of the kinetics need not be a drawback if energy distributions and $k(E)$ functions are known for all participating reactions. Although some progress has been made towards quantitative interpretations of mass spectra of large organic molecules, there are too many unknown, adjustable parameters at present to make any predictive calculations better than semi-empirical. An illustration of semi-empirical calculations of a mass spectrum involving competition reactions may be seen below in section 4-3.

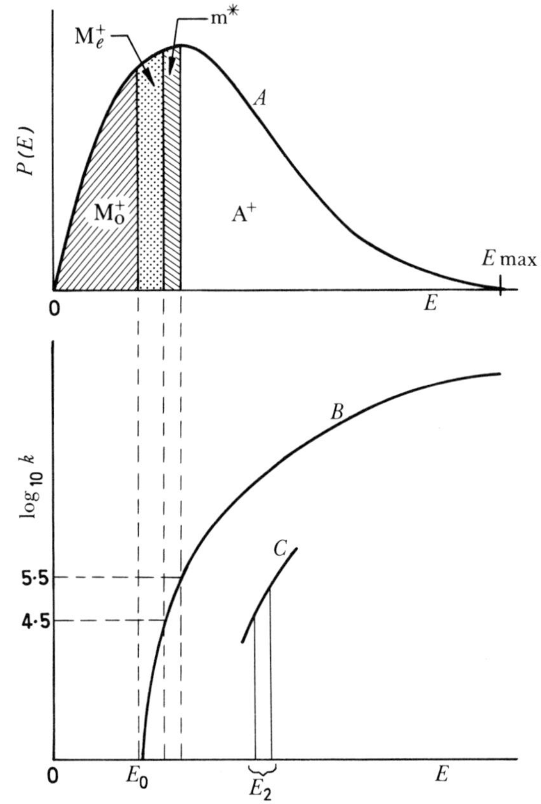

Figure 4-2

At this stage, it is convenient for purposes of exposition to simplify our earlier considerations, and make the approximation (for purposes of illustration) that a specific k-value is sufficient to define an ion as a molecular, metastable or daughter ion. Thus, as an expedient, we temporarily ignore the fact that a specific k does not result in a single lifetime, and make the approximation that all ions having a specific k possess the mean lifetime, i.e., we ignore exponential decay. This allows us to illustrate, to a good approximation, the relative portions of the energy distribution A which give rise to each type of ion (Fig. 4-2) for a single reaction $M^+ \rightarrow A^+$ having the k vs. E curve B. The effect of including exponential decay merely results in some overlapping of the areas of the energy distribution giving rise to M_E^+, m* and A^+, as shown in Fig. 4-3. Incidentally, the simplified treatment in Fig. 4-2 does not account for the so-called 'missing metastables', resulting from decompositions in the analyser tubes.

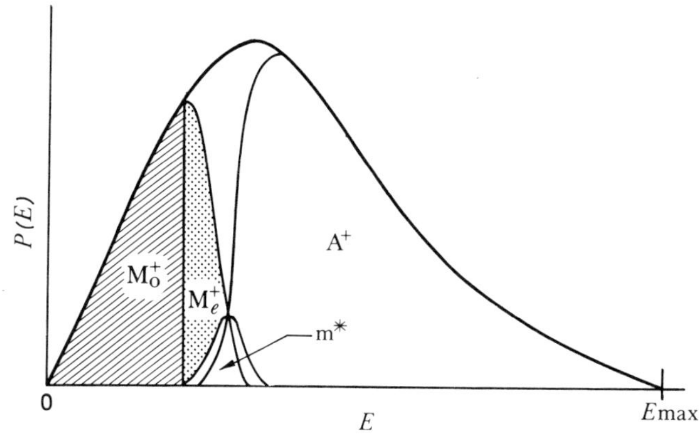

Figure 4-3

The most important factors influencing normal-ion and metastable-ion abundances may now be illustrated with reference to Fig. 4-2. Since the molecular-ion abundance ($M_0^+ + M_E^+$) corresponds largely to the area under the $P(E)$ vs. E curve between the ionization potential and the appearance potential of A^+, the over-riding importance of the difference A.P. − I.P. in determining the molecular ion abundance becomes apparent. Given similar shapes of energy distributions, A.P. − I.P. should be a dominant factor in determining molecular-ion abundances for different organic compounds at similar electron-beam energies. Table 4-3 illustrates this trend in some 20 eV mass spectra. The fact that competitive and consecutive reactions occur from the molecular ion in no way invalidates the illustration and the lowest appearance potential is employed in each case.

Table 4-3
Molecular ion abundances in relation to activation energy

Compound	I.P.	Lowest A.P.	A.P. − I.P.	$[M]/[\Sigma D^+]$
p-$NO_2 \cdot C_6H_4OCH_2C_6H_5$	9·1	10·0	0·9	0·03
$C_6H_5OCH_2C_6H_5$	8·4	9·7	1·3	0·28
p-$NH_2 \cdot C_6H_4OCH_2C_6H_5$	7·6	9·7	2·1	0·33
$CH_3(CH_2)_3CO_2Me$	10·4	10·9	0·5	0·007
p-$MeO \cdot C_6H_4 \cdot (CH_2)_3CO_2Me$	8·0	10·6	2·6	0·17

Since the initial rise of the rate constant with internal energy is rapid, the segments of the energy distribution which yield M_E^+ and m* are relatively narrow. The area of these segments will be increased by factors which decrease the slope of the curve *B* (Fig. 4-2) between $\log_{10} k = 3$ and 6. That is, an increase in E_0 and molecular size and a decrease in ν will in general tend to increase the relative abundances of M_E^+ and m*.

It is a property of mass-spectral processes that the most abundant metastable peak for a decomposition from a given ion will correspond to the lowest activation-energy reaction from that ion (see section 4-1). It is obvious that a reaction having a *k* vs. *E* curve *C* (Fig. 4-2) would not give a metastable ion, since the energy segment E_2 giving rise to decomposition at the appropriate rate would cause reaction *B* to occur several orders of magnitude faster (i.e., in the source) and ions of energy E_2 would have decomposed long before reaching the metastable drift region.

It was pointed out in section 2-2 that the energy distributions of molecular ions are not smoothed but have areas of low and high probability; and it is clear that the abundance of metastable ions depends on the probability distribution in the 'metastable-energy window'—see Fig. 4-2. The importance of the internal energy probability in the metastable region may be illustrated with reference to the energy distributions of some substituted bibenzyls.[8] An energy distribution for bibenzyl itself at various source temperatures may be derived from photo-electron spectra (see Figs. 2-9 and 2-10). The thermal-energy distributions of the neutral bibenzyl at different temperatures were convoluted with the derived low-temperature energy distribution to achieve the overall distribution. At 60°C the molecular ion energies (ca. 11–I.P. eV) giving rise to the metastable transition $M^+ \rightarrow$ *m/e* 91 correspond to a 'valley' in the distribution curve. When the temperature increases (which has the effect of smoothing out the distribution and

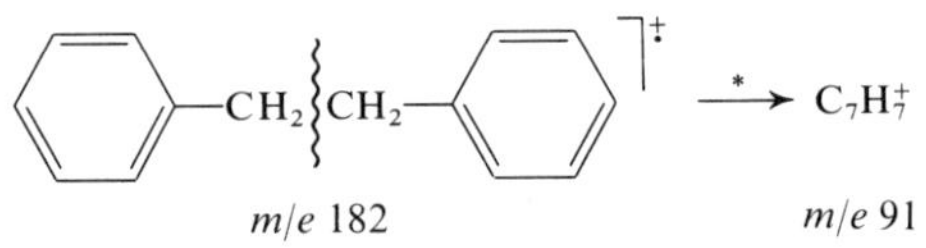

shifting it to higher energies) the metastable ion intensity increases dramatically as the valley is 'filled in'—see Table 4-4.

Table 4-4

Effect of temperature on metastable-ion abundance for the reaction $M^+ \rightarrow C_7H_7^+$ *of bibenzyl*

Temperature	60°	200°	280°
$[m^*]/[M^+]$	7×10^{-4}	4×10^{-3}	$8{\cdot}9 \times 10^{-3}$

The situation regarding the higher-energy molecular ions (forming A^+) is obviously not as simple in the majority of cases as depicted in Fig. 4-2. In the first place, competition may occur with other reactions from M^+ and the relative abundances of the daughter ions would depend on the relative k vs. E curves for the various processes (see section 4-3). In addition, molecular ions above a certain energy may yield primary daughter ions which have the energy to decompose further (e.g., $M^+ \rightarrow A^+ \rightarrow C^+$). As the electron beam energy is increased, the energy distribution becomes more weighted to higher energies and there is a corresponding increase in the abundance of secondary daughter ions. The overall ion abundances of A^+ and C^+ (and the corresponding metastable ions) are determined not only by the $P(E)$ and $k(E)$ curves for the molecular ion, but also by the same parameters for A^+. It is noteworthy that the energy distribution of A^+ formed in the source is influenced by the energy imparted to the neutral fragment lost in its formation.

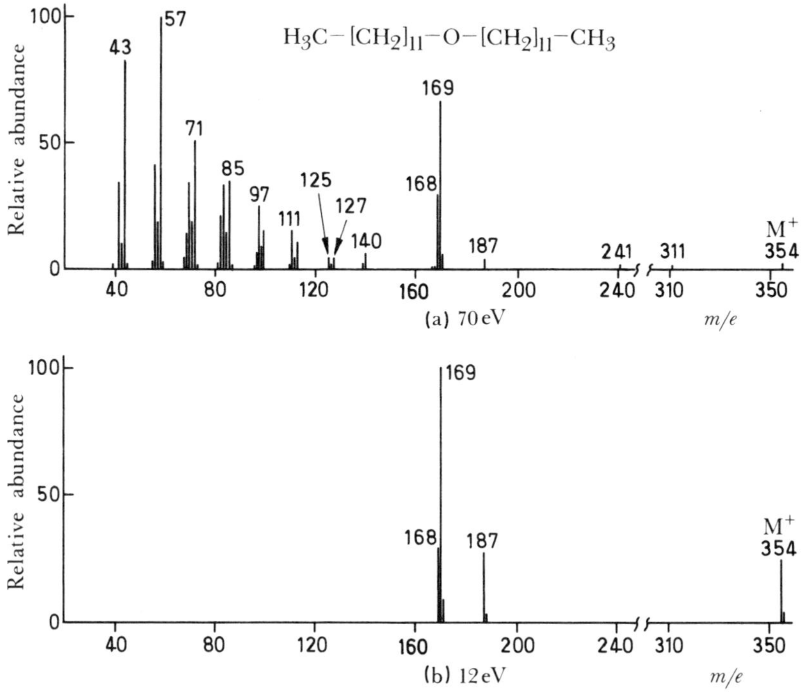

Figure 4-4

The differences between high- and low-voltage spectra may also be understood in terms of Fig. 4-2. As the beam energy is lowered, the energy distribution of the molecular ion becomes weighted to lower energies and E_{max} is reduced. The consequence is that ions formed from reactions requiring more energy (whether produced in one or more steps from the molecular ion) diminish in intensity relative to the intensity of the molecular ion, or disappear altogether. Low eV spectra are consequently much simpler than 70 eV spectra. This is illustrated in Fig. 4-4, where the 70 eV spectrum of di-*n*-dodecylether is compared with that at 12 eV.[9] In the 12 eV spectrum, all ions but those of low appearance potential (generated by the reactions shown below) have disappeared.

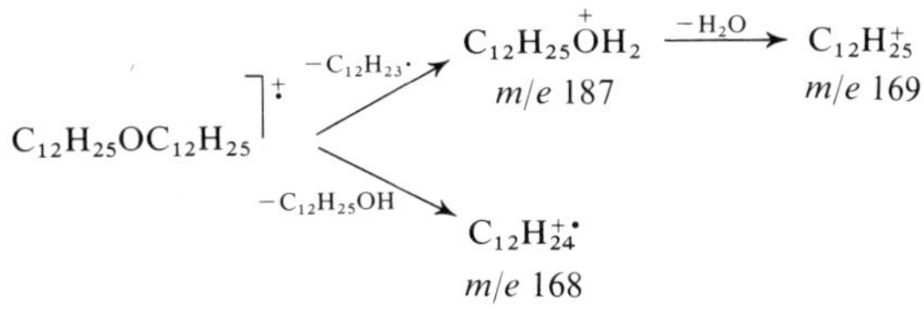

4-3. Competitive reactions

The preceding arguments have illustrated, in a qualitative fashion, the application of the QET to the estimation of ion abundances in mass spectra. It was shown in section 4-1 that the QET is particularly useful in explaining the variation with electron energy in ion yields from competitive reactions; and reasonable quantitative calculations of spectra may be carried out in these circumstances.

Consider the mass spectrum of methyl salicylate (9). The increase, with increasing electron energy in the relative abundance of $M^+ - OMe$ (direct cleavage) compared with that of $M^+ - MeOH$ (rearrangement) would be explained on the basis of a lower activation energy and frequency factor for the rearrangement reaction.

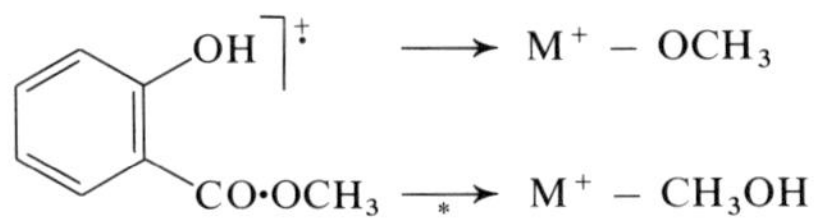

9

It is possible to put this argument on a quantitative basis by assuming realistic values for the $P(E)$ and $k(E)$ functions (see section 4-2), and computing ion abundances using these functions.[10] The energy distribution employed is shown in Fig. 4-5 and a k vs. E function is adopted of the form:

$$k = \nu\left(\frac{E - E_0}{E}\right)^{(s-1)x}, \qquad (4\text{-}10)$$

where the parameter x is varied from 0·2 at threshold to 0·5 at the maximum internal energy of the molecular ion. For each of the competitive reactions, the activation energy E_0 is approximated by the difference between the ionization potential and the appearance potential appropriate to the particular reaction.

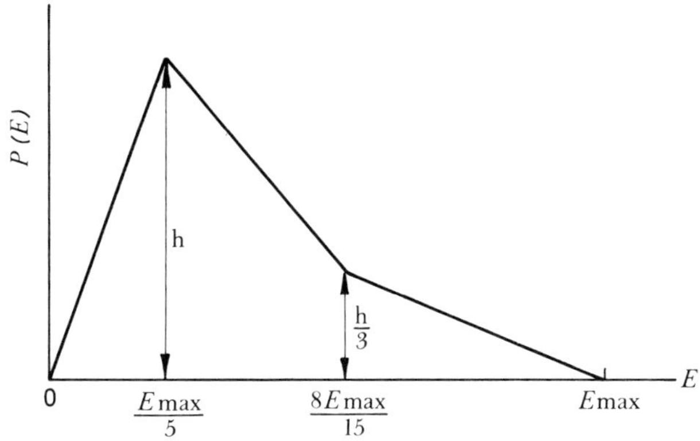

Figure 4-5

Since further breakdown of the $M^+ - OCH_3$ and $M^+ - CH_3OH$ ions is neglected, the only remaining critical requirements for computation of the ion abundances of M^+, $M^+ - OCH_3$ and $M^+ - CH_3OH$ are the frequency factors for the two reactions. If a value of $4 \times 10^{13}\ s^{-1}$ is adopted for the

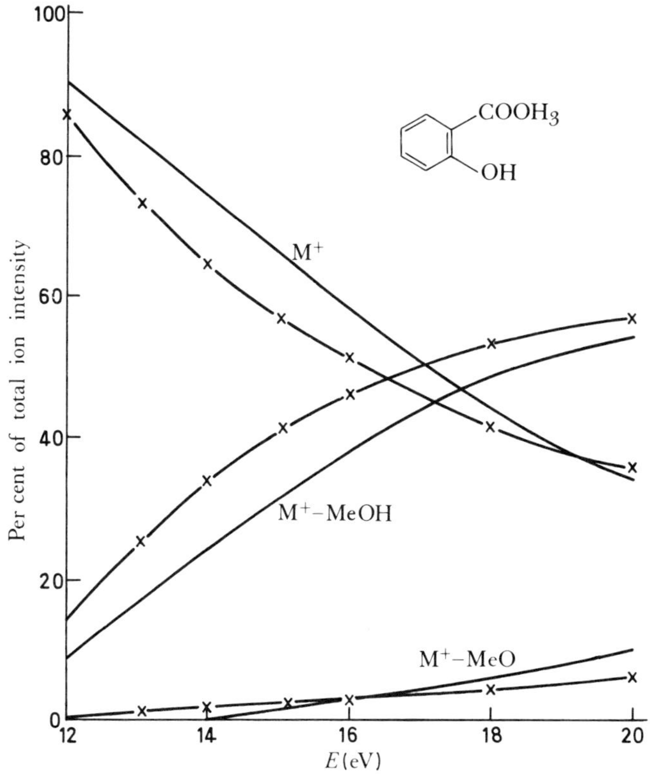

Figure 4-6

directed cleavage reaction and a value of $6 \times 10^{11}\ s^{-1}$ for the rearrangement, a good agreement is found between calculated (continuous) and observed (crossed) spectra for electron beam energies between 12 and 20 eV (see Fig. 4-6). Although certain arbitrary assumptions are necessary in the calculations, the treatment *is* sensitive to the frequency factors chosen for these competitive reactions.

It is evident from much of the discussion in this chapter that relative appearance potentials are an important factor in determining which fragmentations will predominate in bifunctional or polyfunctional compounds. Provided that the substituents do not interact directly through space in the decomposition reactions, or strongly through bonds (e.g., resonance interaction), the predominant fragment will be predictable from the appearance potential of the *characteristic fragment ion* of each substituent. Thus for example, in order to provide a reference series for the *relative ease of fragmentation* of different groups in disubstituted benzenes, the appearance potentials have been measured[11] for the characteristic fragmentations of X from monosubstituted benzenes $C_6H_5X^{+\cdot}$. The order of appearance potentials is shown in Table 4-5.

Table 4-5

Appearance potentials for fragmentation of some C_6H_5X *compounds*

X	Neutral fragment	A.P. (eV)
$COCH_3$	CH_3	9·99
$C(CH_3)_3$	CH_3	10·26
$CH(CH_3)_2$	CH_3	10·65
CO_2CH_3	OCH_3	10·80
$N(CH_3)_2$	H	10·80
CHO	H	10·99
C_2H_5	CH_3	11·25
OCH_3	CH_2O	11·30
I	I	11·46
OH	CO	11·67
CH_3	H	11·80
Br	Br	12·02
NO_2	NO_2	12·16
NH_2	HCN	12·50
Cl	Cl	13·20
CN	HCN	14·60
F	C_2H_2	14·73

Since the reactions of lowest appearance potential in the series of monosubstituted benzenes (Table 4-5) also have high frequency factors (e.g., CH_3 loss from $C_6H_5COCH_3$, CH_3 loss from $C_6H_5C(CH_3)_3$, OCH_3 loss from $C_6H_5COOCH_3$), these reactions will *always* go faster than, say, CH_2O loss from $C_6H_5OCH_3$, Cl loss from C_6H_5Cl, CO loss from C_6H_5OH or HCN loss from C_6H_5CN, when two such very different groups are put into the same benzene ring. It is an experimental fact that the above generalization holds when the substituents are chosen with very different activation energies (see, for example, the 18 eV spectrum of *p*-cyano-*t*-butylbenzene in Fig. 4-7). Even the difference of 0·48 eV in A.P.s for fragmentation of Br and NH_2 substituents appears still to be reflected in the 18 eV spectrum of *p*-bromoaniline (Fig. 4-8), but obviously when substituents fall so near to each other in Table 4-5 modification of the relative activation energies must be considered because competition occurs between loss of Cl and HCN in the 18 eV spectrum of *p*-chloroaniline (Fig. 4-9).

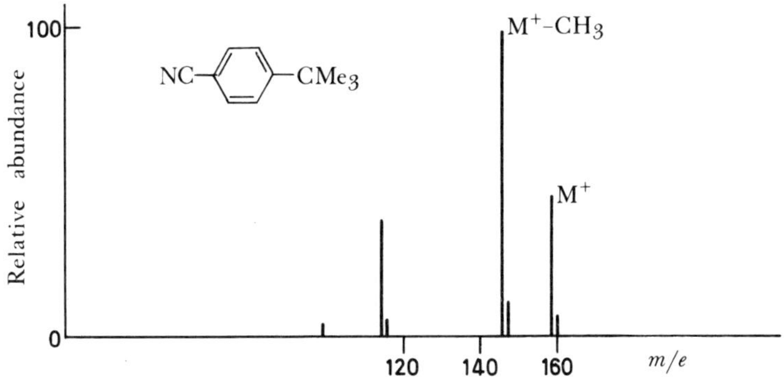

Figure 4-7

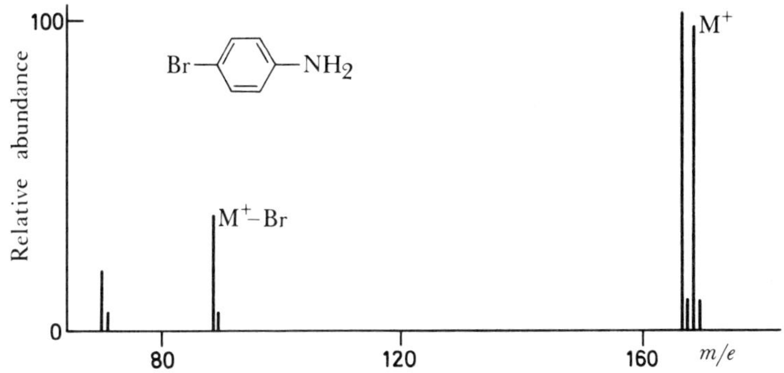

Figure 4-8

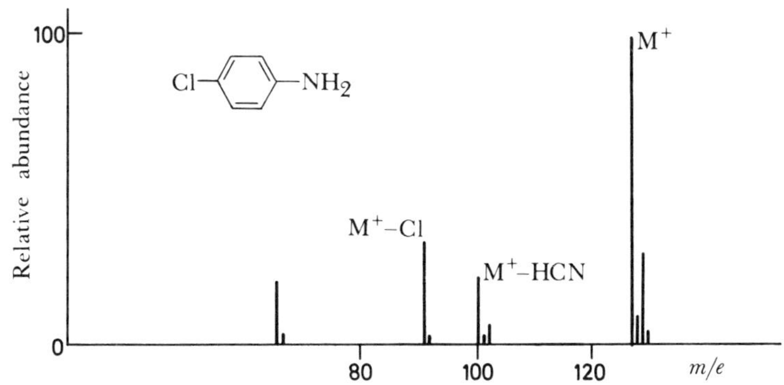

Figure 4-9

Such results show that (a) the appearance potentials for decomposition in the monosubstituted benzenes are not *drastically* altered in *meta*- and *para*-disubstituted benzenes; they will however be slightly modified by interaction and therefore the mass spectrum of a disubstituted benzene cannot confidently be predicted from the relative order given in Table 4-5 if the two substituents fall near together (e.g., CH_3 and Br); (b) the fragmentations involving the two substituents are truly competing, at least in most cases, at internal energies at which fragmentation of both substituents is, in principle, possible. Thus the spectra reproduced in Figs. 4-7 to 4-9 support the basic tenets of the quasi-equilibrium theory.

No predictions can be made concerning the mass spectra of *ortho*-disubstituted benzenes since activation energies may be drastically changed by direct interaction of the two substituents. For *para*- and *meta*-disubstituted benzenes, and other aromatic disubstituted systems where the two substituents cannot interact directly through space, Table 4-5 constitutes a useful 'fragmentation probability' table.

The effect of substituent groups in an aromatic system on the relative abundance of daughter and molecular ions constitutes an important section of organic mass spectrometry; and the QET again provides an explanation for the 'substituent effects' observed. Much of the work on substituent effects has been reviewed[12] and the care necessary in the interpretation of results emphasized.[13, 14]

It has been empirically demonstrated in certain cases that if a common fragment ion A^+ is formed from a precursor $M_Y^{+\cdot}$ containing a *meta* or *para* substituent Y and from a precursor $M_H^{+\cdot}$ lacking the substituent, then

$$\log\left(\frac{[A^+]_Y}{[M_Y^{+\cdot}]}\bigg/\frac{[A^+]_H}{[M_H^{+\cdot}]}\right) \propto \sigma_Y,$$

where $[A^+]_Y$ and $[A^+]_H$ are the recorded ion abundances of the common ion A^+ in the spectra of M_Y and M_H, respectively, $[M_Y^{+\cdot}]$ and $[M_H^{+\cdot}]$ are the corresponding molecular ion abundances, and σ_Y is the Hammett σ-constant for the substituent. An instance where a very good correlation is found is in the formation of benzoyl ions a from substituted benzophenones.

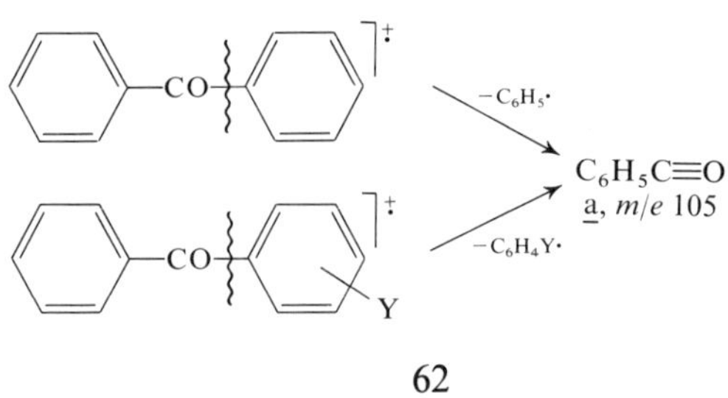

If we define Z as $[C_6H_5CO^+]/[C_6H_5COC_6H_4Y^{+\cdot}]$ and Z_0 as $[C_6H_5CO^+]/[C_6H_5COC_6H_5{}^{+\cdot}]$, then the plot of log (Z/Z_0) against σ_Y appears as shown in Fig. 4-10.[15]

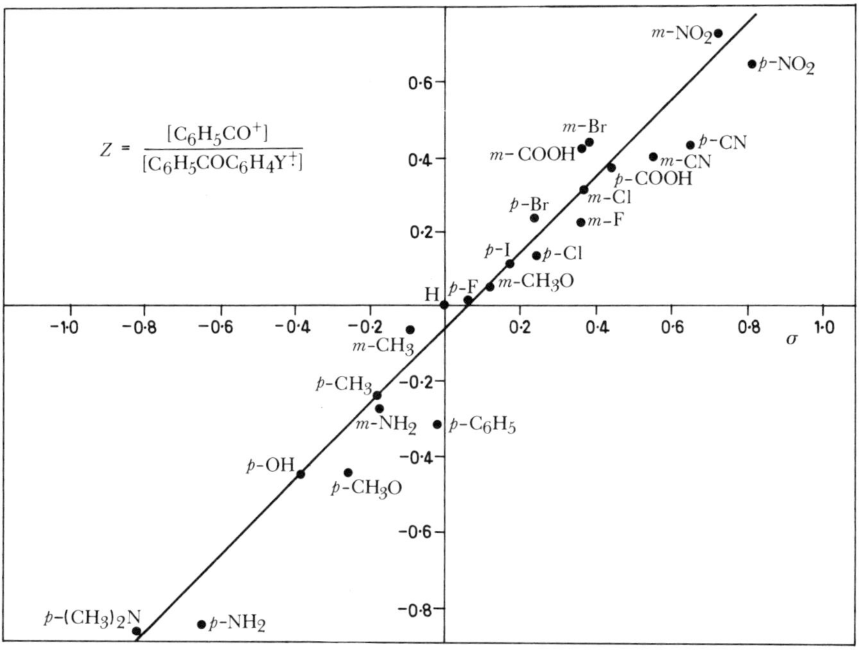

Figure 4-10

To understand the significance of plots such as this it is necessary to recall some of the factors influencing ion abundances. We have already seen (section 4-2) that A.P. − I.P. (where A.P. is the lowest appearance potential for reactions from M^+) is an important factor in influencing the observed abundance of M^+. In general the result of introducing an electron-donating substituent into an aromatic ring is to lower the I.P. (and vice versa for an electron-withdrawing substituent). There are many examples in the literature to verify this and the I.P. values frequently correlate with σ^+ (see Fig. 2-5). However, the effect of substituents on A.P. is frequently much smaller than that on I.P.; and the overall effect in the majority of cases is for electron-donating substituents to increase the difference (A.P. − I.P.) between the I.P. and the lowest A.P. Hence it is a common situation that an electron-donating substituent will increase the fraction of ions which do not have sufficient energy to undergo any dissociation process. For example, a correlation exists[16] between the fraction of total ion current carried by the molecular ion of 10 and σ for the substituent X (see Fig. 4-11).

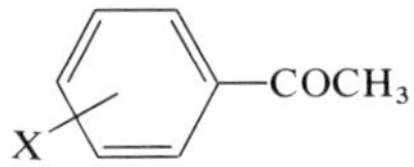

10

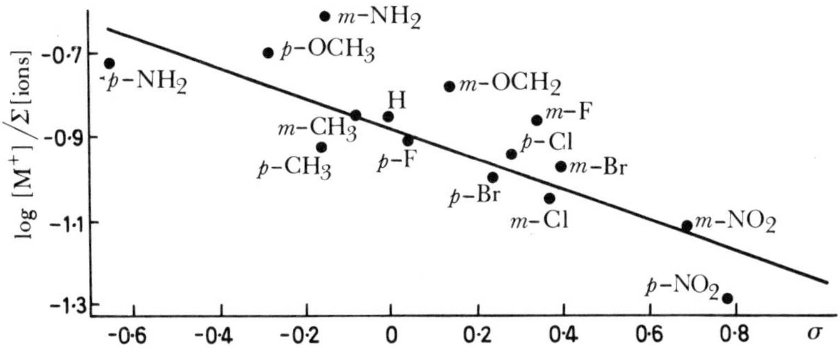

Figure 4-11

Effects such as this will make a contribution to a change in $[A^+]/[M^+]$ (for the reaction $M^+ \rightarrow A^+$). In the few cases where the lowest A.P. is markedly *decreased* by an electron-donating substituent and if A.P. − I.P. consequently decreases, then a contribution to an *increase* in $[A^+]/[M^+]$ will result.

Competitive reactions also constitute an important influencing factor on $[A^+]/[M^+]$ ratios. A change in substituent may implement a change in the appearance potential and rate for a competitive fragmentation from M^+, other than the one under study; or new fragmentations may be created by decomposition of the variable substituent.

The importance of competition reactions may be understood by reference to Table 4-5. If a series of disubstituted benzenes is investigated in which the fixed substituent has a characteristically low A.P. fragment (e.g., $COCH_3$, CO_2CH_3) then the variable substituent will not in general introduce a competing reaction to the one under study. Hence the variable substituent will exert its effect purely via changes in I.P. and the A.P. of the fragment under investigation; and under these circumstances a correlation of $[A^+]/[M^+]$ with the substituent constant may be anticipated. It is noteworthy that, as expected, the majority of such correlations exhibit positive ρ-values.

However, when the variable substituents introduce reactions which compete with, or even effectively swamp the reaction of the fixed substituent (i.e., for cases where the fixed substituent occurs lower down Table 4-5), irregular substituent effects would be anticipated. For example, regular substituent effects are not found for the loss of CH_2O from anisole and substituted anisoles $YC_6H_4OCH_3$. The loss of formaldehyde is drastically reduced by most Y substituents,[17] but it should not be concluded that these substituents have a retarding effect on the reaction by increasing its activation energy. Rather CH_2O loss is unable to compete with the fragmentation of Y when the Y group appears above CH_3O in Table 4-5.

The correlation in Fig. 4-10 should in effect be influenced by competition (the formation of $YC_6H_4C{\equiv}O^+$), but this is an instance in which it is reasonable to suppose that the competition reaction should influence $C_6H_5C{\equiv}O^+$ formation in accord with the substituent constant.

A simple kinetic treatment has been devised[16] which formulates in a quantitative fashion the effect of substituents on ion abundances.

Consider the following general fragmentation scheme:

$$M^+ \begin{cases} \xrightarrow{k_1} A^+ \longrightarrow \text{products} \\ \xrightarrow{k_f} \text{other products} \end{cases} \tag{4-11}$$

To a good approximation the molecular-ion abundance $[M_0^+]$ recorded at the collector represents the fraction, f, of total molecular ions M^+ produced at initial ionization which have insufficient energy to decompose (see Fig. 4-2). Hence we may write

$$[M_0^+] = f[M^+]. \tag{4-12}$$

If it is assumed that decomposition of M^+ occurs via competing unimolecular reactions, the total abundance, $[A^+]$, of A^+ ions initially formed from M^+ is given by

$$[A^+] = \frac{k_1}{k_t}([M^+] - [M_0^+]), \tag{4-13}$$

where k_1 is the rate constant for formation of A^+ and k_t is the sum of all rate constants for decomposition from the molecular ion ($k_t = k_1 + k_f$). These rate constants are averaged over the appropriate energy distribution and depend on the k vs. E curves for that energy region.

A certain fraction, f', of A^+ ions will not possess sufficient energy to decompose further and will be nearly equal to the total abundance $[A_0^+]$ of ions observed as A^+ at the collector:

$$[A_0^+] = f'[A^+]. \tag{4-14}$$

Combination of eqs. (4-12)–(4-14) gives

$$\frac{[A_0^+]}{[M_0^+]} = f'\frac{k_1}{k_t}\left(\frac{1}{f} - 1\right). \tag{4-15}$$

Equation (4-15) neatly illustrates the factors influencing $[A^+]/[M^+]$. A substituent effect may be exerted by (a) a change in the fraction f of M^+ ions with insufficient energy to decompose, (b) a change in the fraction f' of A^+ ions with insufficient energy to decompose, or (c) the fraction k_1/k_t of fragmenting molecular ions that initially form A^+. Point (c) illustrates that $[A^+]/[M^+]$ is dependent on competing unimolecular decompositions from M^+.

Hence the application of substituent effects is restricted, and in any event, care is always necessary in their interpretation.

4-4. Competing metastable transitions

We have already seen that since the initial rise of the rate constant k with internal energy E is rapid, the segment of the energy distribution which leads to metastable transitions ($\log_{10} k \simeq 5$) is very narrow. For reactions with low activation energies and high frequency factors it will be of the order of 0·1 eV, and for reactions with relatively high activation energies and low frequency factors it will be of the order of 1 eV. There will also be a tendency for the width of this energy segment to be increased as molecular size increases. For reactions to compete effectively near the threshold, and so give metastable ions of similar abundances, the energies of activation for the competing reactions must be very similar (the same within, say, 20 per cent may serve as a useful guide, but the precise limits will vary from case to case). Thus, the requirements for competing metastable transitions are very stringent, and if an ion of given composition generated from different precursors decomposes by two reactions which give metastable transitions in the same ratio, this is an excellent indication that the ions from the two sources have the same structure, or consist of the same mixture of structures (in the energy range appropriate to metastable transitions).

For example, hexyl ions ($C_6H_{13}^+$) have been prepared by electron impact ionization and decomposition of a variety of normal alkanes, n-hexyl bromide, and di-n-hexyl ether.[18] These ions undergo the following metastable transitions

$$C_6H_{13}^+ \xrightarrow[38\cdot2]{m^*} C_4H_9^+ + C_2H_4$$

$$C_6H_{13}^+ \xrightarrow[21\cdot8]{m^*} C_3H_7^+ + C_3H_6.$$

Appearance potential data indicate that the onsets of the two transitions are within 0·2–0·3 eV of each other, and this suggests that the reactions are in competition. In 30 eV spectra the ratio of competing metastable transitions is constant within experimental error for all the compounds investigated, indicating that ions of the same structure(s) are decomposing in each case. This constant ratio occurs despite the probability that the internal energy distributions of the $C_6H_{13}^+$ ions from different sources will initially be somewhat different. The occurrence of the competing reactions in the same ratio further suggests that any initial non-equilibrium distribution of energy has disappeared, i.e., the rate of randomization of internal energy is fast on the mass-spectrometer time scale, in support of the basic assumption of the QET. The same ratio of competing metastable transitions is observed even if the $C_6H_{13}^+$ ions are produced from secondary or tertiary bromides ($C_6H_{13}Br$) or iodides ($C_6H_{13}I$), indicating that the same decomposing ions are produced from these branched chain structures.[19]

The criterion of the ratio of competing metastable transitions has also been used in a classification of the types of decomposing $C_2H_5O^+$ ions generated from compounds CH_3OCH_2Y, $HOCH_2CH_2Y$, $CH_3CH(OH)Y$ and CH_3CH_2OY, where Y is a variable group.[20] The relevant competing reactions are

$$C_2H_5O^+ \xrightarrow[8{\cdot}02]{m^*} H_3O^+ + C_2H_2 \tag{4-16}$$

$$C_2H_5O^+ \xrightarrow[18{\cdot}7]{m^*} CHO^+ + CH_4, \tag{4-17}$$

and the data are summarized in Table 4-6.

Table 4-6

Competing metastable transitions and kinetic-energy release in the decomposition of $C_2H_5O^+$ *ions*

Structure	[*m/e* 8·02] / [*m/e* 18·7]	Kinetic energy (eV) of *m/e* 18·7
CH_3OCH_2Y	<0·01	<0·1
$HOCH_2CH_2Y$	1·8 ± 0·2	0·50 ± 0·04
$CH_3CH(OH)Y$	1·9 ± 0·1	0·44 ± 0·02
CH_3CH_2OY	2·0 ± 0·1	0·52 ± 0·05

$C_2H_5O^+$ ions formed from compounds CH_3OCH_2Y do not undergo the loss of acetylene as a metastable transition to any significant extent, and these ions therefore have a unique structure (probably b).

$$CH_3\overset{+\bullet}{O}CH_2{-}Y \xrightarrow{-Y\cdot} \underset{\underline{b},\ m/e\ 45}{CH_3\overset{+}{O}{=}CH_2} \xrightarrow[*]{-CH_4} \underset{m/e\ 29}{\overset{+}{O}{\equiv}CH}$$

For the remaining compounds (Table 4-6), the ratios of ion abundances at *m/e* 8·02 and 18·7 are approximately the same and hence it is concluded that the same $C_2H_5O^+$ ion (or mixture of $C_2H_5O^+$ ions) is decomposing in each case. In terms of a single structure, there is evidence to support the intermediacy of the protonated oxirane c, especially for $C_2H_5O^+$ ions of long lifetimes.

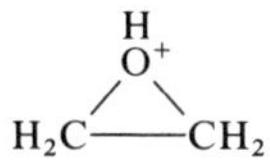

c

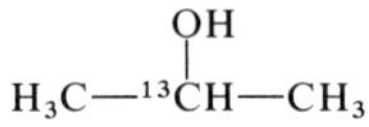

11

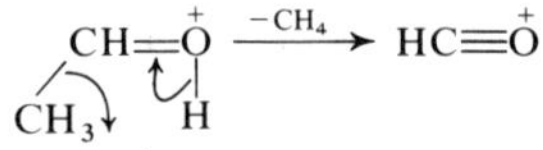

d

Another distinguishing feature is that the reaction leading to CHO^+ from b is different from that leading to CHO^+ from the other $C_2H_5O^+$ ion, since the shape of the metastable peak is different. The ion b decomposes by methane loss without a significant release of kinetic energy (as measured by the width of the metastable peak, section 3-1) whereas the alternative $C_2H_5O^+$ ion releases approximately 0·5 eV of kinetic energy in losing methane (Table 4-6). Since the kinetic energy released in metastable decompositions is dependent on the potential surface over which the reaction occurs (section 3-1), the two reactions must be occurring over different potential surfaces.

When $C_2H_5O^+$ ions are prepared by fragmentation of 2-alkanols $CH_3CH(OH)Y$, the four hydrogens attached to C-1 and C-2 become equivalent prior to further decomposition of $C_2H_5O^+$ to H_3O^+ (as shown by deuterium labelling). The symmetrical structure c accommodates this observation, and also the fact that the hydroxyl hydrogen does not lose its identity to any significant extent.[21] Furthermore, the mass spectrum of 2-propanol-2-^{13}C (11) has been examined,[22] and establishes 74 per cent ^{13}C retention in the CHO^+ ion derived from $^{13}CCH_5O^+$ at 70 eV, 68 per cent ^{13}C retention in the range 12–20 eV, and approximately 50 per cent ^{13}C retention in ions undergoing metastable transitions. One explanation is that methane loss occurs from structure d but increasing numbers of ions at low energies (metastable decompositions) have time to undergo carbon and hydrogen scrambling via structure c (see section 4-5).

In contrast to the ratio of competing metastable transitions, the ratios $[A^+]/[X^+]$, $[A^+]/[m^*]$ and $[X^+]/[m^*]$, where $[A^+]$, $[m^*]$ and $[X^+]$ are

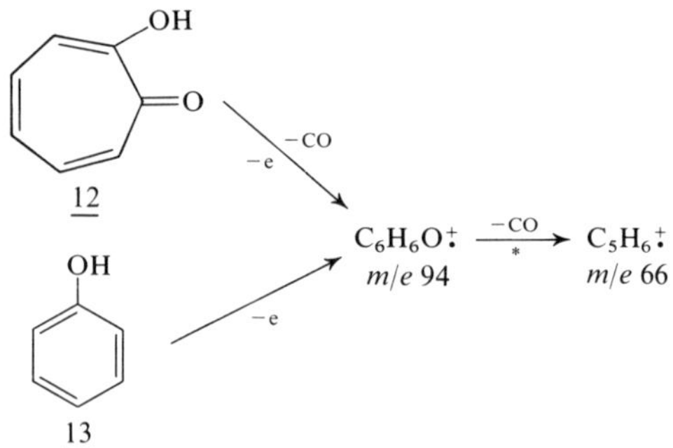

12

13

the ion abundances for a reaction $X^+ \xrightarrow{m^*} A^+$, are not characteristic of a given structure. It is obvious that if one method of producing X^+ weights high internal energies of X^+ more than an alternative method, then this will result in general in an increase of $[A^+]$ relative to $[X^+]$, even though all the X^+ ions produced by the two methods have the same structure. This point may be illustrated by the behaviour of $C_6H_6O^{+\cdot}$ ions produced both by loss of CO from tropolone (12) and by direct ionization of phenol (13). In each case the $C_6H_6O^{+\cdot}$ ions decompose further by loss of CO with a metastable peak of the same shape, indicating that $C_6H_6O^{+\cdot}$ ions decomposing after a few μs have the same structure. However, Figs. 4-12(a) and (b) show quite clearly the greater value of [*m*/*e* 66]/[*m*/*e* 94] when $C_6H_6O^{+\cdot}$ is produced by CO loss from tropolone. Similarly [*m*/*e* 66]/[m*] and [m*]/[*m*/*e* 94] are

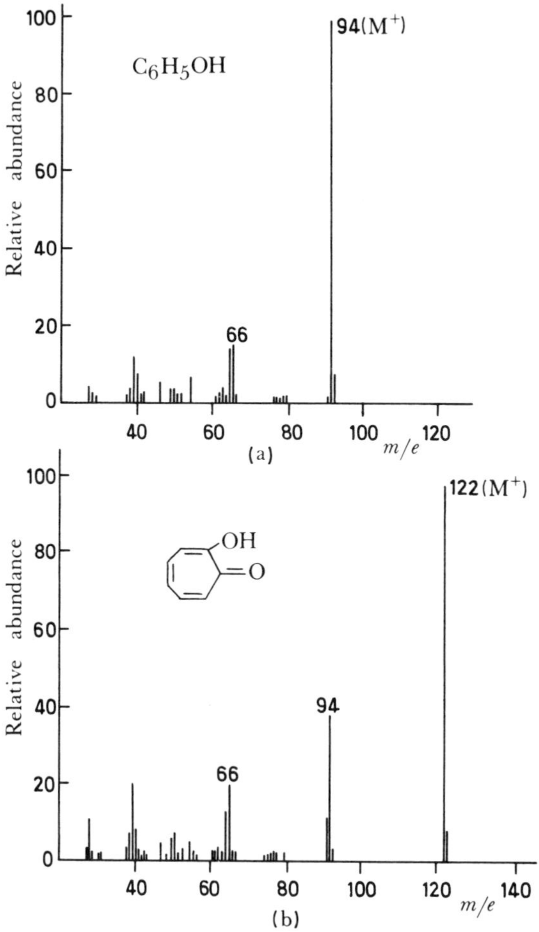

Figure 4-12

greater in the tropolone case, indicating the greater weighting of $C_6H_6O^{+\cdot}$ ions of higher internal energies.[23]

There is evidence to indicate that in compounds of a given functional type (e.g., alcohols or ketones), the distribution of internal energies produced upon electron impact at a given electron-beam energy is very similar in a homologous series (e.g., in $CH_3CH_2CH_2OH$, $CH_3CH_2CH_2CH_2OH$, etc.). The average energy per bond therefore decreases with increasing molecular size and hence if one considers a series of *n*-alkanols (propanol upwards) producing $C_2H_5O^+$ ions, the mean internal energy of $C_2H_5O^+$ will decrease regularly as the size of the ejected alkyl radical increases and carries off more of the internal energy. As a result it is a reasonable expectation that $[m^*]/[C_2H_5O^+]$ will decrease regularly as the size of the precursor alcohol is increased, $[m^*]$ being the abundance of a metastable peak for further decomposition of $C_2H_5O^+$. That is, a regular increase in the fraction of $C_2H_5O^+$ not decomposing should occur (see Fig. 4-13(a), large ejected neutral; cf. Fig. 4-13(b), small ejected neutral). Using the metastable

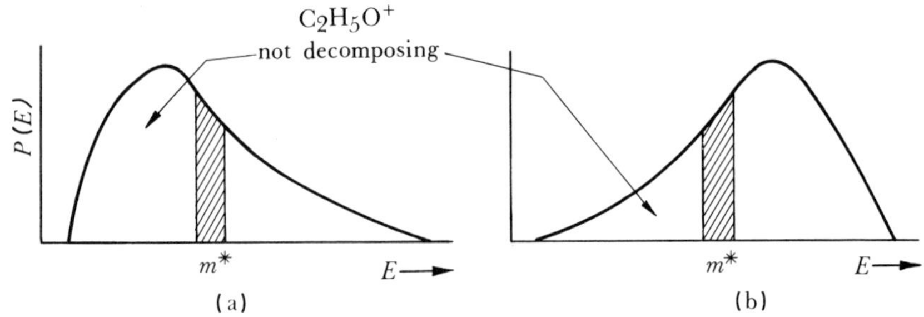

Figure 4-13

abundances for decomposition of $C_2H_5O^+$ given in eqs. (4-16) and (4-17), it is found that plots of $\log [m^*]/[C_2H_5O^+]$ against the reciprocal of number of degrees of freedom in the precursor alcohols give straight lines (Fig. 4-14).[24] In the figure the various plots correspond to homologous series of primary alcohols, secondary alcohols and terminal diols as precursors of $C_2H_5O^+$. The upper line of each pair corresponds to the metastable transition for $C_2H_5O^+ \rightarrow H_3O^+ + C_2H_2$, and the lower to $C_2H_5O^+ \rightarrow CHO^+ + CH_4$. The degree-of-freedom effect has also found theoretical support:[25] calculations employing the QET have shown that trends such as those shown in Fig. 4-14 would be anticipated but the observed linearity was not predicted.

4-5. Scrambling as a function of ion lifetime

Hydrogen (and carbon) scrambling on the time scale of the mass spectrometer may be extensive in organic ions produced by electron impact and it has become apparent that many of these scrambling reactions possess activation energies below that for an elimination reaction. It is a common situation that where scrambling occurs prior to decomposition, it is more extensive in the longer-lifetime ions (e.g., metastables) compared with shorter lifetime ions (e.g., decomposition in the source).

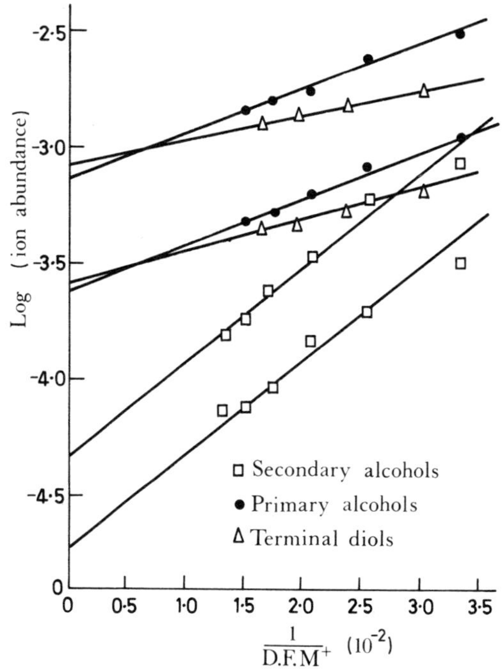

Figure 4-14

Consider, for example, the elimination of propylene from the deuteriated ketone 14† to form the McLafferty rearrangement ion e.[26]

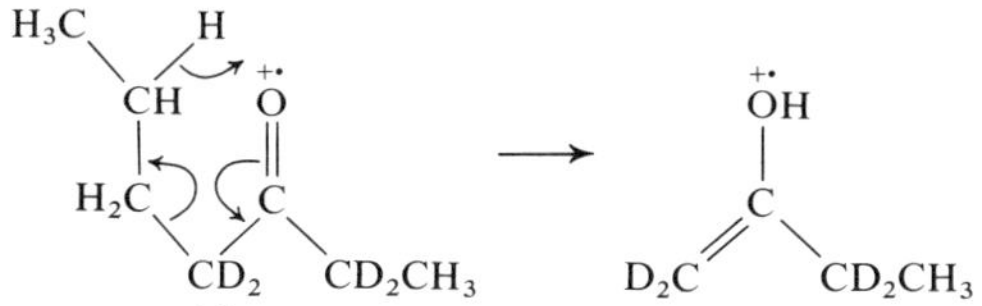

14 e

At 70 eV, 14 eliminates largely C_3H_6 in the source; C_3H_5D is eliminated to a small extent. At 10 eV, loss of $C_3H_4D_2$ also occurs and loss of $C_3H_3D_3$

† It is emphasized that we are depicting the reaction 14 → e in a formal manner and there is no implication that the rearrangement is a concerted process, and indeed may be stepwise. This qualification applies to other rearrangements discussed subsequently.

is detected in the first drift region. It is even possible to detect a weak metastable transition for loss of $C_3H_2D_4$ in the second drift region.

The above results clearly demonstrate an increase in H/D scrambling within the ion 14 as the ion lifetime becomes longer. Although suggestions for the scrambling mechanisms must be speculative at present, 15 is one feasible transition state for H/D exchange:[26]

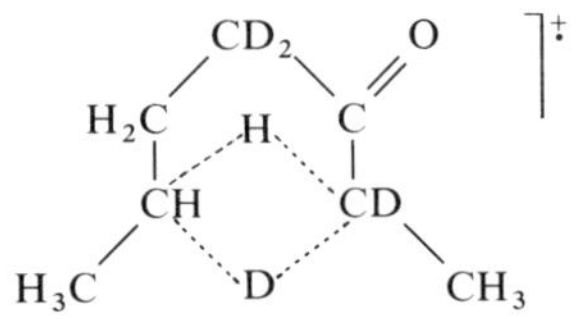

15

The QET can again account for these facts. The scrambling reaction 15 is just another instance of a rearrangement, in which the precursor and product ions have the same elemental composition and no neutral is eliminated. Such rearrangements would be expected to involve very tight transition states and possess k vs. E curves corresponding to curve B in Fig. 4-1. If the decomposition reaction under study has a rate curve corresponding to A, then for ions of low internal energy (long lifetime), scrambling prior to decomposition will be extensive, whereas for higher-energy (short-lifetime) ions scrambling will be only partial or undetectable. If the reaction C corresponds to the decomposition reaction then scrambling will predominate at all ion energies.

The mass spectroscopist now has a number of complementary techniques at his disposal for direct sampling of ions having a wide variety of lifetimes (see later—Fig. 9-6). Indirect techniques may also be devised to sample higher-energy ions than those sampled directly in the electron-impact mass spectrometer.[27]

For example, the molecular ion of 16 has been found to eliminate OD and OH in the ratio 1 : 2 in metastable transitions (consistent with exchange between the hydroxyl- and *ortho*-hydrogens), but in ions produced in the source, the loss of OD is the more prevalent.[28] An energy range above that

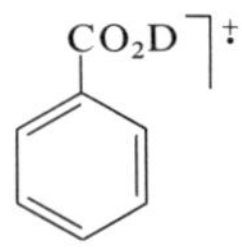

16

for unimolecular metastable formation can be attained by the use of collision-induced metastables (see section 8-3). Molecular ions 16 which have sufficient energy to lose hydroxyl *and* to decompose by further loss of CO provide another energy range. The relative intensities of the metastables for loss of CO from $C_6H_5CO^+$ and $C_6H_4DCO^+$ will represent the relative proportions of molecular ions 16 at an energy ($E + I$) of ca. 16·0 eV (the A.P. of $M^+ - CO_2H$ from benzoic acid) which undergo the initial reaction via loss of OD and OH respectively. Furthermore, the relative intensities of $C_6H_5^+$ and $C_6H_4D^+$ ions formed in the source represent molecular ions from still higher-energy states which decompose via the respective pathways. Five energy segments have thus been defined[27] from which the relative extent of initial decomposition via loss of OH and OD may be investigated (see Fig. 4-15). The relative extent of decomposition via the two pathways is indicated for energy segments (1)–(5) in Table 4-7. The observed increase, with increasing energy, in the reaction course not involving scrambling is completely in accord with the QET.

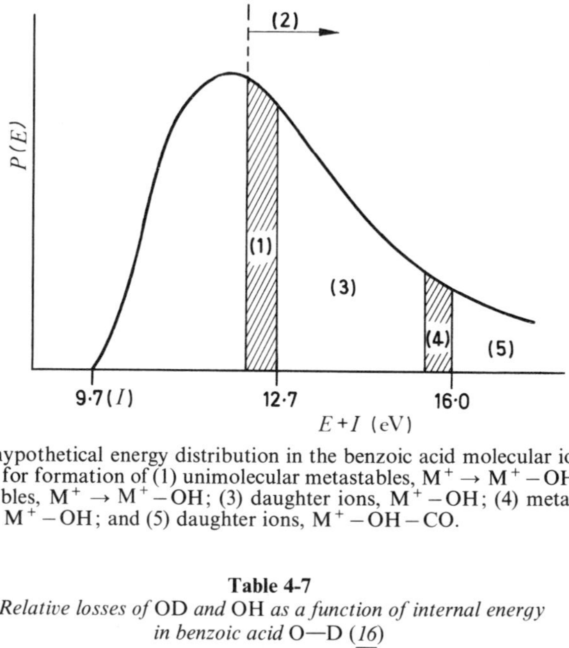

Figure 4-15: A hypothetical energy distribution in the benzoic acid molecular ion showing the energy segments for formation of (1) unimolecular metastables, $M^+ \rightarrow M^+ - OH$; (2) collision-induced metastables, $M^+ \rightarrow M^+ - OH$; (3) daughter ions, $M^+ - OH$; (4) metastable ions for loss of CO from $M^+ - OH$; and (5) daughter ions, $M^+ - OH - CO$.

Table 4-7

Relative losses of OD *and* OH *as a function of internal energy in benzoic acid* O—D (16)

Segment	$[M^+—OD]/[M^+—OH]$
1	0·67 ± 0·07
2	0·96 ± 0·08
3	2·9 ± 0·2
4	12 ± 2
5	>20

References

[1] H. M. Rosenstock, M. B. Wallenstein, A. L. Wahrhaftig and H. Eyring, *Proc. Natl. Acad. Sci. U.S.* **38**, 667 (1952).

[2] H. M. Rosenstock and M. Krauss in *Mass Spectrometry of Organic Ions*, F. W. McLafferty, ed., Academic Press, New York, N.Y., 1963, Chapter 1.

[3] H. M. Rosenstock in *Advances in Mass Spectrometry*, Vol. 4, E. Kendrick, ed., The Institute of Petroleum, London, 1968, pp. 523–545.
[4] W. A. Chupka, *J. Chem. Phys.* **30**, 191 (1959).
[5] F. W. McLafferty and R. B. Fairweather, *J. Am. Chem. Soc.* **90**, 5915 (1968).
[6] D. H. Williams and R. G. Cooks, *Chem. Comm.* 663 (1968).
[7] R. H. Shapiro and T. F. Jenkins, *Org. Mass Spec.* **2**, 771 (1969).
[8] F. W. McLafferty, T. Wachs, C. Lifshitz, G. Innorta and P. Irving, *J. Am. Chem. Soc.* **92**, 6867 (1970).
[9] M. Spiteller-Friedmann and G. Spiteller, *Chem. Ber.* **100**, 79 (1967).
[10] A. N. H. Yeo and D. H. Williams, *J. Am. Chem. Soc.* **92**, 3984 (1970).
[11] I. Howe and D. H. Williams, *J. Am. Chem. Soc.* **91**, 7137 (1969).
[12] M. M. Bursey, *Org. Mass Spec.* **1**, 31 (1968).
[13] R. G. Cooks, I. Howe and D. H. Williams, *Org. Mass Spec.* **2**, 137 (1969).
[14] F. W. McLafferty, *Chem. Comm.* 956 (1968).
[15] M. M. Bursey and F. W. McLafferty, *J. Am. Chem. Soc.* **88**, 529 (1966).
[16] M. S. Chin and A. G. Harrison, *Org. Mass Spec.* **2**, 1073 (1969).
[17] M. M. Bursey and F. W. McLafferty, *J. Org. Chem.* **33**, 124 (1968).
[18] H. M. Rosenstock, V. H. Dibeler and F. N. Harllee, *J. Chem. Phys.* **40**, 591 (1964).
[19] W. G. Cole and D. H. Williams, *Chem. Comm.* 784 (1969).
[20] T. W. Shannon and F. W. McLafferty, *J. Am. Chem. Soc.* **88**, 5021 (1966).
[21] D. Van Raalte and A. G. Harrison, *Can. J. Chem.* **41**, 3118 (1963).
[22] A. G. Harrison and B. G. Keyes, *J. Am. Chem. Soc.* **90**, 5046 (1968).
[23] D. H. Williams, R. G. Cooks and I. Howe, *J. Am. Chem. Soc.* **90**, 6759 (1968).
[24] F. W. McLafferty and W. T. Pike, *J. Am. Chem. Soc.* **89**, 5951 (1967).
[25] Y. N. Lin and B. S. Rabinovitz, *J. Phys. Chem.* **74**, 1769 (1970).
[26] A. N. H. Yeo and D. H. Williams, *J. Am. Chem. Soc.* **91**, 3582 (1969).
[27] I. Howe and F. W. McLafferty, *J. Am. Chem. Soc.* **92**, 3797 (1970).
[28] J. H. Beynon, B. E. Job and A. E. Williams, *Zeitschrift für Naturforschg.* **22a**, 180 (1965).

5. Energetics of ion decompositions

5-1. Appearance potentials

In the discussion of rate processes in the mass spectrometer (chapter 4), for simplicity it was necessary to assume that decompositions were occurring from the ground electronic state of the ion. Experimentally, the absolute ground state of the ion is given by the ionization potential (I.P.) although some slight error may occur here if a vertical transition produces the ion M^+ with some vibrational energy (section 2-1). If we can measure the minimum amount of extra energy which must now be added just to produce daughter ions A^+ in the source for the reaction $M^+ \rightarrow A^+$, then it may be argued that the extra energy will correspond to the vibrational activation energy (E_0) for reaction from the ground state of M^+.

The minimum energy required to produce A^+ ions in the source is known as the appearance potential (A.P.) of A^+ and hence we are assuming E_0 = A.P. − I.P. The methods of determining appearance potentials of organic ions are the same in principle as those for determining ionization potentials (section 2-1). It is sometimes a problem in the determination of A.P.s by the semi-log plot method (section 2-1) that the curve for A^+ is not parallel to that for M^+ or the calibrant inert gases. Although several ingenious methods to minimize such effects have been described,[1] the error limits for A.P.s are quite likely to be in the range 0·2–0·6 eV.

Since energies of activation are extremely important, it is necessary to know how closely A.P. − I.P. represents the activation energy. Here the problem is that A.P. − I.P. does not merely represent the energy necessary to attain the transition state, but rather the energy required to make the reaction occur with a rate constant approaching 10^5 s^{-1}, since the daughter ion must be formed within a few μs in the source to be detected as such. Hence due to this effect, A.P. − I.P. will in general be larger than the true activation energy, and by an amount which corresponds to the energy required to raise the rate constant from zero to approximately 10^5 s^{-1}. This extra energy has been designated the kinetic shift. Using the classical approximation (eq. (4-1), now repeated as eq. (5-1)) for the relationship between k and E, generalizations can be made concerning the magnitude of the kinetic shift.

$$k = \nu\left(\frac{E - E_0}{E}\right)^{s-1} \simeq 10^5\ \mathrm{s}^{-1}. \qquad (5\text{-}1)$$

(a) For a given v, E_0, the kinetic shift will be larger for larger molecules.
(b) For a given v, s, the kinetic shift will increase with an increase in E_0.
(c) For a given E_0, s, the kinetic shift will increase with a decrease in v.

In general, the kinetic shift will be larger the slower the rise of k with internal energy close to threshold, and a particularly slow rise of k with E would be brought about by a combination of a high energy of activation, low frequency factor, and a large molecule.

However, as we have already seen (section 4-1), eq. (5-1) has serious limitations at the low energies close to threshold. So although the above generalizations are valid, any *absolute* calculations of the kinetic shift employing the unmodified eq. (5-1) would result in gross overestimation of the kinetic shift in some cases. If the improved equation using an exact enumeration of states is employed (section 4-1), realistic estimations of the kinetic shift may be made, as shown in Table 5-1.[2]

Table 5-1
Calculated values of the kinetic shift†

Reaction	E_0(eV)	*Kinetic shift*
$C_3H_8^{+\cdot} \rightarrow C_3H_7^+ + H$	0·94	<0·01
i-$C_4H_{10}^{+\cdot} \rightarrow$ *sec*-$C_3H_7^+ + CH_3$	0·69	0·03
$C_4H_6^{+\cdot} \rightarrow C_3H_3^+ + CH_3$	2·28	0·27
$C_6H_6^{+\cdot} \rightarrow C_6H_5^+ + H$	3·9	1·50
$C_7H_8^{+\cdot} \rightarrow C_7H_7^+ + H$	1·95	0·60
$CH_3COCH_3^{+\cdot} \rightarrow CH_3CO^+ + CH_3$	0·70	<0·01

† $(E - E_0)$ in eV at which $k = 10^5\ s^{-1}$ for dissociation.

The results in Table 5-1 show that the kinetic shift may be considerable in some cases, and therefore A.P. − I.P. may be a poor reflection of the true activation energy. For example, the activation energy for the reaction $C_7H_8^{+\cdot} \rightarrow C_7H_7^+ + H$ from toluene would be ca. 2·55 eV from A.P. − I.P. measurements, whereas the true value would be ca. 1·95 eV, according to the calculations.

For small or medium-sized molecular ions undergoing low activation energy elimination of radicals via direct cleavage, the kinetic shift is negligible. However, in other cases—particularly for large unsaturated ions in which E_0 for the lowest energy reaction is large—the shifts may be substantial.

The kinetic shift (1·50 eV), calculated for the reaction $C_6H_6^{+\cdot} \rightarrow C_6H_5^+ + H$ in benzene, is surprisingly high. The reason for this kinetic shift is the high density of states at an internal energy close to the activation energy. For most molecular ions the density of states is low close to threshold and few extra states are available over which to distribute the excess energy as the energy increases. The rise of the rate constant with energy is consequently rapid. Where the density of states is high, as in the case of benzene, the rapid rate of increase with internal energy of the states available for randomization of the excess energy results in a slow rise of *k* with *E* and a large kinetic shift. The calculated *k* vs. *E* curves for five low-activation energy reactions from benzene are shown in Fig. 5-1.[2] Metastable peaks are observed for all these reactions and originally it was thought that isolated electronic states might be involved, as the appearance potentials of the daughter ions were widely spaced. Figure 5-1 shows that the wide 'metastable-energy window' in this case allows these metastables within the framework of the QET.

A special case of the kinetic shift (termed the *competitive shift*)[3] may arise where *k* vs. *E* curves cross for two or more reactions from the same ion.

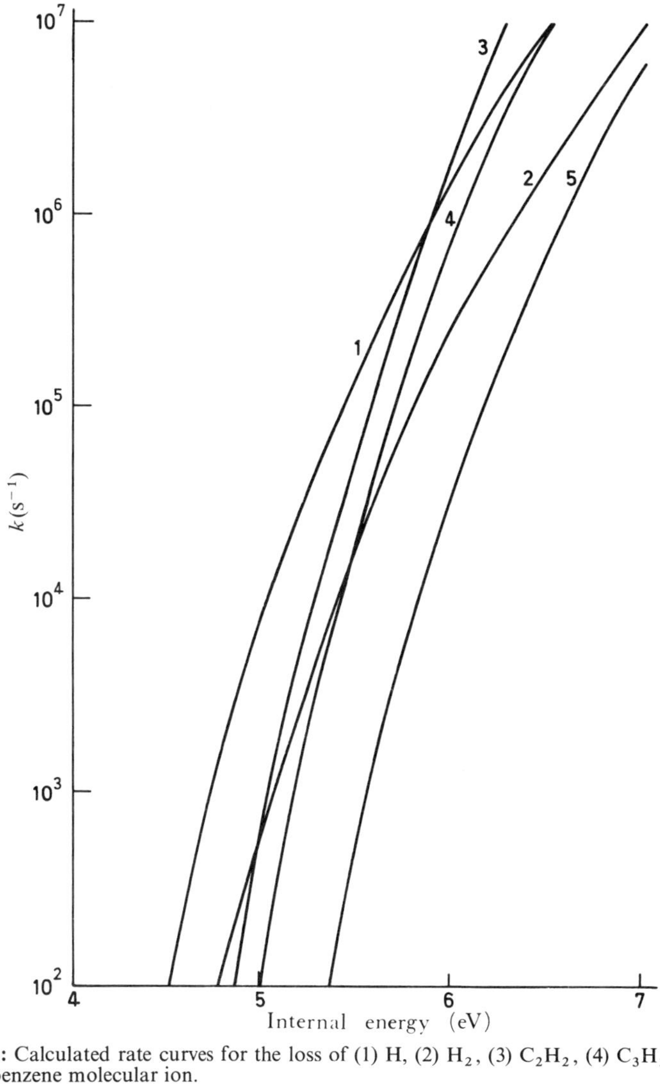

Figure 5-1: Calculated rate curves for the loss of (1) H, (2) H_2, (3) C_2H_2, (4) C_3H_3, (5) CH_3 from the benzene molecular ion.

Consider, for example, the reaction $M^+ \rightarrow B^+$ depicted in Fig. 5-2. Since this reaction has a higher activation energy and frequency factor than the $M^+ \rightarrow A^+$ reaction, a reaction rate considerably faster than 10^6 s^{-1} may be necessary to produce observable quantities of B^+ in the source.

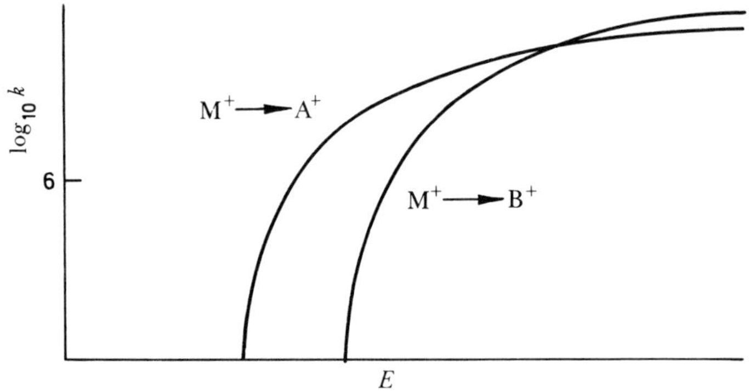

Figure 5-2

Appearance potentials for formation of second- and third- (etc.) generation ions may also give a false representation of the minimum energy necessary for formation of the ion, because partitioning of excess internal energy between charged and neutral fragments must be considered.[4]

Consider the unimolecular fragmentation sequence:

$$\begin{array}{ccl} M^+ & \longrightarrow & A^+ + B \\ & & \downarrow \\ & & C^+ + D \\ & & \downarrow \\ & & E^+ + F \end{array}$$

If E^+ is an ion-source product of a three-step sequence as shown, it is highly probable that the molecular ions which have sufficient energy to undergo the full sequence $M^+ \rightarrow A^+ \rightarrow C^+ \rightarrow E^+$ in the source will decompose with high rate constants (say $> 10^9$ s^{-1}). Considerable excess energy would therefore be available for partitioning between A^+ and B in the initial reaction. Some internal and/or kinetic energy will therefore be lost to B and, by successive reactions, also to D and F. This loss of internal energy to successive neutrals may constitute a considerable kinetic shift, which should be greater for higher activation-energy reactions.

Thermal energy of the incident molecule also has an effect on measured appearance potentials. Even a small molecule such as butane has an average internal energy of 0·18 eV at 400°K, which is of the same order of magnitude as the kinetic shift. The general effect of thermal energy on the kinetic shift is uncertain, however, and depends on the method used for A.P. determination.

5-2. Heats of formation

We have seen above that there are serious potential sources of error in equating A.P. – I.P. to the activation energy. When appearance potential measurements are employed for heat-of-formation determinations, the possibility arises that the ion and neutral products of a decomposition may be formed with excess kinetic and/or internal energy.

For the reaction $A + e \rightarrow B^+ + C + 2e$, we may write

$$\text{A.P.}(B^+) = \Delta H_f(B^+) + \Delta H_f(C) - \Delta H_f(A) + E, \qquad (5\text{-}2)$$

where A.P.(B^+) is the appearance potential of B^+ and the term E accounts for the fact that B^+ and/or the neutral particle C may be produced with excess energy at the threshold, i.e., the ions B^+ and C may not be in their ground electronic states, or not vibrationally cold, or may possess excess K.E. when produced at the A.P. Equation (5-2) is illustrated by the reaction profile shown in Fig. 5-3 and it is evident that E may incorporate both a kinetic shift and an energy of activation for the reverse reaction.

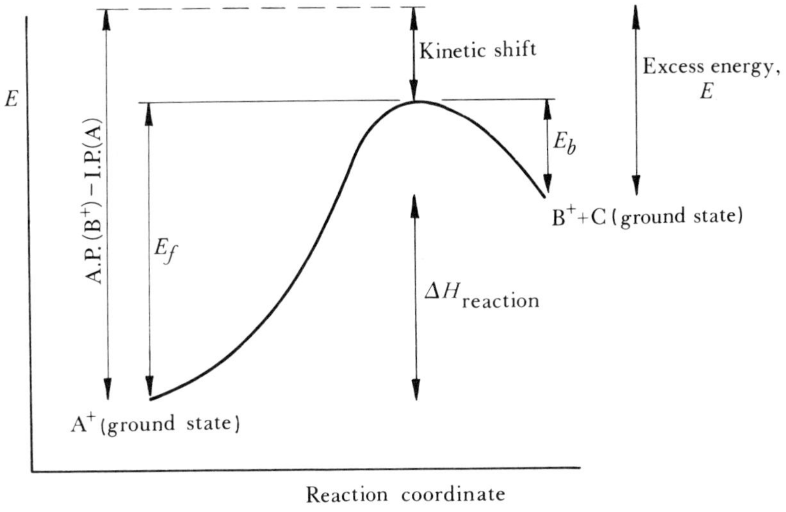

Figure 5-3

If the heats of formation of A and C are known (this requires an assumption of the structure of C), then the heat of formation of B^+ can be calculated if the excess energy can be reliably assumed to be zero, or calculated or estimated. Often, the structure of B^+ is unknown, and indeed the purpose of the measurement may be to obtain information about the structure of B^+ at the threshold. It has been common practice to compare the heat of formation of B^+ obtained in the above manner with that obtained by direct ionization of a molecule (or radical) B_1 of the same elemental composition (see also section 10-2A). For the ionization process of $B_1 \rightarrow B_1^{+}$, we may write

$$\text{I.P.}(B_1) = \Delta H_f(B_1^{+}) - \Delta H_f(B_1) + E_1, \tag{5-3}$$

where E_1 accounts for the fact that B_1^{+} may be produced with excess energy at the threshold. E_1 may be finite because we may be measuring the vertical rather than adiabatic ionization potential (section 2-1). This term is not usually large (0·0–0·2 eV) and may be effectively cancelled in many cases by the initial thermal energy of neutral B_1. Regarding the excess energy E (eq. (5-2)) we have already seen that the kinetic shift may be quite large if the rise of k with E is slow. A second complication in estimating E arises because there may be an energy of activation for the back reaction.

In considering the last problem, it is convenient to divide unimolecular reactions into two broad classes:

(a) reactions which proceed with bond formation in the transition state and expulsion of a neutral molecule and
(b) reactions which proceed via simple bond cleavage and loss of a radical.[5]

As an example of the first class, we can consider the loss of CO_2 from the molecular ion of diphenyl carbonate (1) which might plausibly be assumed to give ionized diphenyl ether (2). Whatever the structure of the product ion, partial bond formation must be occurring in the transition state and there will be a specific geometrical requirement for reaction to occur.

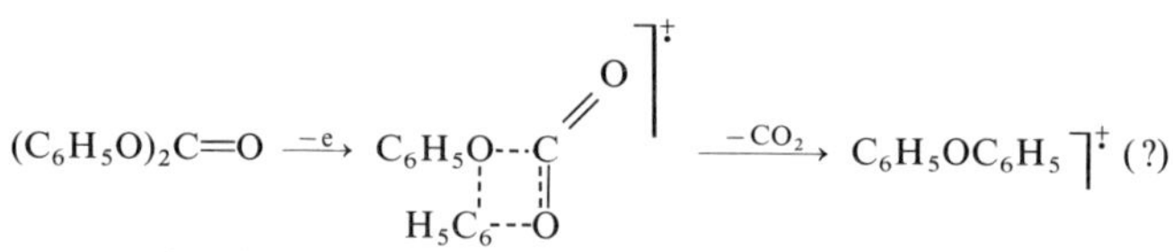

1, m/e 214 2, m/e 170

Thus, the rise of k with E may be anticipated to be slower than for a simple bond cleavage, and for this reason the measured appearance potential will be higher than the true value that should be inserted in eq. (5-2) (to reflect the energy of activation (E_f) for the forward reaction), and this error will give a positive contribution to E. In addition, this same type of reaction may well have an energy of activation (E_b) for the back reaction, since the back reaction involves bringing stable neutral CO_2 and (possibly) ionized diphenyl ether together to attain the transition state. If this is the case then the reaction profile will appear as in Fig. 5-4(a), whereas in general the reaction profile for simple bond cleavage with radical loss will appear more as in Fig. 5-4(b).†

† In support of this, kinetic studies of ion-molecule reactions involving simple bond formation have indicated zero or very small activation energies (see section 8-4).

Even neglecting the kinetic shift, the appearance potential measured for a reaction of the type shown in Fig. 5-4(a) would correspond to the energy needed just to give product ions, and would be too large a value (by an amount E_b) to be appropriate to eq. (5-2). Hence, we see that two positive contributions (kinetic shift and E_b) to the excess-energy term E may well arise for reactions proceeding with bond formation in the transition state and loss of a stable neutral molecule (see also Fig. 5-3).

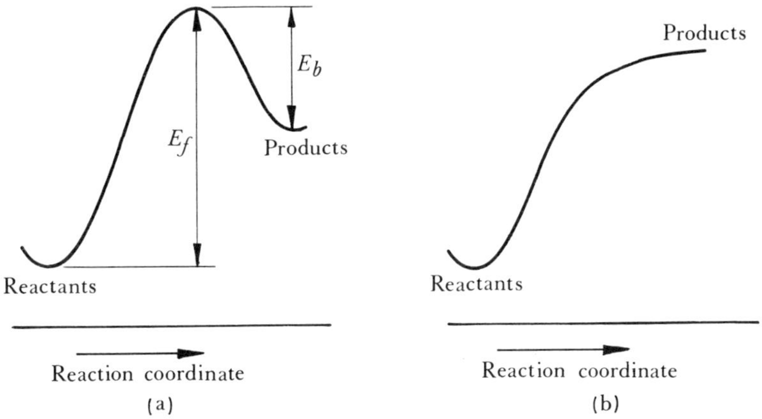

Figure 5-4

The ion formed by loss of CO_2 from the diphenyl carbonate molecular ion has a heat of formation of 257 kcal (1074 kJ)/mole (from its appearance potential).[6] The molecular ion formed by ionization of diphenyl ether has a heat of formation of only 220 kcal (920 kJ)/mole. The difference in these two figures does not allow the conclusion that the $M^+ - CO_2$ ion from diphenyl carbonate and the ion produced by direct ionization of diphenyl ether have different structures. On the contrary, the difference between the two figures is consistent with (but does not prove) structural identity of the two ions, with the sum of the kinetic shift and E_b for CO_2 loss being approximately 37 kcal (155 kJ)/mole (1·6 eV).

5-3. Estimation of kinetic shift and energy of activation of back reaction

It will be evident from the above discussion that heat of formation data will be useless for inferring ion structures where rearrangement reactions are involved, unless the errors due to the kinetic shift and E_b can be reliably measured.†

† In comparing heats of formation of two ions *both* formed by the *same type* of (rearrangement) reaction, the excess-energy terms may of course reasonably be assumed to cancel out.

Since metastable decompositions occur from lower energies than those in the source, the appearance potential of a metastable peak reflects more closely the true activation energy.[7] It is, however, difficult to make generalizations, since the fraction of the kinetic shift which is corrected for in this manner will depend upon the precise shape of the k vs. E curve for reaction, and so will vary from reaction to reaction.

Semilogarithmic plots of ion current against uncorrected electron energy are given for the molecular ion, fragment ion, and metastable ion (first drift region) for the reaction

$$C_6H_6^{+\cdot} \xrightarrow{m^*} C_4H_4^{+\cdot} + C_2H_2$$

occurring in the benzene mass spectrum (Fig. 5-5).[7] The curves are very closely parallel and the appearance potential for the metastable ions is 0·77 eV

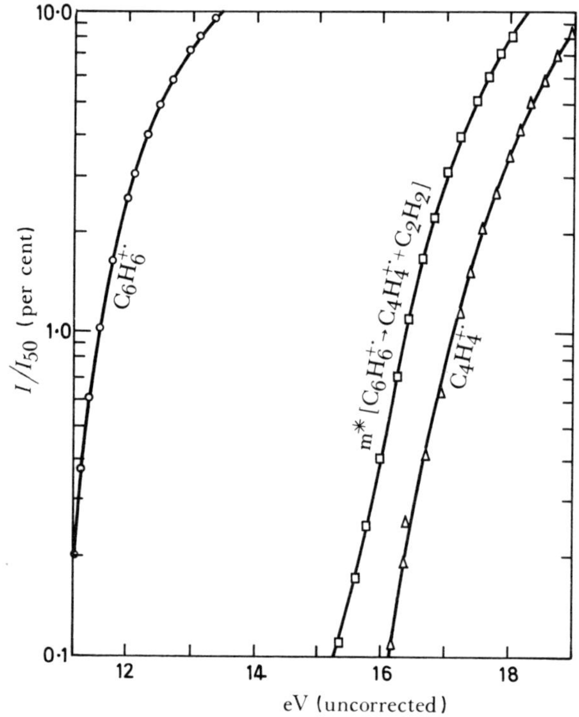

Figure 5-5

below that of the $C_4H_4^{+\cdot}$ daughter ions. Using the subscript α to denote that the fragment ions were formed in the ionization chamber, and β to signify that the ions were formed *via* metastable processes, then:

$$\text{A.P.}(C_4H_4^{+\cdot})_\alpha = 14{\cdot}96 \text{ eV},$$
$$\text{A.P.}(C_4H_4^{+\cdot})_\beta = 14{\cdot}19 \text{ eV}.$$

Other relevant thermochemical data for the calculation of the heat of formation of $C_4H_4^{+\cdot}$ are:

$$\Delta H_f\,(\text{benzene})_g = 19{\cdot}82 \text{ kcal/mole} = 82{\cdot}9 \text{ kJ/mole} = 0{\cdot}86 \text{ eV}$$
$$\Delta H_f(C_2H_2)_g = 54{\cdot}19 \text{ kcal/mole} = 226{\cdot}7 \text{ kJ/mole} = 2{\cdot}35 \text{ eV}.$$

Then we obtain (temporarily neglecting the term E in eq. (5-2))

$$\Delta H_f(C_4H_4^{+\cdot})_\alpha = 13{\cdot}47 \text{ eV} = 310{\cdot}6 \text{ kcal/mole} = 1299 \text{ kJ/mole},$$
$$\Delta H_f(C_4H_4^{+\cdot})_\beta = 12{\cdot}70 \text{ eV} = 292{\cdot}9 \text{ kcal/mole} = 1225 \text{ kJ/mole}.$$

Similar calculations for HCN loss from the pyridine molecular ion give an independent set of values:

$$\Delta H_f(C_4H_4^{+\cdot})_\alpha = 13{\cdot}33 \text{ eV} = 307{\cdot}3 \text{ kcal/mole} = 1285 \text{ kJ/mole}$$
$$\Delta H_f(C_4H_4^{+\cdot})_\beta = 12{\cdot}63 \text{ eV} = 291{\cdot}2 \text{ kcal/mole} = 1217 \text{ kJ/mole}.$$

We see that in the above cases the kinetics shifts are $\geqslant 0{\cdot}77$ eV and $0{\cdot}70$ eV for formation of $C_4H_4^{+\cdot}$ from benzene and pyridine, respectively. The heats of formation of $C_4H_4^{+\cdot}$ calculated using appearance potentials of metastable ions are in good agreement with the heat of formation of $C_4H_4^{+\cdot}$ (294 kcal (1229 kJ)/mole)[8] obtained by direct ionization of vinylacetylene. Other examples of heat of formation calculations in ion structure investigations may be seen in section 10-2.

It is of passing interest that appearance potential measurements on metastable peaks may uncover other phenomena. A general reaction in the mass spectra of many aromatic nitro-compounds is a rearrangement resulting in the loss of NO as a neutral particle, for example:

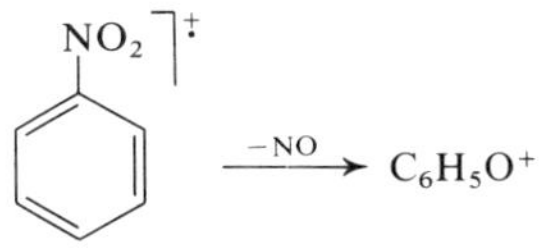

m/e 123 *m/e* 93($M^+ - 30$)

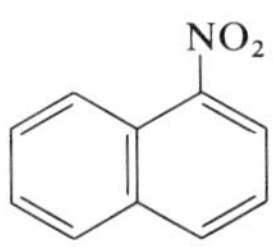

3

In a number of nitro-compounds it is found that the appearance potential of the fragment ion $M^+ - 30$ is *below* that of the metastable peak connecting this ion with the molecular ion (see, for example, the ionization-efficiency curves for 1-nitronaphthalene (3) as precursor in Fig. 5-6). It has been shown[7] that this effect is due to a small amount of catalytic reduction of the nitro-compounds to the corresponding amines on the walls of the ionization chamber. The amines give molecular ions at the same integral mass as the $M^+ - NO$ ions from the nitro compounds. Since the amines have relatively low ionization potentials (section 2-1), valid appearance potential measurements on the $M^+ - NO$ daughter ions are not possible. However, metastable ions give ionization-efficiency curves which are free from this complication (Fig. 5-6).

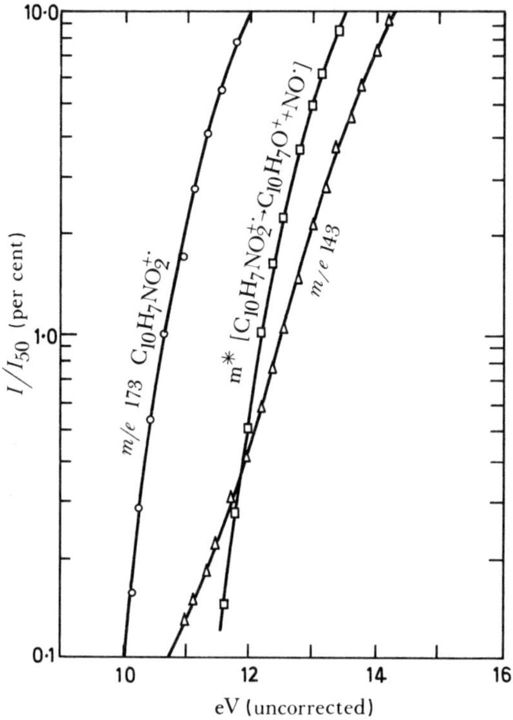

Figure 5-6

The problem of measuring energy of activation of the back reaction, where present, is complicated. It might reasonably be considered that when there is an energy of activation for the back reaction (E_b), at least a part of E_b may appear as translational energy of the products. That is, as the activated complex $(AB\ddagger)^+$ falls down the potential surface to products A^+ and B, the energy released may cause A^+ and B to be pushed apart with a change in velocity of A^+ and B relative to the velocity of $(AB\ddagger)^+$ during the dissociation. It was shown in section 3-1 that the kinetic energy released can be measured from the width of a flat-topped metastable peak. However, we now see a possible origin of this kinetic energy. Indeed, flat-topped metastable peaks are usually observed for reactions (of singly charged ions) in which it is intuitively reasonable that there might be an energy of activation for the back reaction. For example, broad or flat-topped metastable peaks are sometimes observed for loss of H_2, but not for loss of $CH_3\cdot$. Yet rearrangement reactions with loss of stable neutral molecules do also occur to produce Gaussian-shaped metastable peaks, indicating either that (a) some such reactions can have $E_b \simeq 0$ and/or (b) E_b does not necessarily have to appear as translational energy of the products but may appear (totally or in part) as vibrational energy of the products. In support of the latter possibility, Krauss and Dibeler point out[9] that the shape of the potential surface will determine the extent to which the reverse heat of activation and other excess vibrational energy will find its way into relative translational energy rather than into rotational energy or vibrational excitation of the products.

The problem of partition of reverse activation energy between translational, vibrational, and rotational energies is of course the same as that of the relative effectiveness of these three types of energy in supplying the activation energy when passing over the potential surface in the opposite direction. For the reaction $H + Cl_2 \rightarrow HCl + Cl$,† it has been shown that initial rotational or vibrational energy may be more effective than translational energy in increasing reaction probability.[10] Looking at the problem from the alternative viewpoint, it is known that exothermic reactions often produce highly vibrationally excited products.[11] In the case of larger molecules, plausible inferences are sometimes possible *post facto* from experimental data. 2,3-Diazabicyclo[2,2,1]hept-2-ene (4) has been photolysed with monochromatic light (334 nm) to give initially vibrationally excited bicyclo [2,1,0]-pentane (5) which undergoes further unimolecular reactions while 'hot' (prior to collisional deactivation).[12]

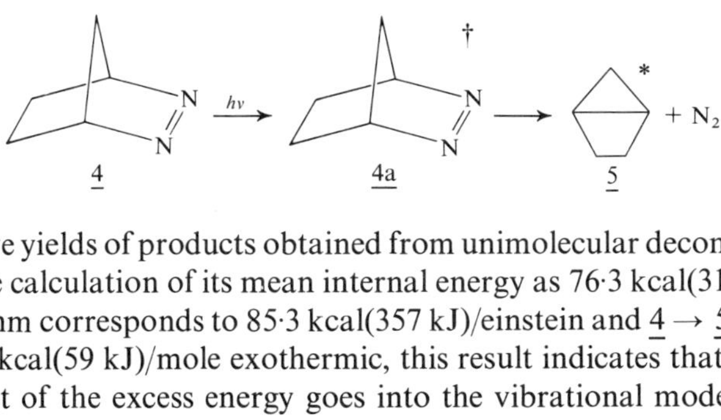

The relative yields of products obtained from unimolecular decomposition of 5 allow the calculation of its mean internal energy as 76·3 kcal(319 kJ)/mole. Since 334 nm corresponds to 85·3 kcal(357 kJ)/einstein and 4 → 5 is approximately 14 kcal(59 kJ)/mole exothermic, this result indicates that on average 77 per cent of the excess energy goes into the vibrational modes of 5. The remaining 23 per cent enters the N_2 vibrational mode, rotational modes of either product, or translational motion of the two products with respect to each other.

The above examples from neutral chemistry serve to illustrate that a considerable fraction of the reverse activation energy may appear as internal excitation energy of the products. As regards the mass spectrometric situation, it may be concluded that

(a) a flat-topped metastable peak indicates an energy of activation for the back reaction, but that the computed translational energy release will

† This problem of energy partition is not of course, peculiar to reactions of positive ions, and useful data may be taken from reactions of neutral species.

only represent some (at present usually unknown) fraction of the total reverse activation energy, and

(b) since the fraction of reverse activation energy released as translation may vary widely, the appearance of a Gaussian-shaped metastable peak does not preclude the occurrence of a reverse activation energy for the reaction in question.

5-4. Examples of kinetic-energy release

It is found that the rearrangement process resulting in loss of NO from *o*-nitrophenol (6) gives rise to a flat-topped metastable peak, the width of which increases as the accelerating voltage is lowered (Fig. 5-7),[13] in accord

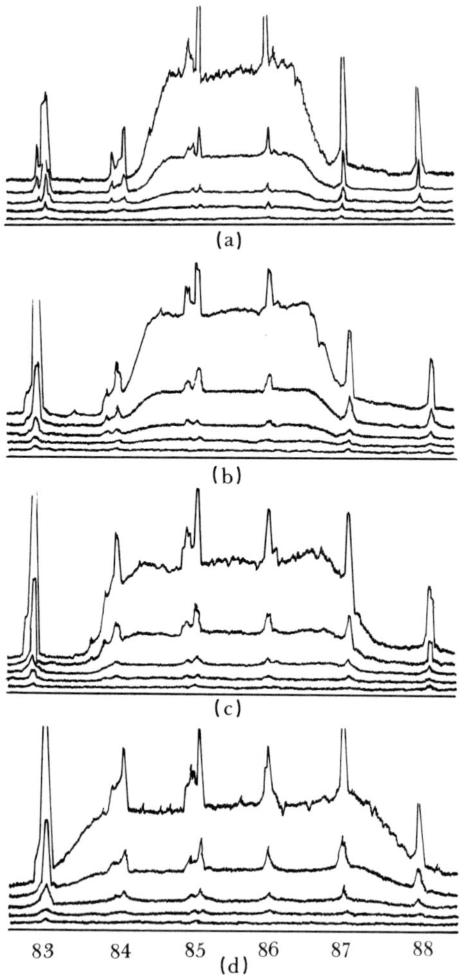

Figure 5-7

with eq. (3-3). Traces (a), (b), (c) and (d) refer to measurements made at 8, 6, 4 and 2 kv respectively.

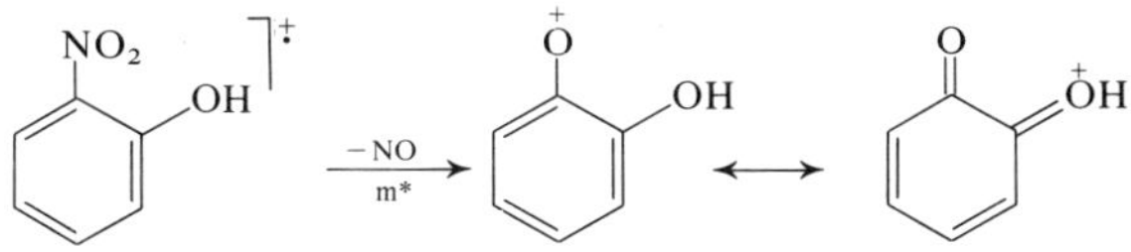

6, M^+, *m/e* 139 — 7, $M^+ - 30$, *m/e* 109

A similar flat-topped metastable peak is also detected for the analogous transition in the mass spectrum of *p*-nitrophenol, but the metastable peak for NO loss from *m*-nitrophenol (8) has the conventional Gaussian shape (Fig. 5-8). Thus kinetic energy is released only in the cases of the *o*- and *p*-

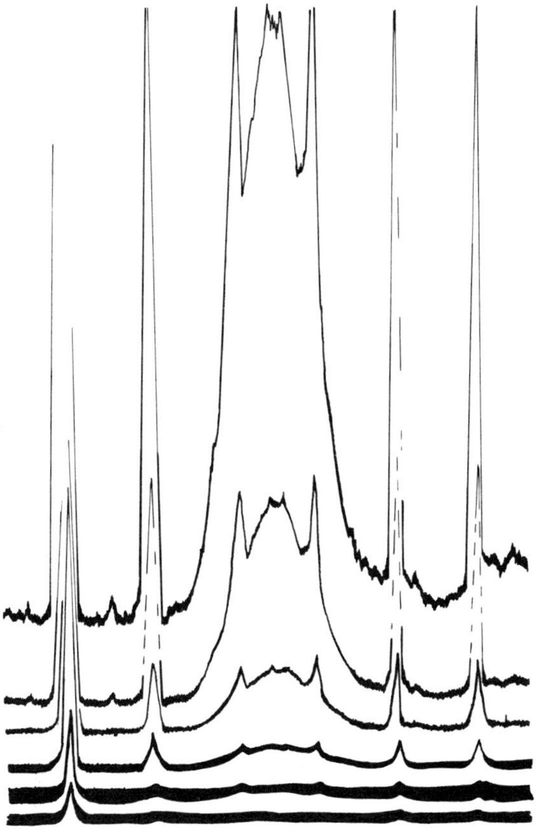

Figure 5-8

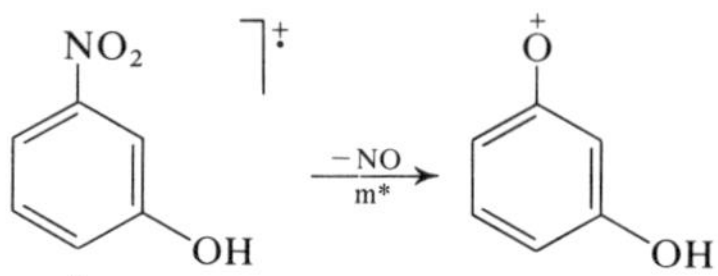

8

isomers (0·25 eV in each case) and it is noteworthy that product stabilization by a classical resonance effect (7) is possible only in these isomers.

The kinetic energy released in ions undergoing metastable transitions has been studied in a wide variety of *p*-substituted nitrobenzenes.[14] The substituents ranged from strongly electron-withdrawing (e.g., $COCH_3$, CHO, CN) to strongly electron-donating (through resonance effects, e.g. $N(CH_3)_2$ and NH_2). In thirteen compounds studied, the metastable peak was flat-topped only in five cases, all where the *p*-substituent could donate electrons through a resonance effect. The data are summarized in Table 5-2,[14] in which the σ_R^0 values of the substituents are included.

Table 5-2
Kinetic energy released in metastable transitions for NO *loss in some p-substituted nitrobenzenes*

Substituent	T (eV)	σ_R^0
$N(CH_3)_2$	0·84 ± 0·02	−0·55
NH_2	0·85 ± 0·04	−0·49
OH	0·74 ± 0·06	−0·43
OCH_3	0·56 ± 0·06	−0·43
C_6H_5	0·38 ± 0·06	−0·09

It may be seen that in general there is an increase in the kinetic energy released with an increased resonance effect. *Meta*-substituted nitrobenzenes do not lose NO with kinetic-energy release.

When kinetic-energy release occurs in mass-spectral transitions, spin-forbidden predissociation is always a possibility (see Fig. 5-9). A species is formed initially in an attractive state (see curve *a*), but subsequently passes onto the repulsive surface of electronic state *b* from which dissociation occurs in one vibration. The transfer from curve $a \rightarrow b$ may well be slow if the transfer is spin forbidden; if the probability of transfer is such that it requires microseconds, the fast dissociation from curve *b* will yield a metastable peak.

The need to consider such a mechanism is evident from studies on a wide metastable peak ($T = 0{\cdot}53$ eV) due to the loss of a nitrogen atom from the molecular ion of isocyanic acid.[15] First, it is noteworthy that the occurrence of a metastable peak for the decomposition of such a small parent ion (4 atoms) is unusual. Vestal[16] has pointed out that the improved form of the QET actually predicts a minimum non-zero rate when the internal energy is equal to the activation energy. This minimum rate is largest in the cases of very small molecules. Thus in the mass spectra of many triatomic molecules the minimum rate for the lowest activation energy reaction is greater than $10^6\ s^{-1}$ and consequently no metastable transitions are observed. Hence, although one would normally expect at least one metastable transition for the decomposition of a parent ion containing, say, 20 or more atoms, the observation of the metastable peak for formation of HCO^+ from $HNCO^{+\cdot}$ is

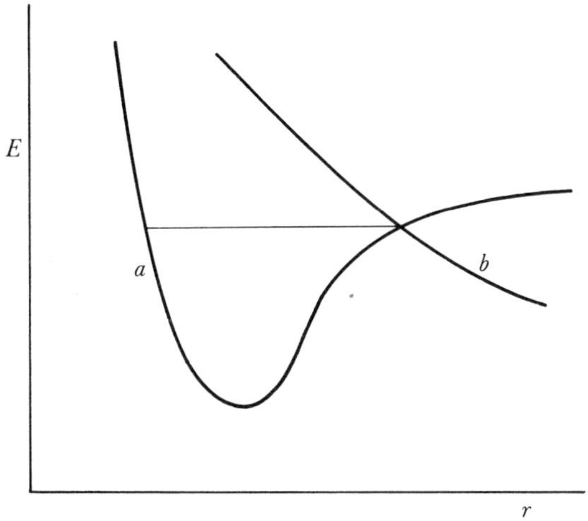

Figure 5-9

somewhat surprising. It has therefore been suggested[15] that this reaction occurs via a predissociation onto a repulsive energy surface. The reason the transfer to the repulsive surface is slow is that the predissociation is spin-forbidden if HCO^+ and N are to be formed in their ground states. The initially produced molecular ion $HNCO^{+\cdot}$ will have one unpaired electron and therefore is a doublet ($2S + 1$, where $S = \frac{1}{2}$); but HCO^+ and N in their ground states respectively contain zero and three unpaired electrons, and therefore correspond to singlet and quartet states. Overall, reaction therefore occurs with $\Delta S = 1$.

$$HNCO^{+\cdot}\ (\text{doublet}) \xrightarrow{\Delta S = 1} HNCO^+(^4A'') \longrightarrow HCO^+(^1\Sigma^+) + N(^4S).$$

The slowness of the forbidden predissociation explains the appearance of a metastable for such a small molecule and repulsion between the closed-shell formyl ion and the quartet nitrogen atom will lead to kinetic-energy release.

If two positively charged ions are approaching each other on a collision path, then their relative translational energy will decrease as the distance between them decreases because work must be done to overcome the Coulombic repulsion between the charges. Conversely, if a doubly charged ion decomposes into two singly charged ions, the particles will fly apart with relative translational energy due to the repulsion between the charges. In the case of the doubly charged benzene molecular ion, it has been shown[17] that the decomposition

$$C_6H_6^{++} \longrightarrow C_5H_3^+ + CH_3^+$$

gives rise to two 'dished-topped metastable peaks' (cf. Fig. 3-4(b) in section 3-1A), since both reaction products are positive ions. From the widths of these metastable peaks it was possible to calculate a kinetic-energy release of 2·8 eV during the decomposition. If all this energy is the result of the separation of the two positive charges in the metastable doubly charged ion from their initial distance apart to infinity, it can be shown from the relation $E = e^2/r$ that the initial inter-charge distance must have been 5·14 Å (0·51 nm).[18] If only part of the energy released arises from this charge separation, then the initial inter-charge distance must be larger than this. Since the distance between two *para*-carbon atoms in doubly charged benzene (9) is only 0·28 nm, $C_6H_6^{++}$ must decompose from an open configuration such as 10 (0·51 nm), 11 (0·56 nm) or 12 (0·61 nm) to account for this observation.

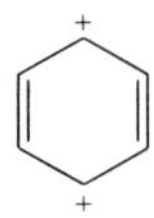

9

10

11

12

In most cases so far investigated, kinetic energy release in the decomposition of singly charged ions lies in the range 0–2 eV. Values found for the decomposition of doubly charged ions into two singly charged ions are frequently larger, as indicated by the values given in Table 5-3.[19]

Table 5-3

Kinetic-energy release in the decomposition of some doubly charged ions

Compound	Metastable transition	T (eV) (± 10 per cent)
n-Butane But-1-ene *cis*-But-2-ene Butadiene	$C_4H_4^{++} \rightarrow C_3H^+ + CH_3^+$	3·0
Pyridine	$C_5H_5N^{++} \rightarrow 51^+ + 28^+$	2·6
Fluorobenzene	$C_6H_5F^{++} \rightarrow C_5H_3^+ + CH_2F^+$	3·2
Toluene	$C_7H_7^{++\cdot} \rightarrow C_4H_4^{+\cdot} + C_3H_3^+$	3·5
	$C_7H_7^{++\cdot} \rightarrow C_5H_5^+ + C_2H_2^{+\cdot}$	2·8
	$C_7H_6^{++} \rightarrow C_5H_3^+ + C_2H_3^+$	2·5

5-5. The sloping sides of metastable peaks

It is very striking that metastable peaks of the conventional Gaussian shape have less steeply sloping sides than normal peaks (see, for example, Fig. 5-8). One of the narrowest metastable peaks yet reported is that for the transition $(M^+ - HCN) \rightarrow C_2H_2^{+\cdot} + HCN$ in the mass spectrum of pyrazine (13). This peak occurs at m/e 12·75 and is reproduced in Fig. 5-10.[20]

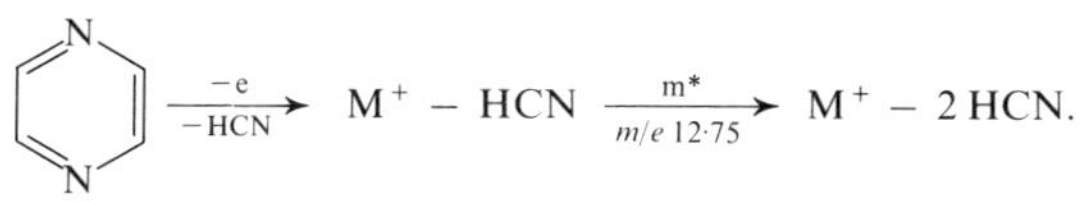

13

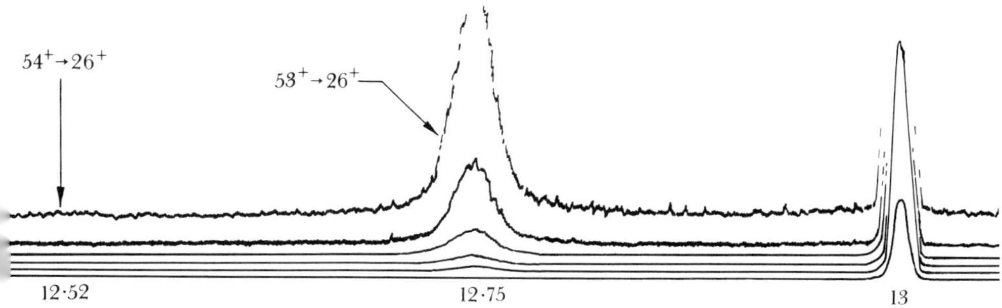

Figure 5-10

Yet even here the peak is broader than the normal peak at m/e 13 and corresponds to an average release of kinetic energy of 2×10^{-3} eV.[20]

Normal daughter-ion peaks are sharp because if any dissociations in the source happen to occur with kinetic-energy release, the change in translational energy which results is a much greater fraction of the total translational energy than it would be after acceleration (repeller voltage of the order of a few volts; accelerating voltages of the order of kV). Consequently, ions undergoing such energetic dissociation in the source have a relatively low probability of passing through the source exit slit. However, ions undergoing metastable transitions will on occasions happen to gather more than E_0 in the reaction coordinate. If the excess energy which may accumulate in the reaction coordinate is designated ΔE, then larger ΔE-values will be found with decreasing probabilities. Thus even in the absence of a *specific* energy release as discussed previously in this section, there will be a *range* of kinetic energies released due to the statistical fluctuation of excess energy in the reaction coordinate. Thus, if a metastable ion species always decomposed with release of the same amount of kinetic energy, then at infinite resolving power the sides of the metastable peak would be as steep as those of a normal mass peak. The fact that most metastable peaks are more diffuse than this suggests that there is release of a small range of kinetic energies in all such decompositions.[21]

The kinetic energy which arises as a result of the statistical fluctuation of excess energy into the reaction coordinate is sometimes referred to as the statistical kinetic energy (E_{st}).[2] Some reactions where the energies of activation (E_0), heat of reaction (ΔH at 0°K), energy of activation of the back reaction (E_b), and experimental kinetic energy released (K.E.^{exp}) are known are listed in Table 5-4.[2] The quoted statistical kinetic energies (E_{st}^{calc}) are calculated values and the portion of E_b appearing as kinetic energy is therefore given by the difference (K.E.^{exp}) − (E_{st}^{calc}).

Table 5-4
Some reactions occurring with excess energies (eV)

Reaction	E_0	ΔH	E_b	K.E.^{exp}	E_{st}^{calc}	E_b	
						Kinetic	Internal
i-$C_4H_{10}^{+\cdot} \rightarrow C_3H_6^{+\cdot} + CH_4$	0·69	0·39	0·30	0·09	0·07	0	0·30
$C_3H_8^{+\cdot} \rightarrow C_3H_6^{+\cdot} + H_2$	1·27	0·31	0·94	0·65	0·10	0·55	0·29
$C_3H_8^{+\cdot} \rightarrow C_2H_4^{+\cdot} + CH_4$	1·00	0·66	0·34	0·09	0·08	0	0·34
$C_2H_6^{+\cdot} \rightarrow C_2H_4^{+\cdot} + H_2$	1·00	0·65	0·35	0·45	0·10	0·35	0
sec-$C_3H_7^+ \rightarrow C_3H_5^+ + H_2$	2·06	1·44	0·62	0·43	0·03	0·4	0·2
$C_2H_5^+ \rightarrow C_2H_3^+ + H_2$	2·4	2·1	0·3	0·03	0·03	0	0·3

These limited data suggest that in the loss of neutral methane, the excess energy remains as internal energy of one or both fragments, while in the loss of a hydrogen molecule the excess energy may appear as kinetic energy or as both kinetic and internal energy.

At the beginning of this section we have seen that the sloping sides of a metastable peak are due to the finite resolving power of the instrument and release of a small range of kinetic energies in decomposition. For a metastable decomposition occurring in the first field-free region with a small energy release, part of the slope in the sides of the resulting peak originates in the limited resolving power associated with the (normally wide) slit (0·200 inch) placed between the electrostatic and magnetic sectors. If this slit width is reduced (0·020 inch), there is greater discrimination against ions formed with components of kinetic energy perpendicular to the direction of the ion beam. Under these conditions of enhanced resolution for metastable peaks,[22] the flat-topped peak for loss of NO from *o*-nitrophenol (Fig. 5-7) becomes concave, and the metastable transition for the reaction $C_3H_7^+ \rightarrow C_3H_5^+ + H_2$ in the spectrum of *n*-butane is clearly resolved into two components (Fig. 5-11). Thus the sloping sides of the metastable peak recorded under the usual conditions (Fig. 5-11, 0·200 inch) are largely associated with limited resolving power, and the steeply sloping sides of the peak at a slit width of 0·020 inch indicate that the decompositions of specific ions take place with the release of a relatively narrow range of kinetic energies. This behaviour seems fairly general, providing a few tenths of a volt or more is released as kinetic energy during the fragmentation, i.e., the total kinetic energy released is appreciably larger than the statistical kinetic energy.

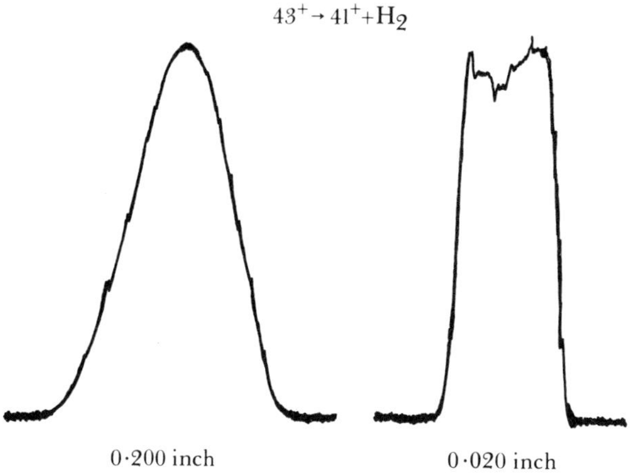

Figure 5-11

At slit widths of 0·020 inch, the peaks indicated in Fig. 5-12[22] are observed for the metastable transitions in the spectra of fluorobenzene and ethyl fluoride respectively.

$$\begin{array}{ccc} C_6H_5F^{+\cdot} & \longrightarrow & C_6H_4^{+\cdot} + HF \\ m/e\ 96 & & m/e\ 76 \\ C_2H_4F^+ & \longrightarrow & C_2H_3^+ + HF \\ m/e\ 47 & & m/e\ 27 \end{array}$$

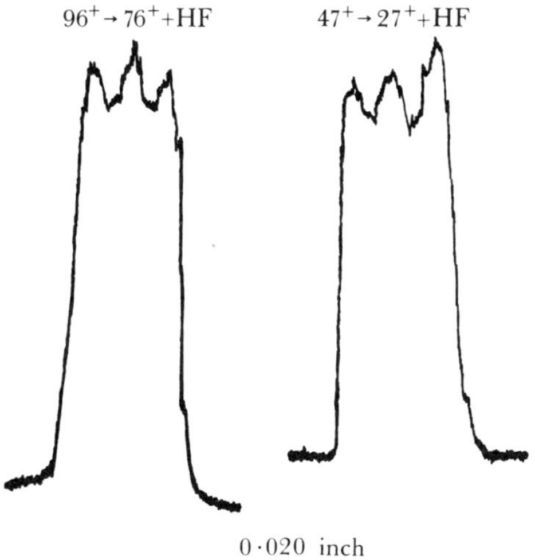

Figure 5-12

The occurrence of fine structure in the peaks has at least two possible explanations.[22] (a) two isomeric ions lose the same neutral fragment, one with the release of appreciable kinetic energy, and the other with release of very little kinetic energy, and (b) the same ion decomposes from two different electronic states, again with the release of different amounts of kinetic energy.

A novel case of fine structure in a metastable peak has been reported for the collision-induced reaction $H_3^+ \rightarrow H^+ + H_2$.[23] Since the product ion of this reaction cannot incorporate vibrational energy, the only way to distribute any excess energy is between kinetic energy of the products and vibrational energy of H_2. A plausible interpretation of the shape of the observed metastable peak (see Fig. 5-13) is that H_2 is formed in a number of vibrational states. The centre of the peak corresponds to product-ion formation with minimum kinetic energy and therefore maximum vibrational energy of H_2. Each notch on the peak corresponds to H_2 formed in a successively lower vibrational state. The energy differences observed are consistent with formation of the vibrational states $v = 14, 13, 12$ and 11 of H_2.

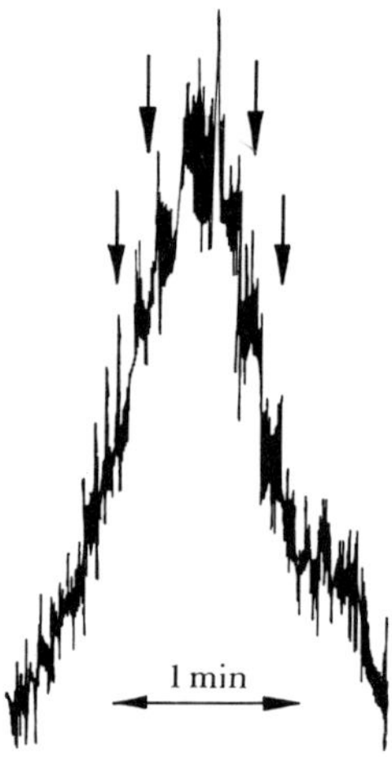

Figure 5-13

5-6. Stevenson's rule

It is relevant to the energetics of ion formation, and also a suitable prelude to the relationship between fragmentation and structure (chapter 6), to discuss a useful rule concerned with the prediction of preferred fragmentations. In connection with the fragmentation of alkanes in the mass spectrometer, Stevenson noted that the positive charge remained on the more substituted fragment,[24] i.e., the one with the lower ionization potential. This principle may be extended to all mass-spectral fragmentations and it may be generally stated that the positive charge will remain on the fragment of lower ionization potential.[25]

This principle may be simply explained by reference to the energy diagram shown in Fig. 5-14.

Figure 5-14 illustrates the energy changes involved in production of A^+ and B^+ from a molecule AB (where A and B are any radicals). The symbols *I* and A.P. have the usual meaning and *D*(A—B) refers to the dissociation energy of the bond connecting the two radicals. Any excess energies or reverse activation energies are neglected.

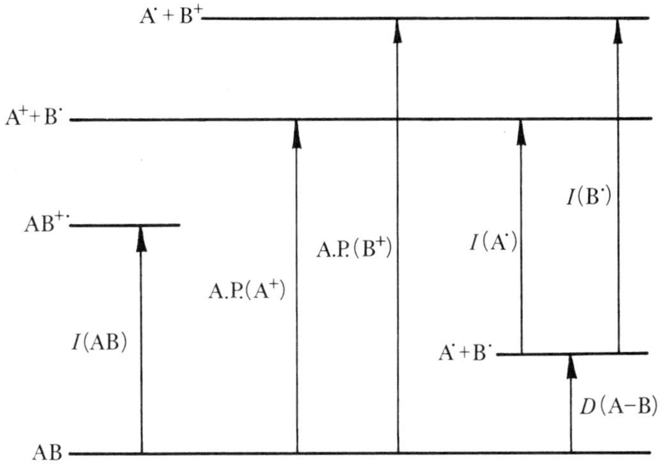

Figure 5-14

It is evident from Fig. 5-14 that the lower A.P. reaction (formation of A^+ in this case) corresponds to the process where the charge resides on the fragment of lower I.P. ($I(A\cdot) < I(B\cdot)$). The rule does not predict which of all the possible fragment ions will be the most prevalent, but tells us which way the charge will go for fragmentation at a given bond.

Consider the ionization potentials of some hydrocarbon radicals reproduced in Table 5-5.[26] Two trends are evident from these figures. First, there is a tendency for the I.P. to decrease as molecular size increases for radicals of the same type (e.g., primary radicals). Thus, for example, formation of $C_8H_{17}^+$ would be predicted to be a lower-energy process than formation of $C_2H_5^+$ from the molecular ion of *n*-decane ($C_{10}H_{22}^{+\cdot}$).† This is borne

Table 5-5
Ionization potentials of hydrocarbon radicals

Radical	*I.P.* (eV)	*Radical*	*I.P.*
$CH_3\cdot$	9·84	$CH_3CH_2CH_2CH_2\cdot$	8·01
$CH_3CH_2\cdot$	8·38	$(CH_3)_2CHCH_2\cdot$	8·01
$CH_3CH_2CH_2\cdot$	8·10	$CH_3CH_2CH\cdot CH_3$	7·41
$(CH_3)_2CH\cdot$	7·55	$(CH_3)_3C\cdot$	6·93

† Especially if hydrogen rearrangement can occur—see 'Heats of Formation', section 10-2A.

out in practice; and a metastable peak is observed for the former but not for the latter process.

The second important tendency is the lowering of the ionization potential as the radical becomes more highly substituted. The reason for the preferential formation of secondary or tertiary hydrocarbon ions in the spectra of branched hydrocarbons is therefore evident.

Figure 5-14 may be modified to account for rearrangement reactions of molecular ions with elimination of stable neutrals (see Fig. 5-15). The sum $\sum D(M)$ represents the difference between the sums of the heats of bond rupture and formation in the rearrangement reaction $M \rightarrow C + D$.

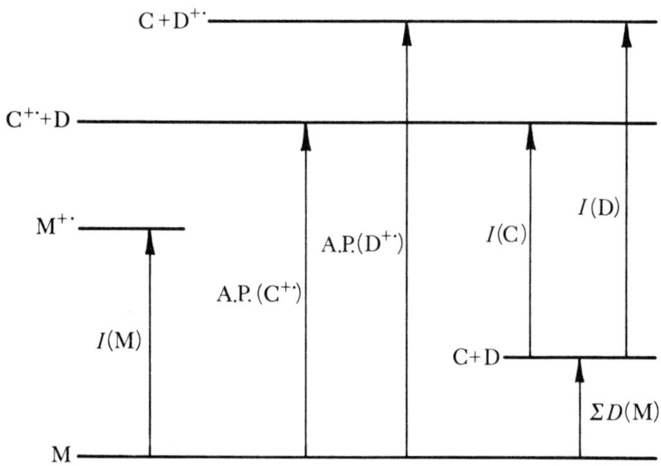

Figure 5-15

The retro Diels–Alder reaction provides a suitable illustration of the application of Stevenson's rule to rearrangements. For example, the molecular ion of α-terpineol (14) can decompose via a retro Diels–Alder reaction, to form the diene 15 and the allylic alcohol 16. The relative ionization potentials of the two even-electron species (15, 9·0 eV; 16, 9·4 eV) lead to the prediction that the peak due to the charged diene will be the more prevalent in the spectrum, and this is borne out in practice.

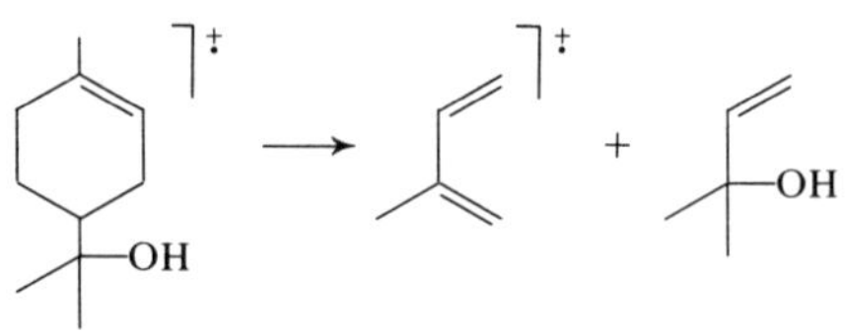

14 15 16

Trans-Δ²-octalin (17) however yields a mass spectrum having peaks at *m/e* 54 and *m/e* 82 of almost equal intensity. Cyclohexene (18) and butadiene (19) both possess ionization potentials close to 9·1 eV.

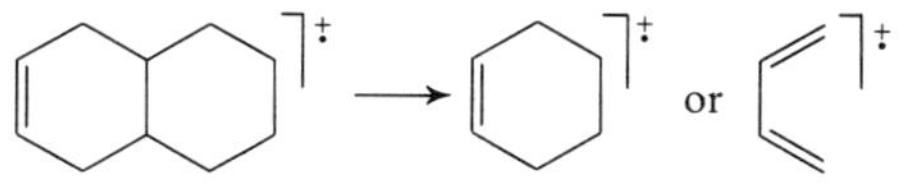

17 18 19

Examples are also forthcoming from the field of substituent effects.[27] A γ-hydrogen (McLafferty) rearrangement from the γ-substituted methyl butyrate 20 may yield either the $M^+ - 74$ ion (21) or the *m/e* 74 ion (22).

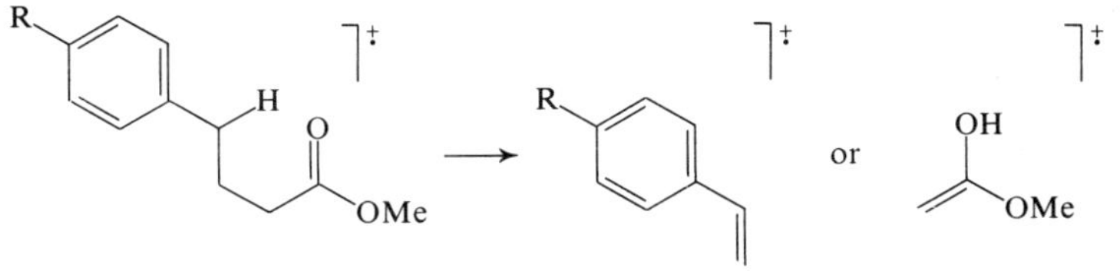

20 21 22

When R = NO_2, ion 22 is the base peak of the spectrum, whereas the abundance of ion 21 is low. When R = OMe the situation is reversed. When R = H, 21 and 22 have similar abundances. This effect is merely a consequence of the lowering of the ionization potential of the neutral compound 21 as the substituent R becomes more electron donating.

References

[1] For a summary of such methods, see R. W. Kiser, *Introduction to Mass Spectrometry and its Applications*, Prentice-Hall, 1965, pp. 166–176.
[2] M. L. Vestal in *Fundamental Processes in Radiation Chemistry*, P. Ausloos, ed., Wiley-Interscience, 1968, chapter 2.
[3] F. W. McLafferty, T. Wachs, C. Lifshitz, G. Innorta and P. Irving, *J. Am. Chem. Soc.* **92**, 6867 (1970).
[4] A. G. Harrison in *Topics in Organic Mass Spectrometry*, A. L. Burlingame, ed., Wiley-Interscience, New York, 1970.
[5] B. S. Rabinovitz and D. W. Setser, *Advan. Photochem.* **3**, 1 (1964).
[6] P. Natalis and J. L. Franklin, *J. Phys. Chem.* **69**, 2943 (1965).
[7] J. H. Beynon, J. A. Hopkinson and G. R. Lester, *J. Mass Spec. and Ion Phys.* **2**, 291 (1969).
[8] F. H. Field and J. L. Franklin, *Electron Impact Phenomena*, Academic Press, New York, 1957, p. 261.
[9] M. Krauss and V. H. Dibeler in *Mass Spectrometry of Organic Ions*, F. W. McLafferty, ed., Academic Press, New York, 1963, p. 131.
[10] J. Lin and J. Light, *J. Chem. Phys.* **45**, 2545 (1966).

[11] K. J. Laidler and J. C. Polanyi, *Progr. Reaction Kinetics* **3**, 1 (1965).
[12] T. F. Thomas, C. I. Sutin and C. Steel, *J. Am. Chem. Soc.* **89**, 5107 (1967).
[13] J. H. Beynon, R. A. Saunders and A. E. Williams, *Ind. Chim. Belge.* **29**, 311 (1964).
[14] M. M. Bursey and F. W. McLafferty, *J. Am. Chem. Soc.* **88**, 5023 (1966).
[15] C. G. Rowland, J. H. D. Eland and C. J. Danby, *Chem. Comm.* 1535 (1968).
[16] M. L. Vestal, *J. Chem. Phys.* **41**, 3997 (1964).
[17] W. Higgins and K. R. Jennings, *Trans. Faraday Soc.* **62**, 97 (1966).
[18] J. H. Beynon and A. E. Fontaine, *Chem. Comm.* 717 (1966).
[19] M. Barber, K. R. Jennings and R. Rhodes, *Zeitschrift für Naturforschg.* **22a**, 15 (1967).
[20] J. H. Beynon in *Advances in Mass Spectrometry*, E. Kendrick, ed., Vol. 4, The Institute of Petroleum, London, 1968, p. 129.
[21] J. H. Beynon, J. A. Hopkinson and G. R. Lester, *Int. J. Mass Spec. and Ion Phys.* **2**, 343 (1968).
[22] R. M. Elliott, S. Jones and K. R. Jennings, private communication; see also J. H. Beynon, W. E. Baitinger, J. W. Amy and T. Komatsu, *J. Mass Spec. Ion Phys.* **3**, 47 (1969).
[23] R. G. Cooks and J. H. Beynon, *Chem. Comm.* 1282 (1971).
[24] D. P. Stevenson, *Disc. Faraday Soc.* **10**, 35 (1951).
[25] See, for example, H. E. Audier, *Org. Mass Spec.* **2**, 283 (1969).
[26] F. P. Lossing and G. P. Semeluk, *Can. J. Chem.* **48**, 955 (1970).
[27] I. Howe, D. H. Williams, D. G. I. Kingston and H. P. Tannenbaum, *J. Chem. Soc.* (**B**), 439 (1969).

6. The relationship between fragmentation pattern and structure

The subject matter of this chapter has been given in detail in several other texts,[1–5] and hence the present treatment will concentrate on (a) general principles, (b) concise presentation, and (c) the most common classes of compounds.

6-1. Molecular ions. Isotope abundances

The great advantage of mass spectrometry, relative to other analytical techniques which are employed in chemistry, is determination of the integral molecular weight of the compound being investigated if a low-resolution instrument is used, and determination of the molecular formula if a high-resolution instrument is used.

The above generalizations need some qualification, since not all compounds give molecular ions, but in a representative cross section of organic compounds up to molecular weight 600, somewhere around 80 per cent may be expected to do so. Molecular ion abundance can be a very useful criterion in assessing the type of compound under investigation. The fraction of ions collected as molecular ions will be roughly proportional to the activation energy for the most facile fragmentation, all other things being equal (see also Table 4-3). Compounds which do not give molecular ions may not do so because (a) the energy of activation for decomposition is very low or zero and no M^+ ions survive to reach the collector, or (b) the sample decomposes thermally prior to ionization. Table 6-1 gives a few loose classifications which may serve as a useful guide to molecular ion abundances in electron-impact mass spectra determined in the range 20–70 eV. 'Strong' implies the molecular ion as the largest peak (base peak) or carrying more than, say, 30 per cent of the total ion current; 'weak' implies a molecular ion of only a few per cent of the abundance of the base peak, and 'medium' applies to intermediate situations.

While it is possible to say that aromatic hydrocarbons will always give strong molecular ion peaks, and in tertiary alcohols the peaks will be negligible or absent, it is difficult to be precise about behaviour between these extremes. For example, in compounds ArCOR or $ArCH_2R$, the molecular

Table 6-1

Molecular ion abundances in relation to molecular structure

Strong	Medium†	Weak or absent
Aromatic hydrocarbons (ArH)	Conjugated olefins	Long-chain aliphatic compounds
	Ar⫶Br	
ArF	Ar⫶I	Branched alkanes
ArCl	ArCO⫶R	Tertiary aliphatic alcohols
ArCN	$ArCH_2$⫶R	Tertiary aliphatic bromides
$ArNH_2$	$ArCH_2$⫶Cl	Tertiary aliphatic iodides

† In this column, wavy lines indicate a relatively weak bond.

ion abundance will decrease quite rapidly with increasing size of the alkyl group R. Lengthening the size of an alkyl chain will reduce the M^+ abundance because (a) a larger R group may interact with Ar— or ArCO— to give a new, low-energy fragmentation, and (b) larger, more-branched radicals R tend to be lost more readily because of their greater stability; thus H· is classed as an unstable radical and CH_3·, C_2H_5·, $(CH_3)_2CH$·, $(CH_3)_3C$· have progressively increasing stabilities. Effects (a) and (b) usually more than offset the opposing effect which results in less average energy per bond as molecular size increases in a homologous series.

All singly charged ions in the mass spectrum which contain carbon will also give rise to a peak at one mass unit higher. This arises because of the natural abundance of ^{13}C (1·1 per cent). For an ion containing n carbon atoms, the abundance of the isotope peak will be $n \times 1{\cdot}1$ per cent of the ^{12}C-containing peak. Thus $C_5H_{12}^{+\cdot}$, $C_{40}H_{70}^{+\cdot}$ and $C_{100}H_{170}^{+\cdot}$ would give isotope peaks at one mass unit higher of approximate abundances 5·5 per cent, 44 per cent and 110 per cent of the abundance of the ions containing ^{12}C only. These isotope peaks are due to $^{13}C^{12}C_4H_{12}^{+\cdot}$ etc. Since most of the compounds which are encountered will contain considerably less than 100 carbon atoms, the ^{13}C isotope peaks are usually appreciably less abundant than the molecular ion. Obviously, the probability of finding two ^{13}C

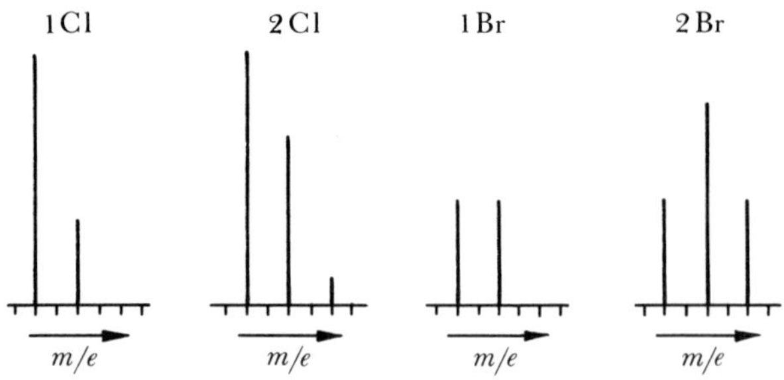

Figure 6-1

atoms in an ion is very low and M + 2 peaks are accordingly of very low abundance in most cases, as may be seen by reference to some of the spectra reproduced in this chapter.

Although iodine and fluorine are monoisotopic, chlorine consists of ^{35}Cl and ^{37}Cl in the ratio of approximately 3:1 and bromine of ^{79}Br and ^{81}Br in the ratio of approximately 1:1. Molecular ions (or fragment ions) containing 1 or 2 chlorine or bromine atoms therefore give rise to the patterns shown in Fig. 6-1.

Obviously, the isotope patterns to be expected from any combination of elements can readily be calculated, and provide a useful test of ion composition in those cases where polyisotopic elements are involved (see also section 6-6A). Most of the remaining elements normally found in organic compounds are essentially monoisotopic, with the exception of sulphur and silicon (Table 6-2).

Table 6-2

Atomic weights and approximate natural abundance of some isotopes

Isotope	Atomic weight (^{12}C = 12·000000)	Natural abundance (per cent)
^{1}H	1·007825	99·985
^{2}H	2·014102	0·015
^{12}C	12·000000	98·9
^{13}C	13·003354	1·1
^{14}N	14·003074	99·64
^{15}N	15·000108	0·36
^{16}O	15·994915	99·8
^{17}O	16·999133	0·04
^{18}O	17·999160	0·2
^{19}F	18·998405	100
^{28}Si	27·976927	92·2
^{29}Si	28·976491	4·7
^{30}Si	29·973761	3·1
^{31}P	30·973763	100
^{32}S	31·972074	95·0
^{33}S	32·971461	0·76
^{34}S	33·967865	4·2
^{35}Cl	34·968855	75·8
^{37}Cl	36·965896	24·2
^{79}Br	78·918348	50·5
^{81}Br	80·916344	49·5
^{127}I	126·904352	100

In relation to isotope patterns in mass spectra, it is worthy of mention that simple chemical reagents of mixed, but known, isotope content may be used to determine the number of given functionalities in a molecule. For example, esterification with a 50:50 mixture of CH_3OH and CD_3OH of a compound containing an unknown number of carboxylic acid groups, followed by observation of the mass spectrum in the molecular ion region, would yield information on the number of ester groupings. For one ester group, the molecular ion would consist of two equally abundant peaks three mass units apart (due to CO_2CH_3 and CO_2CD_3), whilst two ester groups would yield a molecular abundance ion pattern of 1 : 2 : 1 at respective masses of M, $M + 3$ and $M + 6$ where M is the mass of the undeuteriated ester. The isotope patterns for larger numbers of functionalities are a matter of simple permutations and mixtures of other isotope reagents may be designed to fit different systems, e.g., the use of $(CH_3CO)_2O/(CD_3CO)_2O$ to determine the number of groups which can be acetylated in a given molecule.

It is very useful to remember that in compounds containing C, H, O only, or C, H, O plus an even number of nitrogen atoms, the molecular weight must be *even*. If a molecule contains C, H, O and an odd number of nitrogen atoms only, then its molecular weight will be *odd*. This generalization applies to all stable even-electron molecules, and a substance such as NO constitutes an exception only because it is a stable radical. The ion at highest mass (except for the ^{13}C isotope peak) will usually be the molecular ion, but in the case of an unknown compound this is an unwarranted assumption because (a) the molecular ion may be absent or (b) the peak at highest mass may be associated with an impurity. The latter possibility may be checked by testing whether the ions at lower mass are related to the supposed molecular ion by the expulsion of plausible neutral particles. For example, differences such as 1, 15 and/or 18 mass units between a supposed molecular ion and the nearest fragment ions encourage the belief that the ion at highest mass is indeed a molecular ion since the loss of H, CH_3 and/or H_2O is relatively common from molecular ions containing C, H, O (although there are of course many C, H, O compounds from which these losses do not occur). An extensive list of 'common losses' observed from molecular ions in mass spectra is given in Table 6-3; plausible structural inferences for each type of loss are given, but it is emphasized that these are only plausible and should preferably be checked by other techniques.

In addition, many other inferences are possible for the larger masses lost which are listed in Table 6-3; the number of entries is limited only to keep the table fairly concise, and only the more common of the possible inferences are included. The structural information which can securely be inferred from mass spectra alone is frequently limited. Fortunately the information often complements that available from infrared, ultraviolet and nuclear magnetic resonance spectroscopy, techniques which should always be used in conjunction with mass spectrometry if possible. In relation to the recognition of molecular ions, it is important to note that the 'loss' of 14 mass units should always make one suspect the presence of a homologue, differing in formula by a CH_2— unit. The direct loss of methylene from M^+ is almost never observed since methylene is such a high-energy neutral species. In compounds containing only C, H, O, N, the loss of 5–13 units is virtually impossible, since loss of many hydrogen atoms (or molecules) would have too high an energy requirement. Loss of 3–5 hydrogens is rarely observed, and usually occurs when a compound is being dehydrogenated in the inlet system. In

Table 6-3
Some common losses from molecular ions

Ion	Groups commonly associated with the mass lost	*Possible* inference
M − 1	H	—
M − 2	H_2	—
M − 14	—	Homologue?
M − 15	CH_3	—
M − 16	O	Ar—NO_2, $\geqslant \overset{+}{N}—\overset{-}{O}$, sulphoxide
M − 16	NH_2	$ArSO_2NH_2$, —$CONH_2$
M − 17	OH	—
M − 17	NH_3	—
M − 18	H_2O	Alcohol, aldehyde, ketone, etc.
M − 19	F	Fluorides
M − 20	HF	Fluorides
M − 26	C_2H_2	Aromatic hydrocarbon
M − 27	HCN	Aromatic nitriles; Nitrogen heterocycles
M − 28	CO	Quinones
M − 28	C_2H_4	Aromatic ethyl ethers; Ethyl esters, *n*-propyl ketones
M − 29	CHO	—
M − 29	C_2H_5	Ethyl ketones, Ar—*n*-C_3H_7
M − 30	C_2H_6	—
M − 30	CH_2O	Aromatic methyl ether
M − 30	NO	Ar—NO_2
M − 31	OCH_3	Methyl ester
M − 32	CH_3OH	Methyl ester
M − 32	S	—
M − 33	$H_2O + CH_3$	—
M − 33	HS	Thiols
M − 34	H_2S	Thiols
M − 41	C_3H_5	Propyl ester
M − 42	CH_2CO	Methyl ketone; Aromatic acetate, $ArNHCOCH_3$
M − 42	C_3H_6	*n*- or *iso*-butyl ketone, Aromatic propyl ether, Ar—*n*-C_4H_9
M − 43	C_3H_7	Propyl ketone, Ar—*n*-C_4H_9
M − 43	CH_3CO	Methyl ketone
M − 44	CO_2	Ester (skel. rearr.); Anhydride
M − 44	C_3H_8	—
M − 45	CO_2H	Carboxylic acid
M − 45	OC_2H_5	Ethyl ester
M − 46	C_2H_5OH	Ethyl ester
M − 46	NO_2	Ar—NO_2
M − 48	SO	Aromatic sulphoxide
M − 55	C_4H_7	Butyl ester
M − 56	C_4H_8	Ar—*n*-C_5H_{11}, ArO—*n*-C_4H_9; Ar—*iso*-C_5H_{11}, ArO—*iso*-C_4H_9; Pentyl ketone
M − 57	C_4H_9	Butyl ketone
M − 57	C_2H_5CO	Ethyl ketone
M − 58	C_4H_{10}	—
M − 60	CH_3COOH	Acetate

such cases, the pattern of ions leading down from M^+ usually shows an ion at each mass (see 1, Fig. 6-2). The pattern shown in 2 might lead one to expect, for example, that A and B were respectively $M^+ - CH_3$ and $M^+ - H_2O$ ions, since the *specific* loss of 3 hydrogens is not found.

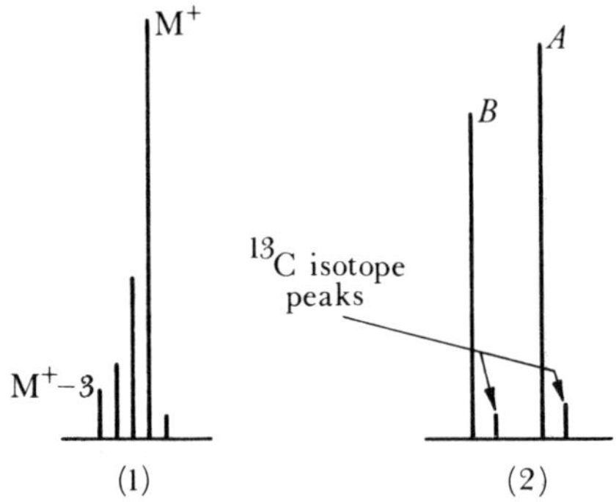

Figure 6-2

Any structural inferences drawn from Table 6-3 will have the highest probability of being significant when the mass loss in question occurs with a fair degree of specificity (rather than the case where the so-called diagnostic mass loss gives rise to a small peak, surrounded by other larger peaks or peaks of similar size). That Table 6-3 can list only a few possibilities is evident from the consideration that although the loss of 15 mass units is very likely to be loss of CH_3, loss of 57 mass units in a C, H, N, O compound can in theory correspond to loss of C_4H_9, C_3H_5O, C_2HO_2, C_3H_7N, $C_2H_5N_2$, CH_3N_3, C_2H_3NO or CHN_2O. It will be seen subsequently that the problem of structure elucidation of an unknown is sometimes best approached with the aid of a computer on-line to a high-resolution mass spectrometer (chapter 12).

6-2. Common fragment ions

The mass spectrum of an unknown can not only be interpreted by considering the masses of the neutral molecules eliminated from molecular ions, but also from a consideration of the masses of fragment ions. Table 6-4 lists the masses of some common fragment ions which may be useful in structure elucidation of simple compounds. Possible inferences should be treated with the kind of caution that was outlined with reference to Table 6-3 in section 6-1.

Inferences drawn from Table 6-4 are most likely to be significant if the peak in question is of great abundance. The Table should be used in conjunction with the information given immediately below and in sections 6-3 and 6-4, since qualifications, not conveniently expressed in the Table, are necessary in some cases.

Table 6-4

Masses and some possible compositions of common fragment ions

m/e	Groups commonly associated with the mass	*Possible* inference
15	CH_3^+	—
18	$H_2O^{+\cdot}$	—
26	$C_2H_2^{+\cdot}$	—
27	$C_2H_3^+$	—
28	$CO^{+\cdot}$, $C_2H_4^{+\cdot}$, $N_2^{+\cdot}$	—
29	CHO^+, $C_2H_5^+$	—
30	$CH_2{=}\overset{+}{N}H_2$	Primary amine?
31	$CH_2{=}\overset{+}{O}H$	Primary alcohol?
36/38(3:1)	$HCl^{+\cdot}$	—
39	$C_3H_3^+$	—
40†	Argon$^{+\cdot}$, $C_3H_4^{+\cdot}$	—
41	$C_3H_5^+$	—
42	$C_2H_2O^{+\cdot}$, $C_3H_6^{+\cdot}$	—
43	CH_3CO^+	CH_3COX
43	$C_3H_7^+$	C_3H_7X
44	$C_2H_6N^+$	Some aliphatic amines
44	$O{=}C{=}\overset{+}{N}H_2$	Primary amides
44	$CO_2^{+\cdot}$, $C_3H_8^{+\cdot}$	—
44	$CH_2{=}CH(OH)^{+\cdot}$	Some aldehydes
45	$CH_2{=}\overset{+}{O}CH_3$ $CH_3CH{=}\overset{+}{O}H$	Some ethers and alcohols
47	$CH_2{=}\overset{+}{S}H$	Aliphatic thiol
49/51(3:1)	CH_2Cl^+	—
50	$C_4H_2^{+\cdot}$	Aromatic compound
51	$C_4H_3^+$	C_6H_5X
55	$C_4H_7^+$	—
56	$C_4H_8^{+\cdot}$	—
57	$C_4H_9^+$	C_4H_9X
57	$C_2H_5CO^+$	Ethyl ketone Propionate ester
58	$CH_2{=}C(OH)CH_3^{+\cdot}$	Some methyl ketones Some dialkyl ketones
58	$C_3H_8N^+$	Some aliphatic amines
59	$COOCH_3^+$	Methyl ester
59	$CH_2{=}C(OH)NH_2^{+\cdot}$	Some primary amides
59	$C_2H_5CH{=}\overset{+}{O}H$	$C_2H_5CH(OH){-}X$
59	$CH_2{=}\overset{+}{O}{-}C_2H_5$ and isomers	Some ethers
60	$CH_2{=}C(OH)OH^{+\cdot}$	Some carboxylic acids
61	$CH_3CO(OH_2)^+$	$CH_3COOC_nH_{2n+1}$ $(n > 1)$
61	$CH_2CH_2SH^+$	Aliphatic thiol
66	$H_2S_2^{+\cdot}$	Dialkyl disulphide
69	CF_3^+	—
68	$CH_2CH_2CH_2CN^+$	—
69	$C_5H_9^+$	—
70	$C_5H_{10}^{+\cdot}$	—

† Appears as a doublet in the presence of argon from air; useful as a reference point in counting the mass spectrum.

Table 6-4 (*cont.*)

m/e	Groups commonly associated with the mass	*Possible* inference
71	$C_5H_{11}^+$	$C_5H_{11}X$
71	$C_3H_7CO^+$	Propyl ketone; Butyrate ester
72	$CH_2{=}C(OH)C_2H_5^{+\cdot}$	Some ethyl alkyl ketones
72	$C_3H_7CH{=}\overset{+}{N}H_2$ and isomers	Some amines
73	$C_4H_9O^+$	—
73	$COOC_2H_5^+$	Ethyl ester
73	$(CH_3)_3Si^+$	$(CH_3)_3SiX$
74	$CH_2{=}C(OH)OCH_3^{+\cdot}$	Some methyl esters
75	$(CH_3)_2Si{=}\overset{+}{O}H$	$(CH_3)_3SiOX$
75	$C_2H_5CO(OH_2)^+$	$C_2H_5COOC_nH_{2n+1}$ ($n > 1$)
76	$C_6H_4^{+\cdot}$	C_6H_5X; XC_6H_4Y
77	$C_6H_5^+$	C_6H_5X
78	$C_6H_6^{+\cdot}$	C_6H_5X
79	$C_6H_7^+$	C_6H_5X
79/81 (1:1)	Br^+	—
80/82 (1:1)	$HBr^{+\cdot}$	—
80	$C_5H_6N^+$	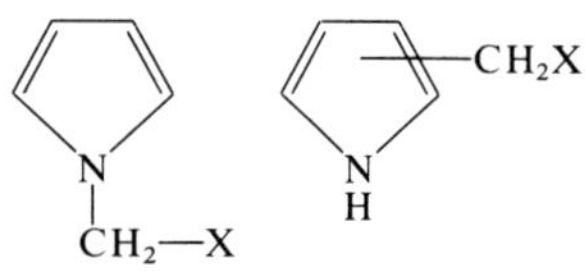
81	$C_5H_5O^+$	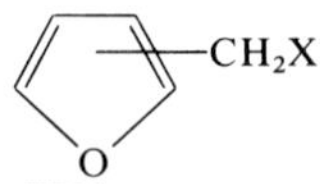
83/85/87 (9:6:1)	$HCCl_2^+$	$CHCl_3$
85	$C_6H_{13}^+$	$C_6H_{13}X$
85	$C_4H_9CO^+$	C_4H_9COX
85	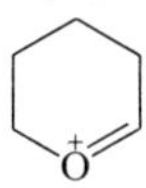	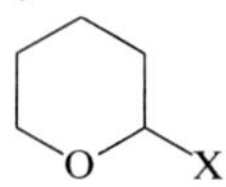
85	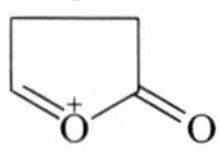	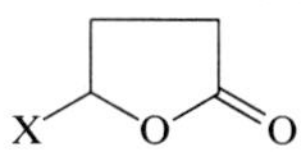
86	$CH_2{=}C(OH)C_3H_7^{+\cdot}$	Some propyl alkyl ketones
86	$C_4H_9CH{=}\overset{+}{N}H_2$ and isomers	Some amines
87	$CH_2{=}CH{-}\overset{+OH}{\overset{\Vert}{C}}{-}OCH_3$	$XCH_2CH_2COOCH_3$
91	$C_7H_7^+$	$C_6H_5CH_2X$
92	$C_7H_8^{+\cdot}$	$C_6H_5CH_2$ alkyl
92	$C_6H_6N^+$	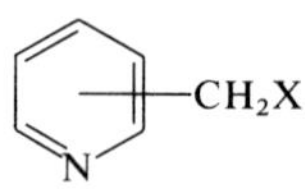
91/93 (3:1)	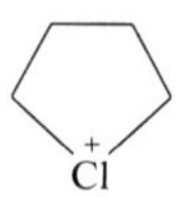	*n*-Alkyl chloride (⩾hexyl)

Table 6-4 *(cont.)*

m/e	Groups commonly associated with the mass	*Possible* inference
93/95 (1:1)	CH_2Br^+	—
94	$C_6H_6O^{+\cdot}$	C_6H_5O-alkyl (alkyl $\neq CH_3$)
94	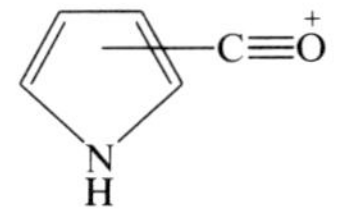	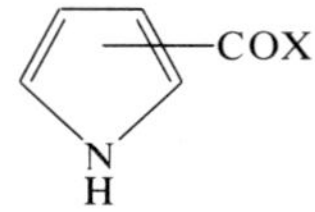
95	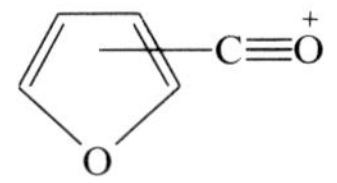	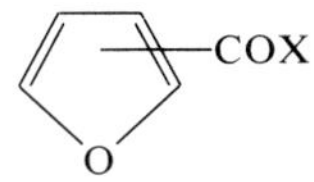
95	$C_6H_7O^+$	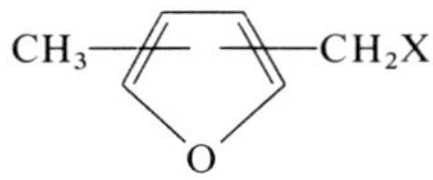
97	$C_5H_5S^+$	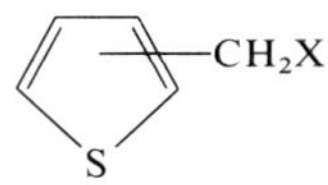
99	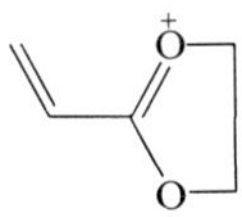	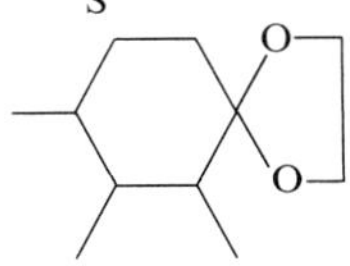
99	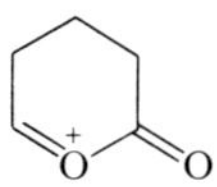	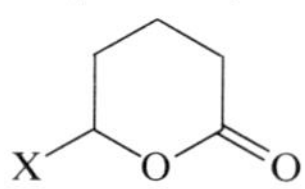
105	$C_6H_5CO^+$	C_6H_5COX
105	$C_8H_9^+$	$CH_3—C_6H_4CH_2X$
106	$C_7H_8N^+$	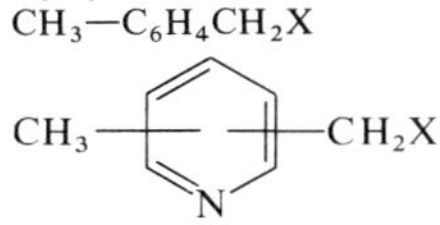
107	$C_7H_7O^+$	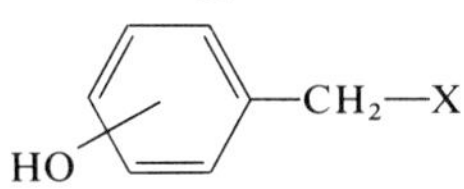
107/109 (1:1)	$C_2H_4Br^+$	—
111	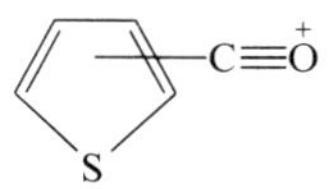	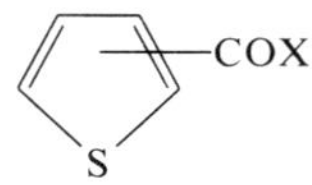
121	$C_8H_9O^+$	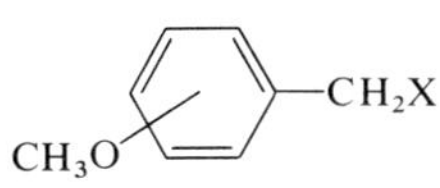
122	C_6H_5COOH	Alkyl benzoates
123	$C_6H_5COOH_2^+$	

Table 6-4 (*cont.*)

m/e	Groups commonly associated with the mass	*Possible* inference
127	I^+	—
128	$HI^{+\cdot}$	—
135/137 (1:1)	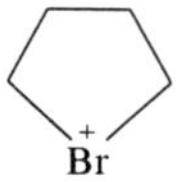	*n*-Alkyl bromide (⩾hexyl)
130	$C_9H_8N^+$	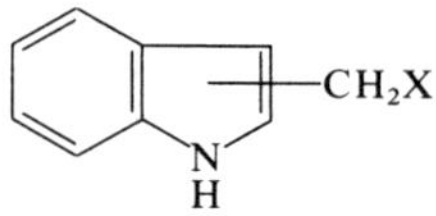
141	CH_2I^+	—
147	$(CH_3)_2Si{=}\overset{+}{O}{-}Si(CH_3)_3$	—
149	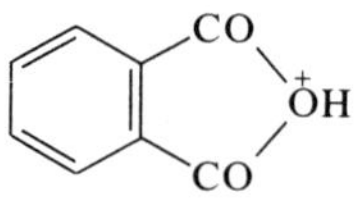	Dialkyl phthalate
160	$C_{10}H_{10}NO^+$	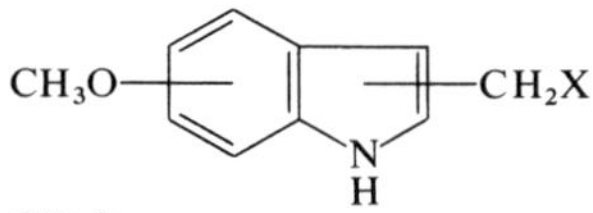
190	$C_{11}H_{12}NO_2^+$	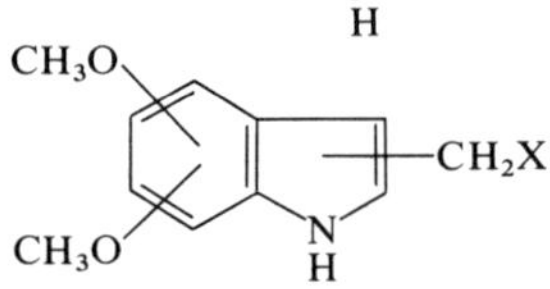

Almost no fragmentation reaction can be said to be always characteristic of a certain structural unit, since the so-called characteristic process may be pre-empted by a lower activation-energy reaction upon the introduction of a substituent grouping which fragments readily. Therefore, an appreciation of the relative ease of fragmentation of common structural units is extremely important in the interpretation of mass spectra. The relative ease of fragmentation of common aromatic substituents has already been summarized in Table 4-5. Reference to that table shows that loss of a methyl radical to give an abundant $M^+ - 15$ species should occur in virtually all structures $ArCOCH_3$ (where Ar is any aromatic nucleus, substituted or unsubstituted). In contrast, all substituents but fluorine will prevent the so-called characteristic loss of C_2H_2 from an aromatic ring.

The division of Table 6-4 into areas separated by horizontal dotted lines is useful since mass spectra usually divide into peak groups. For C-, H-, N-, O-containing compounds, the region from *m/e* 26 to *m/e* 32 covers C_2, CO, CN combinations with varying number of hydrogens, depending on the degree of unsaturation of the ion. Similarly, the region *m/e* 39–45 covers C_3, C_2O, C_2N ions, the region *m/e* 50–61 covers C_4, C_3O, C_2O_2, C_2ON, etc.

ions, and so on. Ions containing heteroatoms such as fluorine, chlorine and sulphur often fall outside these characteristic regions.

6-3. Aliphatic compounds

A. Saturated hydrocarbons. Saturated normal hydrocarbons undergo cleavage of C—C bonds with a fairly low activation energy. The molecular ion abundance is large in the mass spectrum of ethane, but decreases regularly as the size of the hydrocarbon is increased (Fig. 6-3(a)). The most abundant fragment ions are members of the series of saturated carbonium ions $C_nH_{2n+1}^+$. The *m/e*-values of these are 29, 43, 57, 71, 85, 99, etc., the most abundant ones being the lower members of the series (Fig. 6-3(a)). These ions, situated 14 mass units apart, are isobaric with the acyl ions RCO^+ (section 6-3B) so the presence of an abundant *m/e* 85 ion in a spectrum, in the presence of lower members of the series, suggests the presence of $C_6H_{13}^+$ or $C_4H_9CO^+$ units.

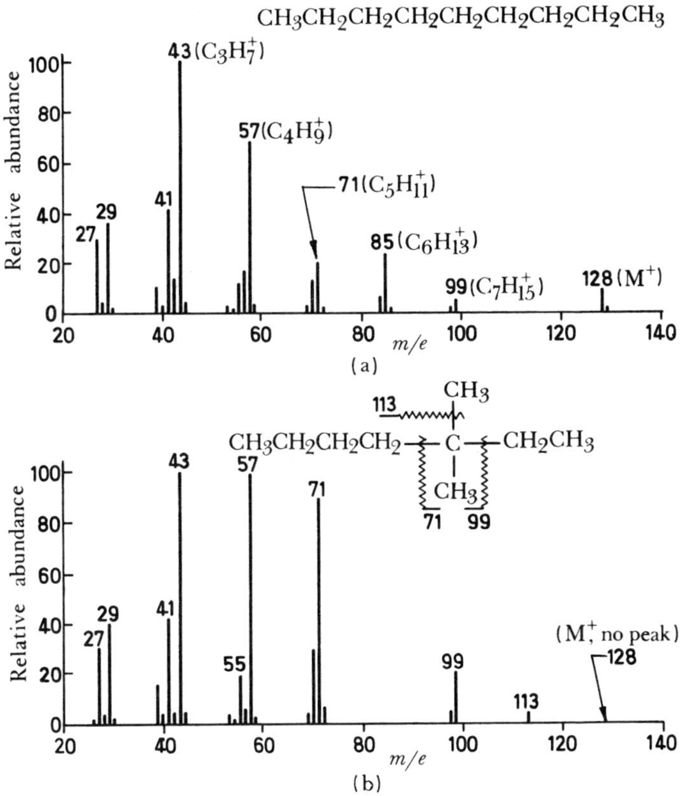

Figure 6-3

These carbonium ions decompose preferentially by loss of even-electron neutral molecules. The lowest activation energy reactions, which occur with observation of the appropriate metastable transitions, are listed below.

$$C_2H_5^+ \longrightarrow C_2H_3^+ + H_2$$

$$\underset{m/e\ 43}{C_3H_7^+} \longrightarrow \underset{m/e\ 41}{C_3H_5^+} + H_2$$

$$\underset{m/e\ 57}{C_4H_9^+} \longrightarrow \underset{m/e\ 41}{C_3H_5^+} + CH_4$$

$$\underset{m/e\ 71}{C_5H_{11}^+} \longrightarrow \underset{m/e\ 43}{C_3H_7^+} + C_2H_4$$

$$\underset{m/e\ 85}{C_6H_{13}^+} \begin{cases} \nearrow \underset{m/e\ 57}{C_4H_9^+} + C_2H_4 \\ \searrow \underset{m/e\ 43}{C_3H_7^+} + C_3H_6 \end{cases}$$

$$\underset{m/e\ 99}{C_7H_{15}^+} \longrightarrow \underset{m/e\ 57}{C_4H_9^+} + C_3H_6$$

$$\underset{m/e\ 113}{C_8H_{17}^+} \begin{cases} \nearrow \underset{m/e\ 57}{C_4H_9^+} + C_4H_8 \\ \searrow \underset{m/e\ 71}{C_5H_{11}^+} + C_3H_6 \end{cases}$$

$$\underset{m/e\ 127}{C_9H_{19}^+} \begin{cases} \nearrow \underset{m/e\ 57}{C_4H_9^+} + C_5H_{10} \\ \searrow \underset{m/e\ 71}{C_5H_{11}^+} + C_4H_8 \end{cases}$$

The effect of branching in a saturated hydrocarbon (a) reduces the molecular ion abundance and (b) causes preferential primary cleavage at the site of branching (Fig. 6-3(b); see also Table 10-2). Hence, the site of branching may be located by mass spectrometry.

The mass spectra of acyclic mono-olefins are dominated by ions of the general formula $C_nH_{2n-1}^+$, but are very similar even when the double bond is in different positions,[6a] except in those cases where the double bond is tetrasubstituted.[6b]

B. Carbonyl compounds. The even-electron rule. Carbonyl compounds which contain a γ-hydrogen atom undergo the McLafferty rearrangement (3 → a), which is analogous to the photochemical Norrish Type II rearrangement in ketones.

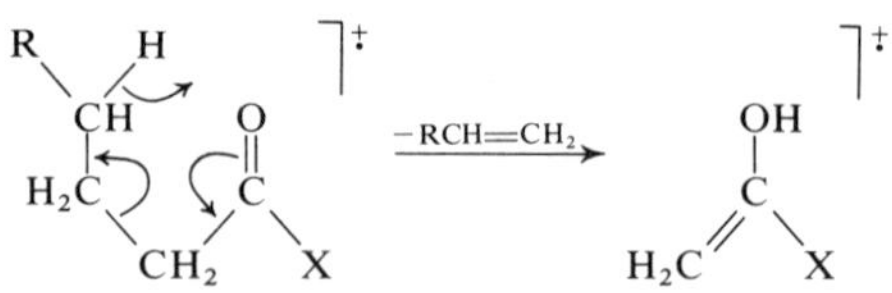

3 a

This reaction has a relatively low activation energy and even in quite complex molecules is rarely completely 'swamped' by other processes. In molecules containing only C, H, (O), ions formed by simple bond rupture, without rearrangement of a hydrogen atom or other group, will occur at *odd* m/e-values. Conversely, fragment ions formed from molecular ions [again containing only C, H, (O)] by bond cleavage with associated rearrangement of a hydrogen atom will occur at *even* masses. Thus in C-, H-, O-compounds, the McLafferty rearrangement ion will always occur at even masses. Ions formed with hydrogen rearrangement are quickly recognized by use of this criterion since examination of the mass spectra of representative C-, H- and C-, H-, O-containing compounds shows that odd m/e-values are more common than even m/e-values, especially at lower masses (see Figs. 6-3 and 6-4). This observation is related to the *even-electron rule*, which states that odd-electron ions decompose by loss of radicals or even-electron molecules, whereas even-electron ions decompose by loss of even-electron molecules. Odd-electron ions are those ions which contain an unpaired electron, and include molecular ions ($M^{+\cdot}$) and fragment ions $A^{+\cdot}$ formed from $M^{+\cdot}$ by loss of an even-electron molecule. Even-electron ions formally do not contain unpaired electrons (possess no radical character). To summarize the rule,

$$M^{+\cdot} \longrightarrow B^{+} + \text{radical}$$

or $$M^{+\cdot} \longrightarrow A^{+\cdot} + \text{even-electron molecule}$$

and $$A^{+\cdot} \longrightarrow C^{+} + \text{radical}$$

or $$A^{+\cdot} \longrightarrow D^{+\cdot} + \text{even-electron molecule}$$

but $$E^{+} \longrightarrow F^{+} + \text{even-electron molecule}$$

and *not* $$E^{+} \longrightarrow G^{+\cdot} + \text{radical.}$$

Exceptions to this rule would constitute the occurrence of the last reaction listed, and are surprisingly rare for decompositions which give metastable peaks. Exceptions to the rule are found for example in some di-iodides, which successively lose iodine radicals from the molecular ion (another way of looking at the rule is that the successive loss of radicals is 'forbidden'). However, note that all the even-electron saturated carbonium ions obey the rule in eliminating even-electron molecules in their lowest energy of activation reactions (section 6-3A). The physical origin of the rule is presumably associated with the greater inherent stability of most even-electron ions relative to ion-radicals. Since the elimination of a stable even-electron molecule (H_2O, $CH_2{=}CH_2$, HCN, CH_3COOH, etc.) is usually possible, even-electron ions will undergo such reactions while preserving their even-electron character; the energy of activation is lowered by partial bond formation (as well as bond stretching) in the transition state. In contrast, odd-electron ions have a choice between eliminating a generally less stable neutral particle (radical loss, no bond formation in the transition state) and forming a generally more stable ion (even electron), or eliminating a stable even-electron molecule and retaining some instability in the resulting ion radical.

The m/e-values of the ionized enols produced by the McLafferty rearrangement in various precursors are listed below. The values are also included in Table 6-4, and we may now add the qualification that the structural inferences require a chain of three saturated carbon atoms adjacent to the carbonyl group and a hydrogen atom in the γ-position.

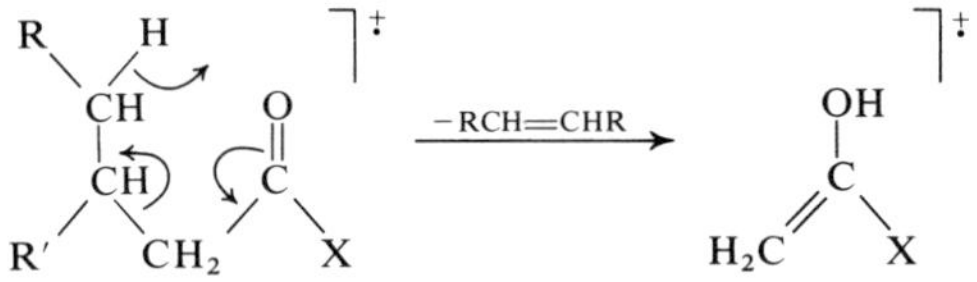

X = H	m/e 44
X = CH_3	m/e 58
X = NH_2	m/e 59
X = OH	m/e 60
X = C_2H_5	m/e 72
X = OCH_3	m/e 74.

Note that only in the case of the amide does the ion occur at an odd mass. A methyl substituent on the α-carbon atom will cause a shift in all cases by 14 mass units to higher m/e-values. Hence, in n-hexanal the McLafferty rearrangement ion occurs at m/e 44 (Fig. 6-4(a)), whereas in 2-methylpentanal it occurs at m/e 58 (Fig. 6-4(b)). The McLafferty rearrangement has a low energy of activation and therefore usually dominates low-voltage spectra more than high-voltage spectra.

Dialkyl ketones containing γ-hydrogens in each alkyl chain can undergo consecutive McLafferty rearrangements and also give rise to m/e 58, for example,

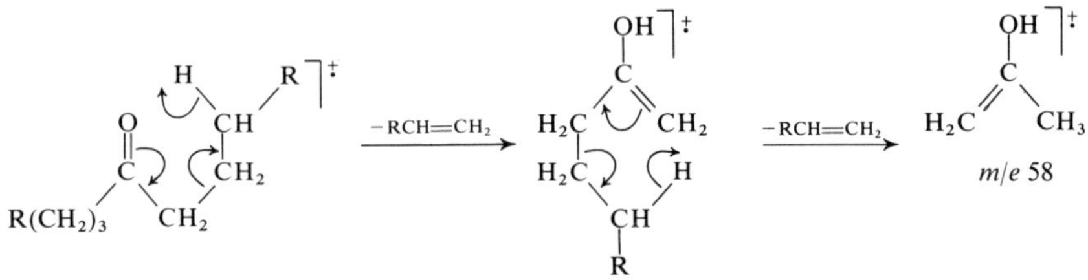

A second favoured reaction type in carbonyl compounds is cleavage of the bond adjacent to carbonyl. This is conventionally known as α-cleavage and is probably triggered by the unpaired electron on oxygen in the ground state of the ionized carbonyl group (4 → b and 4 → c, R = alkyl).

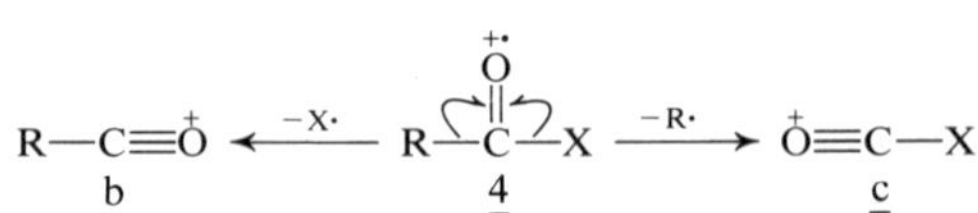

In ketones (X = alkyl) either X or R may be lost, and in methyl esters both $M^+ - R$ ($COOMe^+$, *m/e* 59) and $M^+ - OMe$ ($M^+ - 31$) ions are usually formed. Primary amides afford both $O^+{\equiv}C{-}NH_2$ (*m/e* 44) and $M^+ - NH_2$ ($M^+ - 16$) ions, but carboxylic acids do not in general lose a hydroxyl radical to a significant extent; carboxylic acids are preferably examined as their derived methyl esters (prepared by reaction with diazomethane), both from the viewpoint of volatility and interpretation of the resultant mass spectrum.

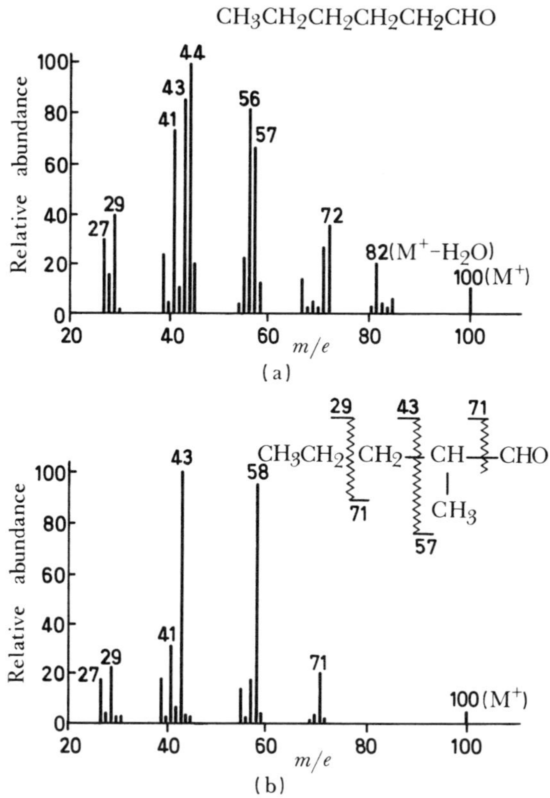

Figure 6-4

In unsymmetrical ketones RCOR′, where R and R′ are different *n*-alkyl groups, loss of the larger radical usually yields the more abundant fragment ion in 70 eV spectra. However, in low-voltage spectra loss of the smaller radical is usually dominant, indicating that this process has the lower activation energy.

The *m/e*-values corresponding to many of the common acyl ions of types b and c are listed in Table 6-4. If R is alkyl they are members of the series 29, 43, 57, 71, 85, etc. In general, they decompose further by loss of CO with an appropriate metastable peak to give alkyl ions which fall in the same series of *m/e*-values but displaced to lower masses by 28 units.

$$RCO^+ \xrightarrow[-CO]{m^*} R^+$$

A third diagnostic fragmentation reaction of carbonyl compounds corresponds to γ-cleavage.

$$\left[RCH_2 \overset{\gamma}{\wr} CH_2CH_2COX\right]^{+\cdot} \longrightarrow CH_2{=}CH{-}\overset{\overset{+}{O}H}{\overset{\|}{C}}{-}X$$

$$X = OMe,\ m/e\ 87$$

Although this ion formally arises by simple bond cleavage, it has been shown[7] by isotopic labelling that cleavage of the γ-bond is associated (often to a large extent) with hydrogen rearrangement both to and from the incipient fragment ion. The mechanistic details need not concern us here, but it may suffice to say that the portion of the chain RCH_2 provides a hydrogen for transfer to the carbonyl oxygen, and this is followed by a reciprocal hydrogen transfer from the carbon atom adjacent to the C=O group to the potential neutral particle. Several reciprocal hydrogen transfer reactions are known to occur upon electron impact and can of course only be uncovered by deuterium labelling experiments, since the fragment ion in question appears initially to have arisen by simple bond scission.

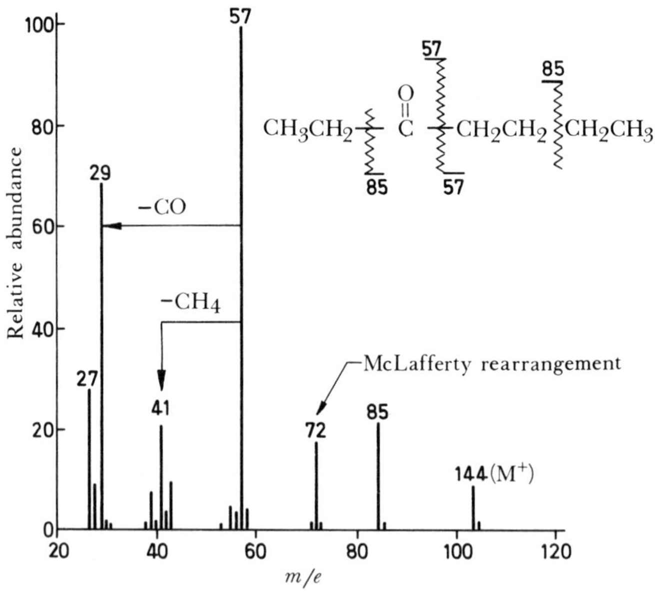

Figure 6-5

γ-Cleavage ions are particularly useful in the mass spectra of methyl esters, where they occur at m/e 87. The mass spectra of an aliphatic ketone and ester are reproduced in Figs. 6-5 and 6-6, to illustrate the abundances of the characteristic cleavage ions. In the spectra of long-chain methyl esters the formation of fragments of the general formula $(CH_2)_nCOOCH_3^+$ where $n = 2, 6, 10$, etc. is favoured. Members of this series fall at m/e 87, 143, 199, 255, 311, 367, etc. The first two members are evident in Fig. 6-6. Notice that although m/e 59 is only of similar abundance to members of the hydrocarbon series 27/29, 41/43, 55/57 in Fig. 6-6, it should be regarded as structurally significant because it appears at a larger m/e-value than the saturated C_4 carbonium ion, and therefore must contain oxygen.

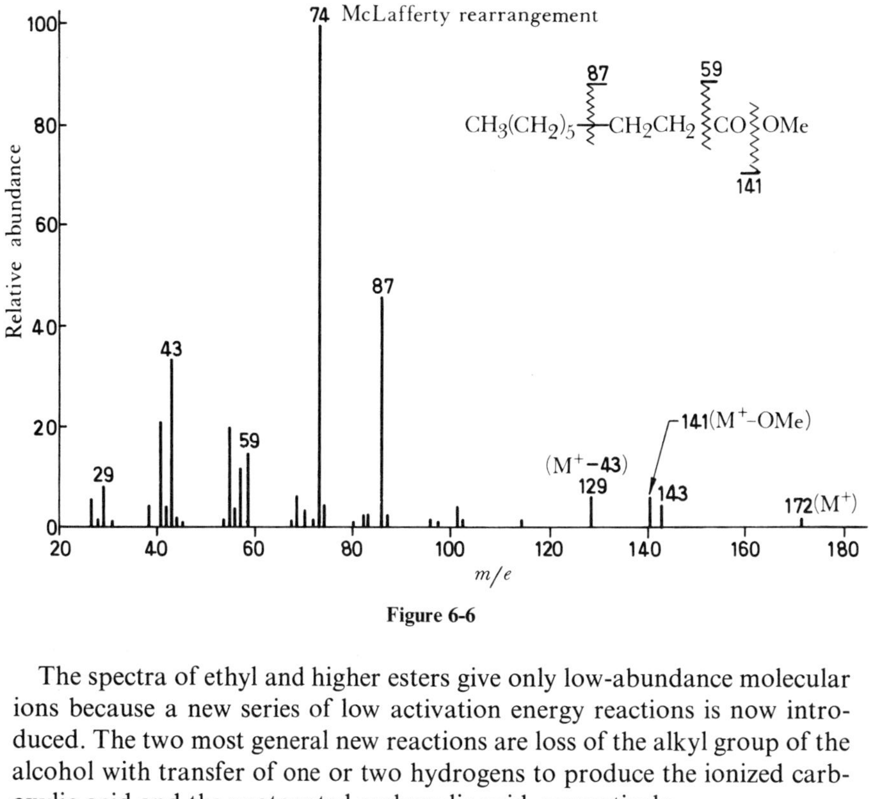

Figure 6-6

The spectra of ethyl and higher esters give only low-abundance molecular ions because a new series of low activation energy reactions is now introduced. The two most general new reactions are loss of the alkyl group of the alcohol with transfer of one or two hydrogens to produce the ionized carboxylic acid and the protonated carboxylic acid, respectively.

$$\left.\mathrm{R\;COOR'}\right]^{+\cdot} \longrightarrow \mathrm{R\;COOH}^{+\cdot} + \mathrm{R'} - \mathrm{H}$$

$$\left.\mathrm{R\;COOR'}\right]^{+\cdot} \longrightarrow \mathrm{R{-}C}\begin{matrix} \overset{+}{\mathrm{O}}\mathrm{H} \\ \mathrm{OH} \end{matrix} + \mathrm{R'} - 2\mathrm{H}.$$

In acetates, these ions fall at m/e 60 and 61, and in propionates at m/e 74 and 75. The reactions also occur in aromatic esters and in benzoates ($R = C_6H_5$) afford ions at m/e 122 and 123. The prevalence of double hydrogen transfer (relative to single hydrogen transfer) increases with increasing size of R′. The double hydrogen transfer has a particularly low activation energy and increases in importance in low-voltage spectra.

C. Alcohols, ethers and amines. The ground ionized state of a saturated aliphatic alcohol corresponds to the species d, which arises by removal of a non-bonding electron from the oxygen atom of the original alcohol. The ion d can in general further decompose to the oxonium ions e, f, and g in the case where R ≠ R′ ≠ R″. Where R, R′ and R″ are H or *n*-alkyl groups, the group with the largest number of carbon atoms is expelled to the largest extent in 70 eV spectra.

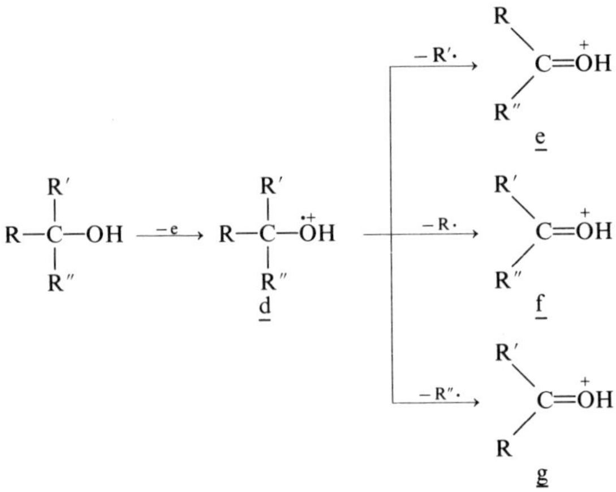

In the case of an *n*-alkanol the resulting oxonium ion ($CH_2{=}\overset{+}{O}H$) is found at m/e 31. Depending on the size of the alkyl groups attached to the sp^2-hybridized carbon atom, a series of ions corresponding to saturated oxonium ions will obviously arise at m/e 31, 45, 59, 73, 87, etc. Ions of the same masses may also be produced from saturated aliphatic ether functions (*vide infra*). Although ions which are members of the above series may be of moderate or large abundance if R, R′ and R″ are all small [e.g., m/e 45 (h) is the base peak in the mass spectrum of butan-2-ol (5) at 70 eV], they fall in abundance as the length of the alkyl chain(s) is increased.

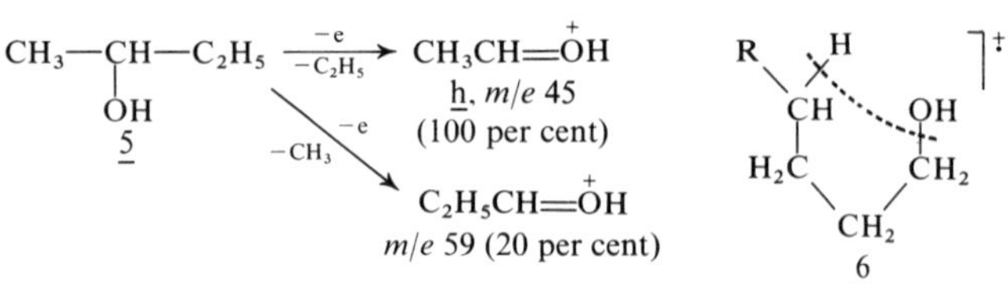

In long-chain aliphatic alcohols, the molecular ion is sometimes completely absent and, if not, it is of very low abundance. In the higher alcohols, the first detectable ion is often $M^+ - H_2O$. Unlike thermal loss of water, which is a 1,2-elimination, electron-impact-induced loss of water is a 1,4-elimination (see 6), as determined by deuterium labelling.

Low molecular weight dialkyl ethers undergo primary cleavages similar in nature to those occurring in low molecular weight alcohols, viz.:

$$CH_2{=}\overset{+}{O}{-}CH_2R' \xleftarrow{R\cdot} R\,\vdots\,CH_2{-}\overset{+\cdot}{O}{-}CH_2\,\vdots\,R' \xrightarrow{R'\cdot} RCH_2{-}\overset{+}{O}{=}CH_2$$

The resulting ions (*m/e* 45, 59, 73, 87, ...) undergo further loss of a neutral olefin if the chain, which was *not* involved in the primary reaction, is ethyl or larger. The eliminated olefin incorporates all the carbon atoms of the alkyl chain in question. Thus, in the case of di-isopropyl ether (7, Fig. 6-7), the primary simple bond cleavage can only lead to an $M^+ - 15$ ion, which then expels propylene to give i (*m/e* 45).

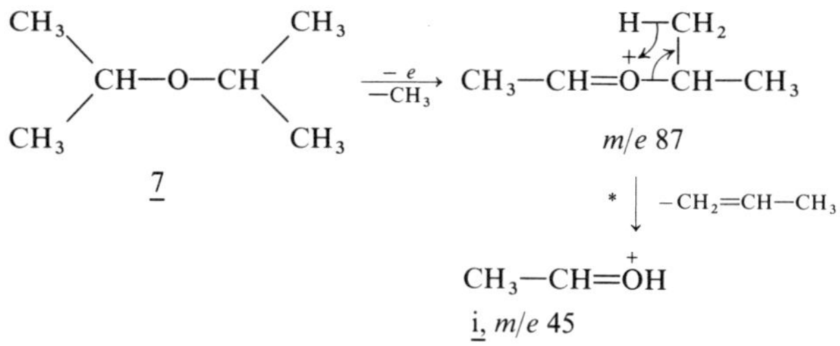

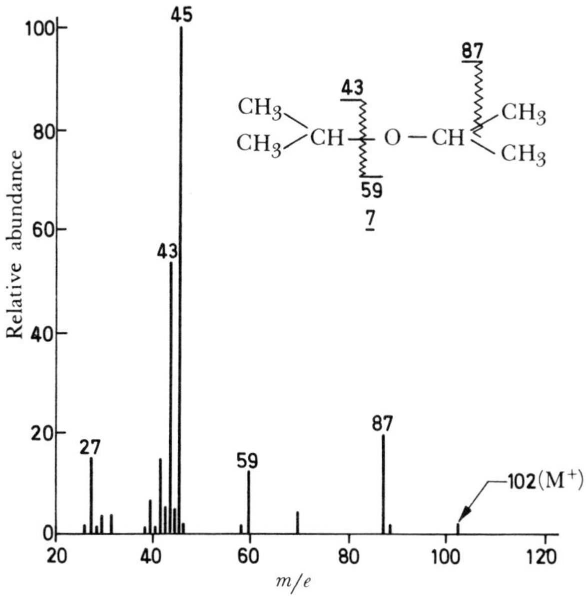

Figure 6-7

Note that the even-electron rule is obeyed, the even-electron oxonium ion decomposing by loss of an even-electron molecule. In general, the final product ions are still members of the series *m/e* 45, 59, 73, 87, ... Members of this series also arise via C—O cleavage with charge retention by the oxygen-containing portion, but this reaction has a relatively high activation energy and occurs only in ions of high internal energy (see *m/e* 59 in Fig. 6-7).

The sequence of events occurring in aliphatic amines is analogous to that occurring in ethers. Initially, a radical will be lost, via α-cleavage, with formation of an immonium ion, and this species then decomposes via olefin elimination when a second chain of two or more carbon atoms is available. Thus in the case of ethyl 1,3-dimethylbutyl amine (8, Fig. 6-8), the largest radical is preferentially lost in the 70 eV spectrum to give *m/e* 72, which then eliminates ethylene to give *m/e* 44 (j). To a lesser extent a methyl radical is expelled to give k and/or l (*m/e* 114).

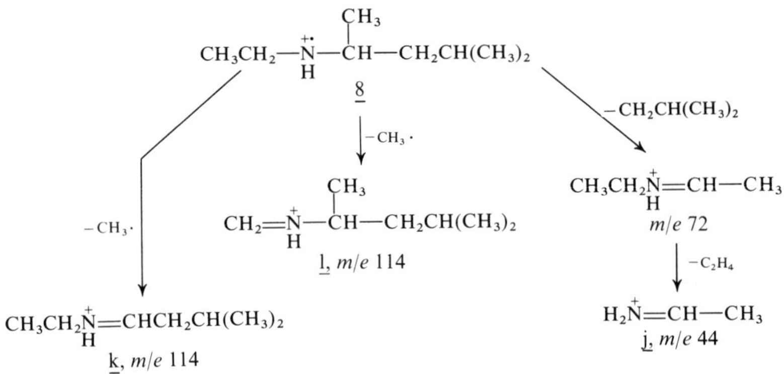

In primary amines the initial radical loss cannot be followed by the type of olefin elimination outlined above. Simple α-cleavage produces the base peak at *m/e* 30.

$$R\text{—}CH_2\text{—}\overset{+\cdot}{N}H_2 \xrightarrow{-R\cdot} CH_2\text{=}\overset{+}{N}H_2 \quad (m/e\ 30)$$

All ions which are characteristic of saturated aliphatic amine groupings occur in the series *m/e* 30, 44, 58, 72, 86, 100, 114, etc. In contrast to the cases of ethers and alcohols, this is an even *m/e*-series.

D. Alkyl halides. The unique advantage of the mass spectrometer in detecting the presence of Cl and Br due to their characteristic isotopic abundances has already been mentioned (section 6-1). Unfortunately, however, the functional groups F, Cl, Br and I do not direct fragmentation strongly. Thus ions of type m (X = F, Cl, Br, I), analogous to those found for amines and ethers, do not have a marked stability and are usually absent or of low abundance in the mass spectra of halides.

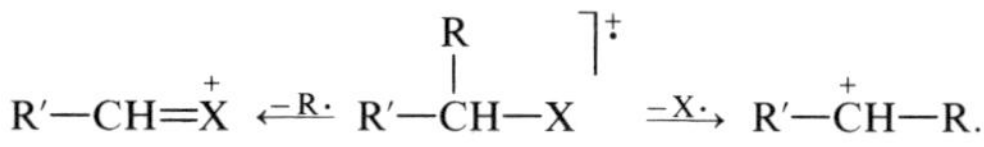

m n

Instead, loss of a halide radical (see n) is frequently preferred in the mass spectra of bromides and iodides (reflecting the relative stability of bromine and iodine radicals), while fluorides and chlorides show a more marked tendency to undergo HF and HCl loss.

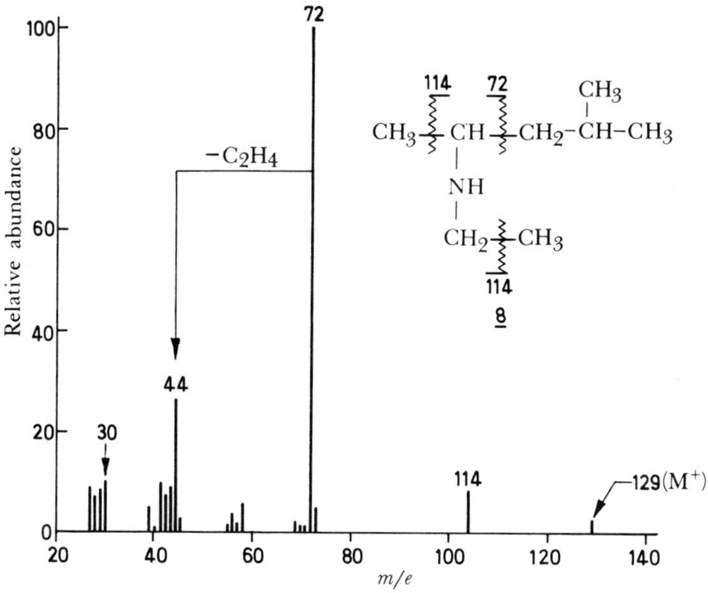

Figure 6-8

All *n*-alkyl chlorides and bromides from hexyl through octadecyl give rise to abundant ions of the formula $C_4H_8X^+$ (*m/e* 91/93 in chlorides and *m/e* 135/137 in bromides).[8] The abundance of these ions may be attributed to formation of a five-membered cyclic ion.

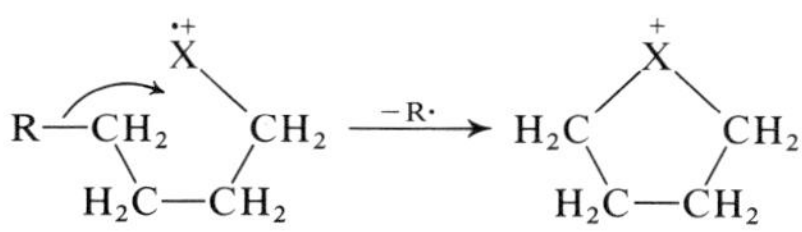

In the case of branched alkyl chlorides and bromides, the abundance of ions corresponding to substituted analogs of the cyclic ion is greatly reduced, since new low activation energy reactions corresponding to rupture of the chain at the site of branching can compete.

As is general in aliphatic compounds, the abundance of the molecular ion decreases as the chain length and/or chain branching are increased. For a given alkyl halide, the molecular ion abundance increases as the electronegativity of the halogen atom is decreased. However, in the higher *n*-alkyl halide spectra, the molecular ion is often absent or of very low abundance.

E. Relative fragment-directing ability of common functional groups in aliphatic compounds. Having indicated in sections 6-3A to 6-3D some of the lower activation energy primary and secondary reactions associated with common functional groups, it is natural to enquire whether the fragmentation associated with one functional group will dominate over another when both functionalities are present in an aliphatic molecule. It is almost impossible to make reliable predictions, since any particular arrangement in a complex molecule may lead to some entirely new (and perhaps unique) rearrangement reaction, thus 'swamping' the so-called characteristic reactions of the functional groups. However, useful guides are available. In particular, Spiteller and co-workers[9] have estimated the relative powers of common functional groups to initiate cleavage by taking a series of compounds 9 and measuring the relative ion yields (i_{OCH_3} and i_X) due to cleavages initiated by the OCH_3-group and the X-group; interpretation of the spectra was aided by reference to those of the monosubstituted 3-X-decane and 3-methoxydecane. The results are summarized in Table 6-5, and refer to 70 eV spectra obtained with a cooled ion source. The use of a cooled ion source ensures that direct thermal excitation of the molecules is minimized (see section 6-8), and hence the fragmentation pattern is relatively specific.[9] The quantitative aspects of the results are nevertheless of limited value, and

Table 6-5

Relative propensities of common functional groups to produce cleavage in compounds of the general formula 9

X	i_X†	X	i_X†
$—COOH$	2	$—SCH_3$	90
$—CH_2OH$	6	$—OCH_3$	100
$—Cl$	8	$—NHCOCH_3$	121
$—COOCH_3$	17	$—I$	131
$—Br$	23	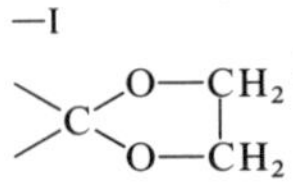 ‡	514
$—OH$	25		
$—SH$	36	$—NH_2$	616
>C=O ‡	41	$—N(CH_3)_2$	1200

† Relative to i_{OCH_3} arbitrarily taken as 100 units.
‡ Data refer to the 3-methoxy-8-ketodecane and 3-methoxy-8-ketodecane ethylene ketal derived from 9.

the quotation of specific numbers serves primarily to indicate where major changes in the fragment-controlling abilities of groups are found.

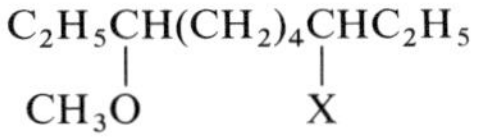

9

As might have been anticipated, groups leading to the most stable onium ions, e.g.,

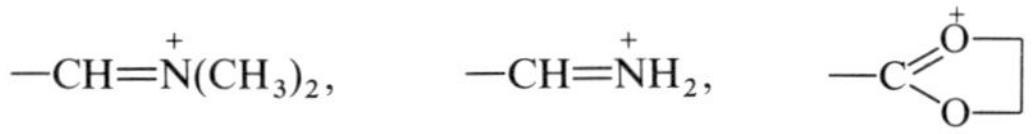

direct cleavage (α-cleavage) the most strongly. Although iodine falls in a part of the table which suggests that it controls fragmentation efficiently, this is not through the formation of stable onium ions, but rather through loss of an iodine radical which does not give information as to the site of iodine substitution.

The qualitative trends may be discerned by reference to the mass spectra[9] of 8-chloro-3-methoxydecane (10), 8-methylmercapto-3-methoxydecane (11), and 3-methoxydecan-8-one ethylene ketal (12) (see Figs. 6-9 to 6-11). In the case of the chloro-compound (10), the spectrum is dominated by α-cleavage adjacent to the methoxyl group with loss of the larger radical to give *m/e* 73. The $M^+ - 29$ ion is associated essentially completely with α-cleavage adjacent to the methoxyl group with loss of the smaller (ethyl) radical to give *m/e* 177/179, which then decomposes further as indicated below.

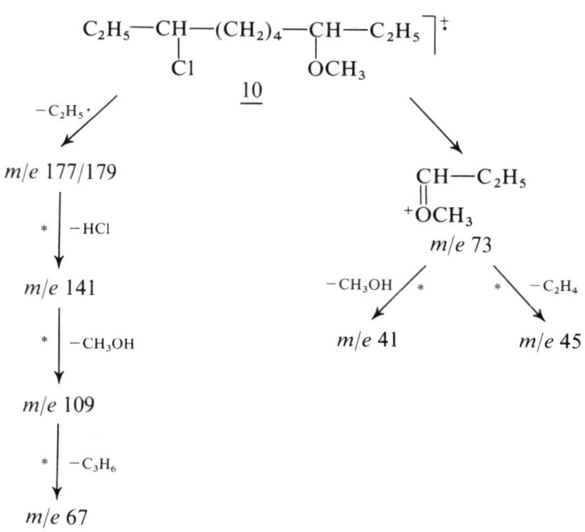

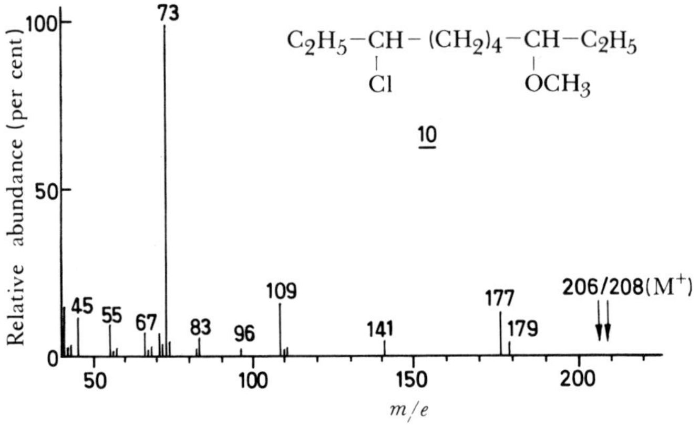

Figure 6-9

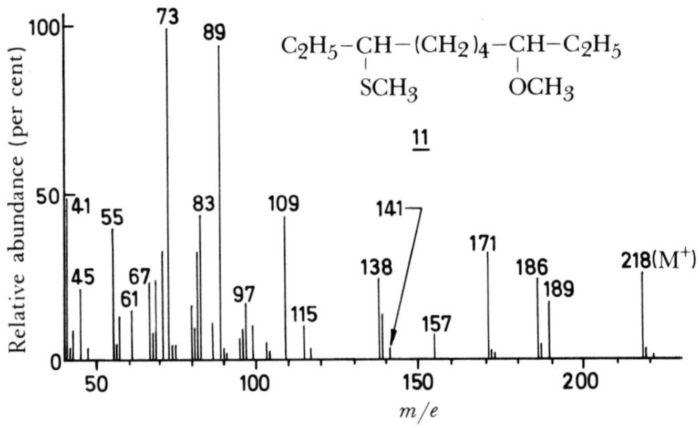

Figure 6-10

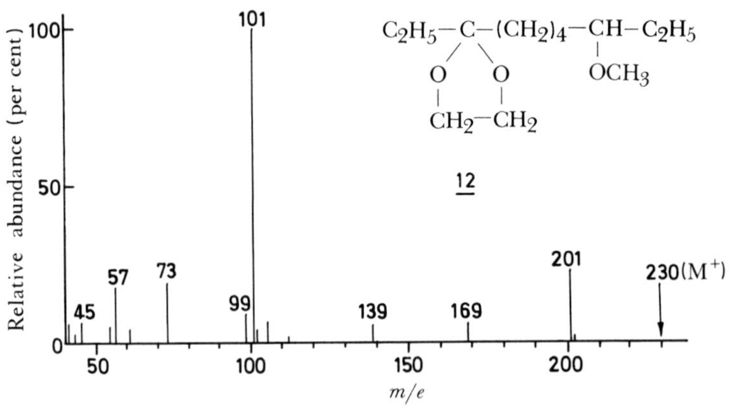

Figure 6-11

In general, where α-cleavage can occur at alternative sides of a functional group with loss of radicals of unequal sizes, all other things being equal (e.g., primary radical loss vs. primary radical loss, or secondary vs. secondary), the loss of the larger radical will be dominant in 70 eV spectra (as in the above case). However in low-voltage spectra (e.g., 12 eV), or in the case of metastable transitions where discernible, the loss of the smaller radical is almost always dominant. Apparently, the loss of the smaller radical occurs with the lower activation energy, but loss of the larger radical becomes the faster reaction in ions of higher internal energy. Obviously, the behaviour could be rationalized in terms of a lower frequency factor for the loss of the smaller radical (section 4-1), but such a rationalization does not appear to have a clear physical basis and the phenomenon is not clearly understood at this time.

In the spectrum (Fig. 6-10) of the thioether (11), the two possible α-cleavage products of differing mass (*m/e* 73 and 89) are of similar abundance, with the former decomposing as indicated above and the latter undergoing losses of C_2H_4 and CH_3SH (to *m/e* 61 and 41, respectively). The primary losses of C_2H_5, CH_3OH and SCH_3 from M^+, lead to *m/e* 189, 186 and 171, with further decomposition as indicated in the reaction scheme.

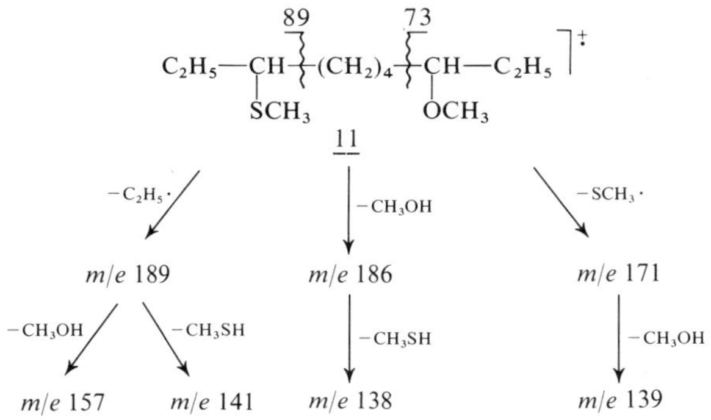

Figure 6-11 needs no comment except to note that the activation energy to form *m/e* 101 is so low that the ion yield of *m/e* 73 is greatly reduced relative to the previous cases.

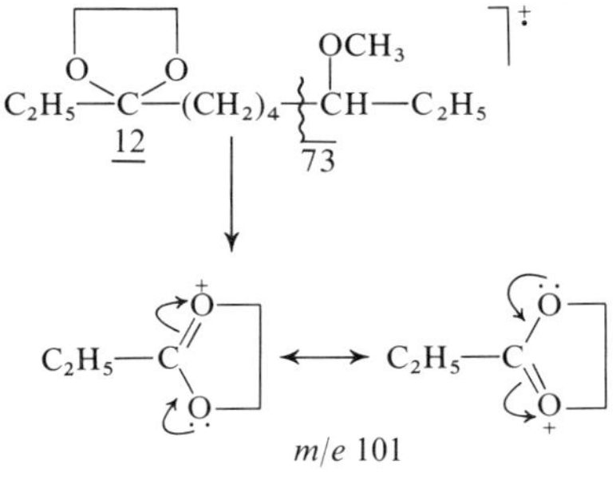

6-4. Aromatic compounds

The major fragmentation paths of compounds C_6H_5X for most common X substituents are given in Table 4-5. In disubstituted aromatics, fragmentation of those substituents X entered near the top of Table 4-5 will in general dominate over, or completely swamp, the cleavages normally associated with X groups entered near the foot of the Table.

In some cases the X groups undergo, at 70 eV, characteristic secondary reactions which follow the primary steps already discussed in chapter 4. These secondary reactions are summarized in Table 6-6.

Table 6-6

Primary and secondary decompositions of some molecular ions C_6H_5X

X	Group expelled in first step	Group expelled in second step
$COCH_3$	CH_3	CO
$C(CH_3)_3$	CH_3	C_2H_4
$COOCH_3$	OCH_3	CO
CHO	H	CO

The above generalizations dispose of most of the main features of the mass spectra of mono- and di-substituted aromatics for small X groups and *any* aromatic nucleus, except where particularly favourable or unfavourable interactions of substituents and/or heteroatoms occur. This point is demonstrated by reference to the 70 eV mass spectrum of 5-bromo-2-carbometh-

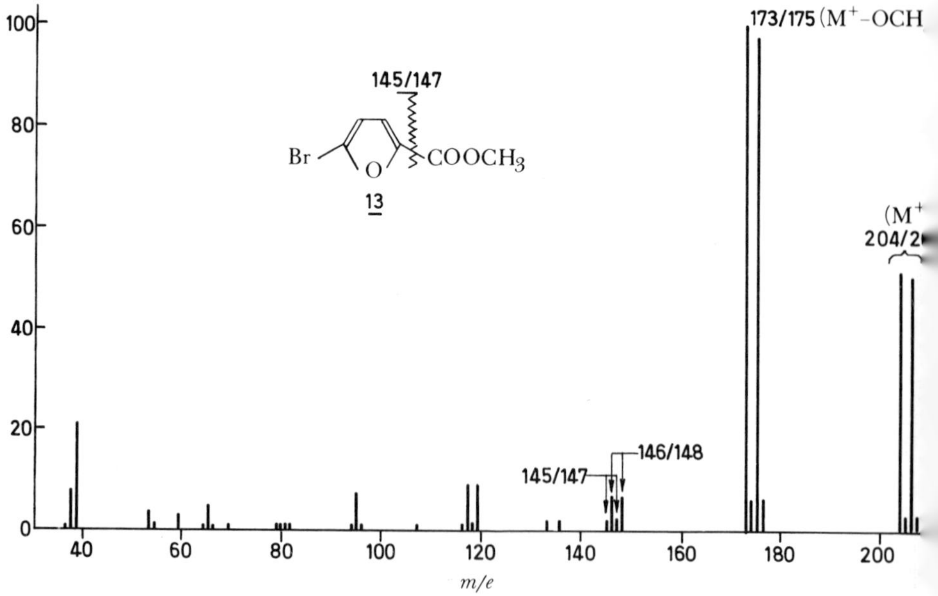

Figure 6-12

oxyfuran (13, Fig. 6-12), where the anticipated OCH_3 loss completely quenches Br loss as a primary fragmentation.

Complicating interactions, which cause changes from the behaviour anticipated on the basis of the spectra of the monosubstituted compounds, may be divided into the two main categories of resonance effects and ortho effects. The former class is easy to understand and even predict. First, those substituents X [$N(CH_3)_2$, $NHCH_3$, NH_2, OCH_3, SCH_3, OH, etc.] which donate electrons to the reaction centre Y—Z through a mesomeric effect enhance loss of Z as a radical (see 14 → 15).

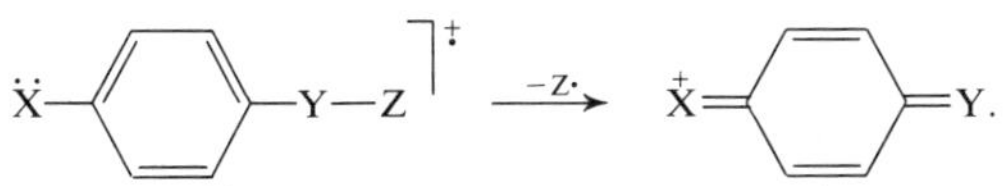

14 15

Such effects can operate optimally only when the two substituents are *o*- or *p*-oriented. An example is found in the 'normal' loss of CH_2O from the OCH_3 group of *m*-methoxyaniline (Fig. 6-13(b)), in contrast to loss of a CH_3 radical as the lowest activation energy reaction of the molecular ion of *o*-methoxyaniline (Fig. 6-13(a)).

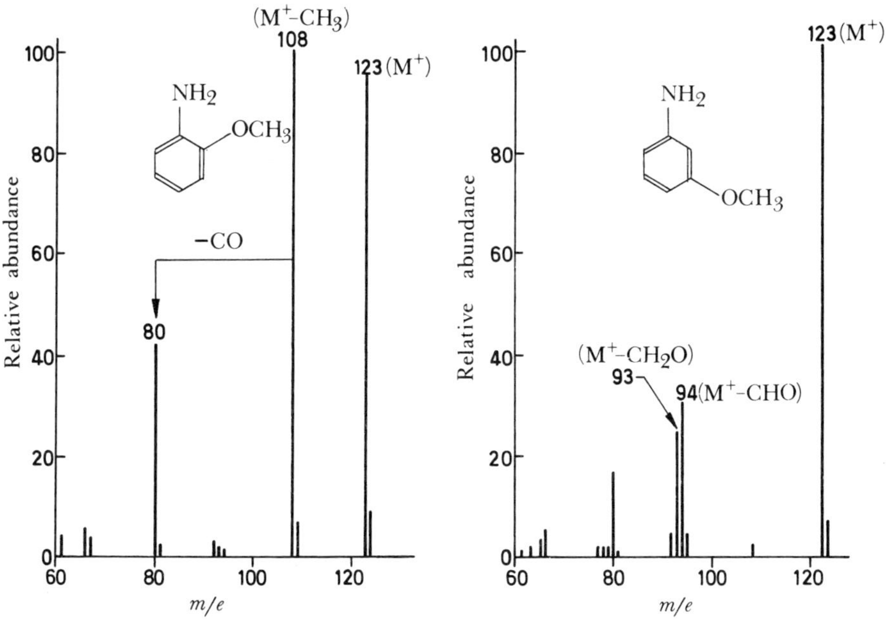

Figure 6-13

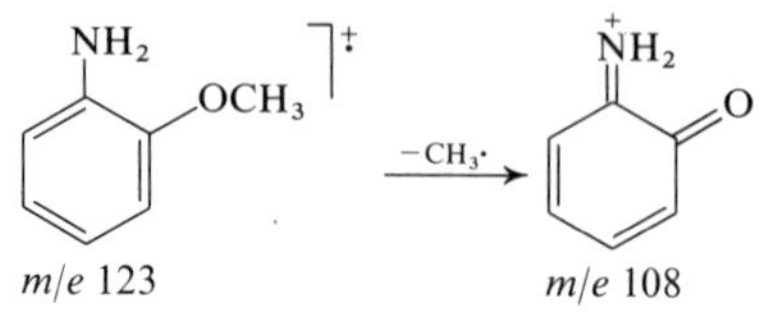

Second, substituents X (NO_2, COR, CHO, COOR, etc.) which withdraw electrons from the reaction centre Y—Z through a mesomeric effect make the loss of Z as a radical less favourable. For example, the 'characteristic' reactions of OCH_3, CH_3 and NO_2 substituents are not grossly different in their energy requirements (A.P.s of 11·30, 11·80 and 12·16 eV for CH_2O, H and NO_2 loss, respectively, in the monosubstituted benzenes). In the absence of interactions between the substituents in events leading up to the transition state, *p*-methoxytoluene would be anticipated to undergo predominantly CH_2O loss, and *p*-nitrotoluene hydrogen atom loss. However, since the energy requirements for all the reactions are similar, resonance effects should be considered. It is anticipated that the energy of activation for H loss from *p*-nitrotoluene will be raised (NO_2 destabilizes the carbonium ion 16) and for H loss from *p*-methoxytoluene will be lowered (17).

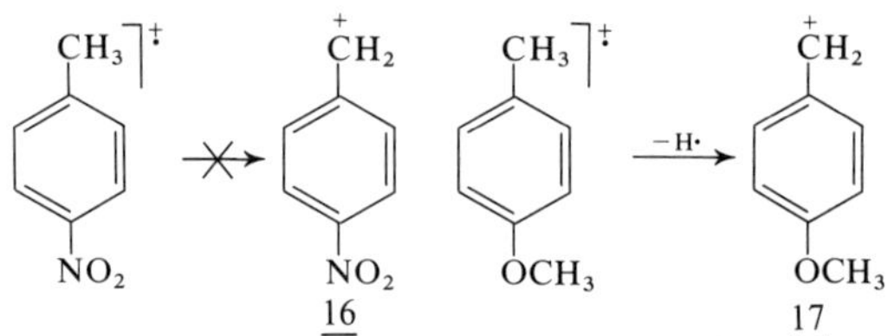

In fact, $M^+ - H$ is the most abundant daughter ion arising from primary decomposition of *p*-methoxytoluene (Fig. 6-14) and hydrogen radical loss from *p*-nitrotoluene is negligible relative to NO and NO_2 loss (Fig. 6-15).

Similarly, since carbonium ions 18 (e.g., X = O, NR) are energetically unfavourable, methyl radical loss from 2-ethylpyridine (19) is almost negligible, while from 3-ethylpyridine (20) it accounts for the base peak.

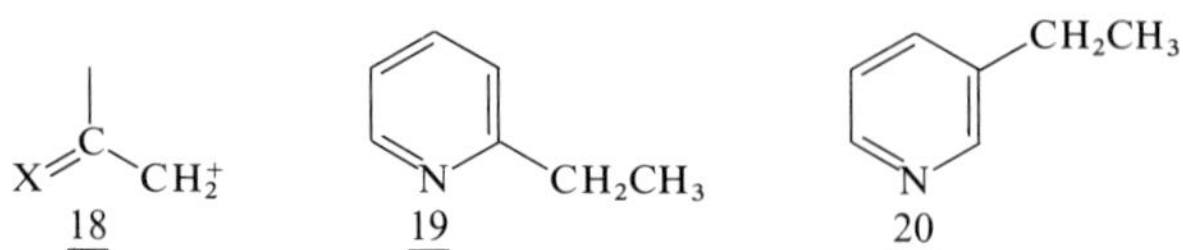

'Ortho effects' are those in which adjacent substituents interact directly through space, so that a neutral particle, whose constituents originate in part from each substituent, may be eliminated with a relatively low activation energy. Thus, the presence of a methyl group *ortho* to the nitro-function in *o*-nitrotoluene introduces a new, low-energy process, namely loss of a hydroxyl radical (Fig. 6-16, contrast Fig. 6-15). One of the most useful *ortho*-effects in structure elucidation is associated with ROH loss (in competition

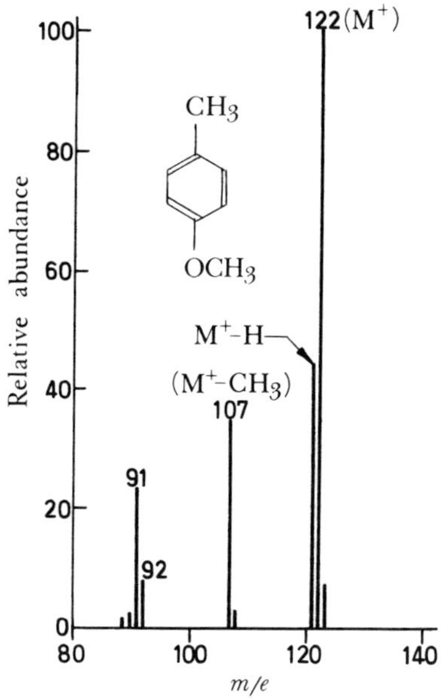

Figure 6-14

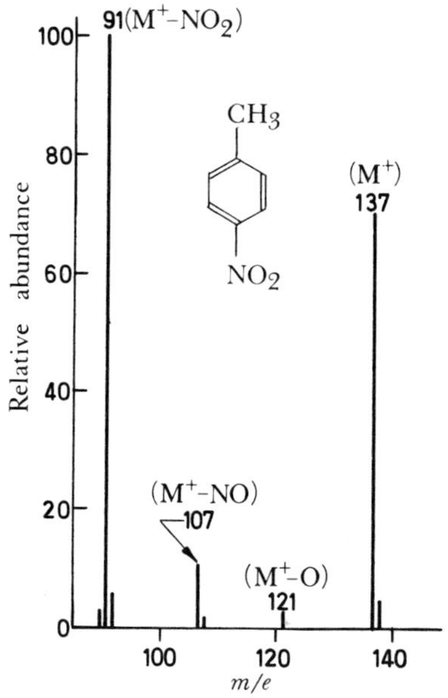

Figure 6-15

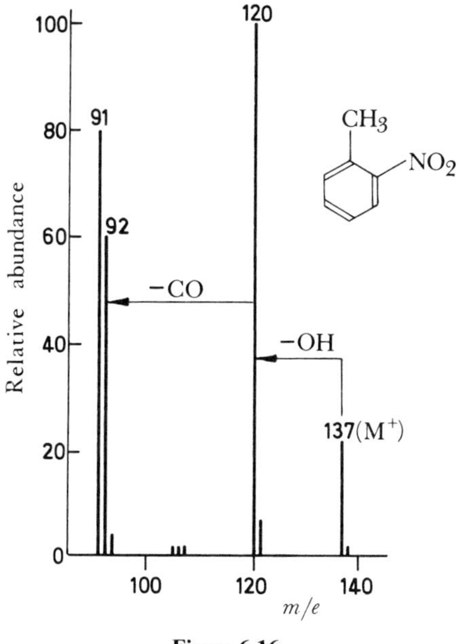

Figure 6-16

with OR loss) from various kinds of aromatic esters carrying XH groups in the *ortho* position.

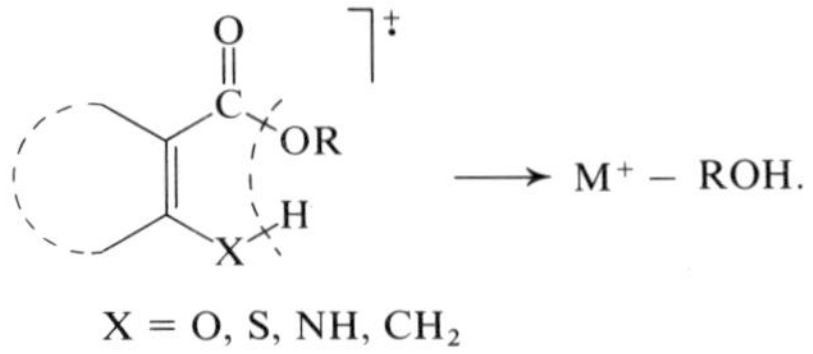

This *ortho*-effect is illustrated in the mass spectrum (Fig. 6-17) of ethyl 3-hydroxythiophen-2-carboxylate, where the base peak arises due to loss of ethanol from the molecular ion.

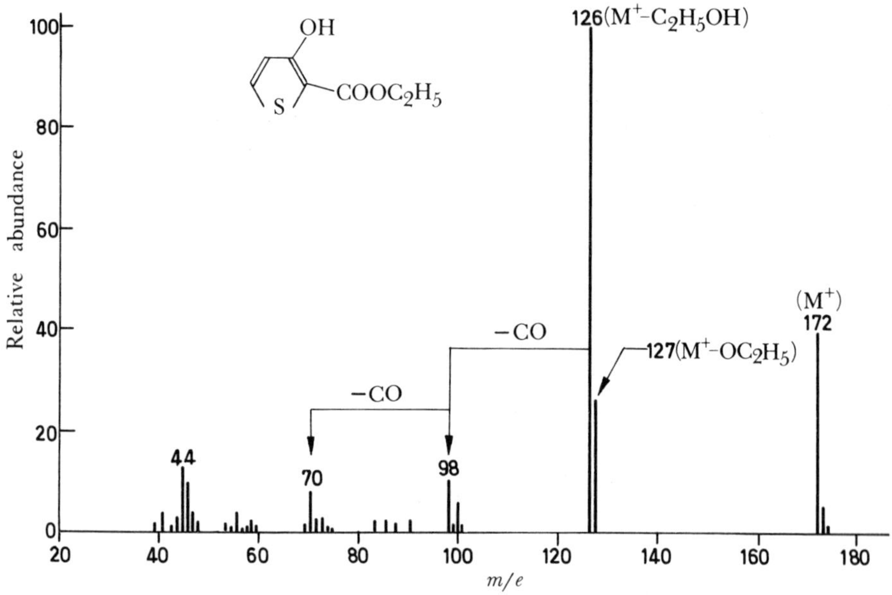

Figure 6-17

6-5. Natural products

A. Molecular formula determination. The most general use of mass spectrometry in natural products chemistry is for the determination of molecular weight (low resolution) or molecular formula (high resolution). A prerequisite of such determinations by electron impact is that the intact molecule can be vapourized prior to ionization. Currently, the practical upper limit of molecular weights of natural products that can be determined by mass spectrometry is around 1500, given favourable volatility and thermal stability. Nevertheless, the advantages of molecular formulae determinations up to this region cannot be overemphasized, especially in the cases of non-crystalline substances available only in microgram quantities, where no alternative approach may be adopted.

It is worthy of mention that care must be taken in the interpretation of mass spectra obtained from small quantities of material after fairly extensive handling (e.g., from tlc plates, or columns) to recognize peaks due to plate or column material. For example, the widely used plasticizer, tri-*n*-butyl acetylcitrate is a common contaminant and yields intense peaks at *m/e* 129, 185, 259 and 329.

Frequently, in the case of compounds such as steroids, triterpenes and alkaloids, microanalysis is insufficient to determine the molecular formula, commonly accepted errors in C and H analyses being up to ±0·3 per cent, which readily accommodates differences in the molecular formula of at least $\pm CH_2$. This point may be illustrated by the example of two sterols where mass spectrometry was required to determine the molecular weight.[10]

Table 6-7
Empirical formula determination of some sterols

	Alcohol A (m.p. 137°)		Alcohol B (m.p. 165°)	
Empirical formula	$C_{32}H_{56}O$	$C_{29}H_{50}O$	$C_{32}H_{54}O$	$C_{30}H_{50}O$
Molecular weight	456	414†	454	426†
Carbon content (per cent)	84·14	83·99	84·51	84·44
Hydrogen content (per cent)	12·36	12·15	11·97	11·81

† Determined by mass spectrometry.

B. The shift rule. A severe limitation of mass spectrometry in structure elucidation is that sometimes there is no simple or predictable relationship between the fragmentation pattern and the structure of the original molecule. This situation may arise if the fragmentation of no like basic skeleton has previously been examined in the mass spectrometer, or if the basic skeleton is so stable to electron impact that only side chains or structurally uninteresting substituents fragment. Conversely, the greatest successes of mass spectrometry have arisen when the basic skeleton (in its natural state, or after suitable derivatization) undergoes such a low-energy fragmentation that this always occurs irrespective of the presence or absence of additional substituents. The first and classical example of the latter type of behaviour was observed in the mass spectra of some indole and dihydroindole alkaloids by Spiteller and Biemann,[11] who formulated their observations in terms of a *shift rule*. This rule may be formulated in terms of a molecule $AB^{+\cdot}$ breaking down into, say, both $A^{+\cdot}$ and $B^{+\cdot}$ (e.g., in two competing reactions from the molecular ion), and a biogenetically related molecule $R{-}AB^{+\cdot}$ similarly breaking into $R{-}A^{+\cdot}$ and $B^{+\cdot}$. The shift of $A^{+\cdot}$ to $R{-}A^{+\cdot}$ indicates the mass, or with high resolution the elemental composition, of the substituent R and its location in a certain part of the alkaloid.

For example, alkaloids possessing the aspidosperma skeleton (21) undergo the following sequence of bond ruptures upon electron impact with a relatively low activation energy. In particular, the ion p (*m/e* 124 or a substituted analogue) is usually the base peak in these spectra.

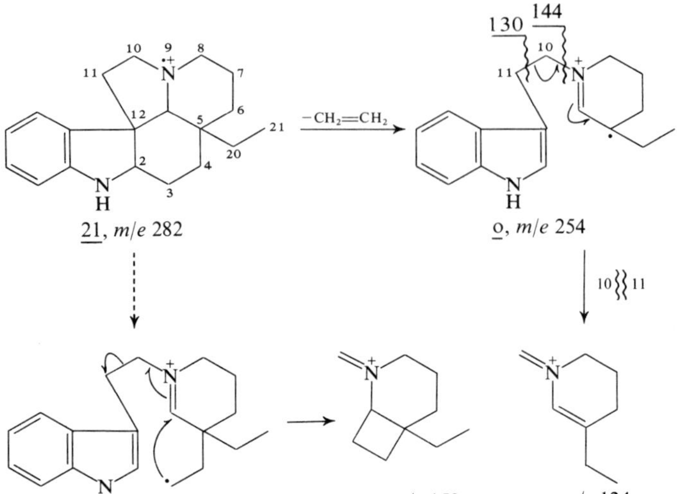

21, *m/e* 282 — o, *m/e* 254 — q, *m/e* 152 — p, *m/e* 124

Clearly, the primary reactions from 21 are favoured since the strain of the polycyclic system is released with simultaneous aromatization of the indole system. The structure assigned to *m/e* 152 (q) is less secure than that assigned to *m/e* 124 (p), but the important point is that sufficient members of the aspidosperma series have been examined to ensure that these ions differ in terms of atomic content only in the additional presence of C-3 and C-4 (and attached hydrogens) in *m/e* 152.

The partial mass spectra (showing only peaks of 10 per cent or greater relative abundance) of two aspidosperma alkaloids, demethoxypalosine (22)[12] and 2,3-dihydrovincadifformine (23) are reproduced (Figs. 6-18 and 6-19). The increments in the molecular weights over the basic aspidosperma skeleton are respectively 56 and 58 mass units, which may be associated with C_2H_5CO and $COOCH_3$ substituents through the use of proton magnetic resonance spectroscopy and high-resolution mass measurements. The loss of 28 mass units ($CH_2{=}CH_2$) from $M^{+\cdot}$ in Fig. 6-18 is replaced by the loss of 86 mass units in Fig. 6-19 ($CH_2{=}CH{-}COOCH_3$). Therefore, the C_2H_5CO substituent of demethoxypalosine (22) cannot be attached to either C-3 or C-4, but the carbomethoxyl substituent of 2,3-dihydrovincadifformine (23) must be located at C-3 or C-4. The occurrence of the base peak at *m/e* 124 (p) in both spectra indicates the absence of substituents on atoms 5–10, 20–21 (see 21), and the occurrence of *m/e* 152 in Fig. 6-18 indicates further the absence of substituents on C-3 and C-4 in 22. The C_2H_5CO substituent of 22

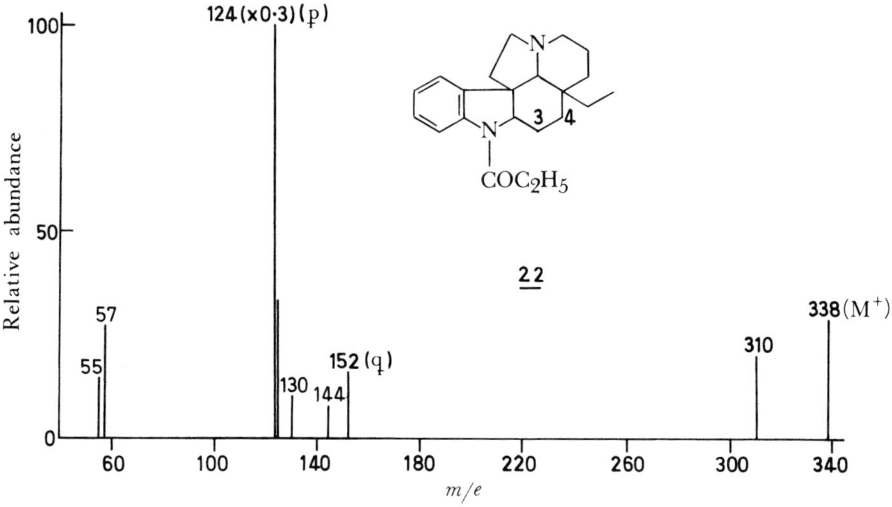

Figure 6-18

must therefore be located in the indolic part of the molecule, as ascertained through the expenditure of only a few μg of sample. Where sufficient material is available, the attachment of the C_2H_5CO group of 22 to the indolic nitrogen may readily be established through the application of complementary techniques such as nuclear magnetic resonance, infrared, and ultraviolet spectroscopy.

It is interesting to note that within a spectrum (Fig. 6-18) where the shift rule can be applied to great advantage, there occurs also an example of a

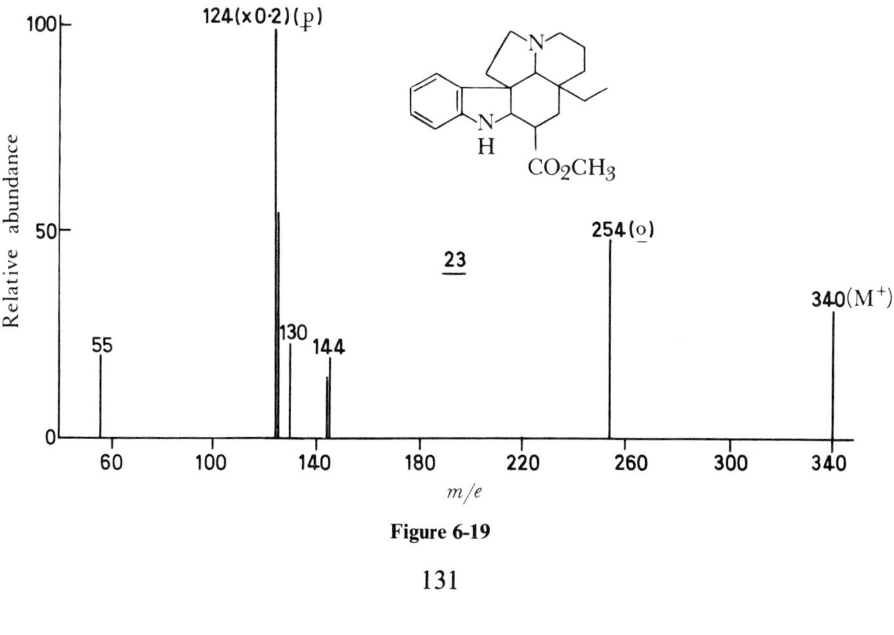

Figure 6-19

substituent being lost with a relatively low activation energy *en route* to ions (*m/e* 130 and 144) which are normally considered to indicate an unsubstituted indole moiety. The ions *m/e* 130 and *m/e* 144 are known to contain the indole nucleus plus carbon atom 11, and carbon atoms 10 and 11, respectively (see o). It is therefore apparent that the C_2H_5CO substituent of demethoxypalosine (22) must be lost as C_2H_4CO with associated hydrogen rearrangement (almost certainly to nitrogen) in the course of the formation of *m/e* 130 and 144. In contrast, the presence of a methoxyl substituent in the benzene ring of the indole system of such alkaloids *is* uncovered by shifts of *m/e* 130 and 144 to *m/e* 160 and 174 respectively. Thus the importance of considering the relative activation energies of various competing reactions before making sweeping generalizations in mass spectrometry is again emphasized.

C. Fragment-directing derivatives. In those natural products where the fragmentation pattern upon electron impact is not very specific, and is prone to change in an unpredictable manner upon changing the substituent groupings, it may be desirable to convert the compound into a derivative which can strongly control the fragmentation. Suitable derivatives are dimethylamines or ethylene ketals,[2, 4] which may be obtained from ketones in the following manner.

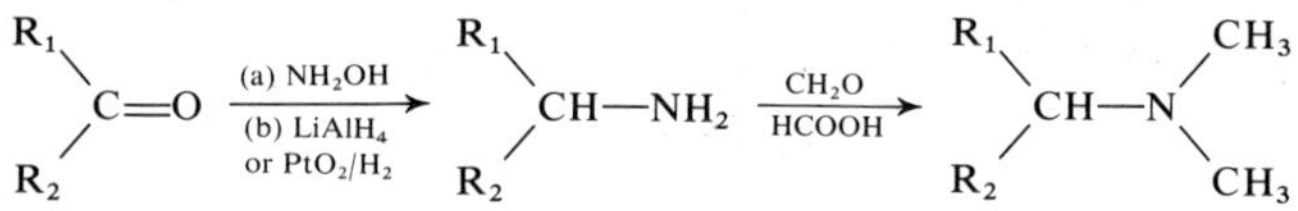

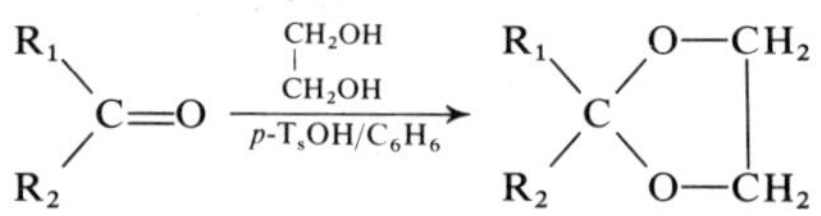

We have already seen (section 6-3E) that such groupings undergo very facile α-cleavage. If limited amounts of material are available, it is evident that the

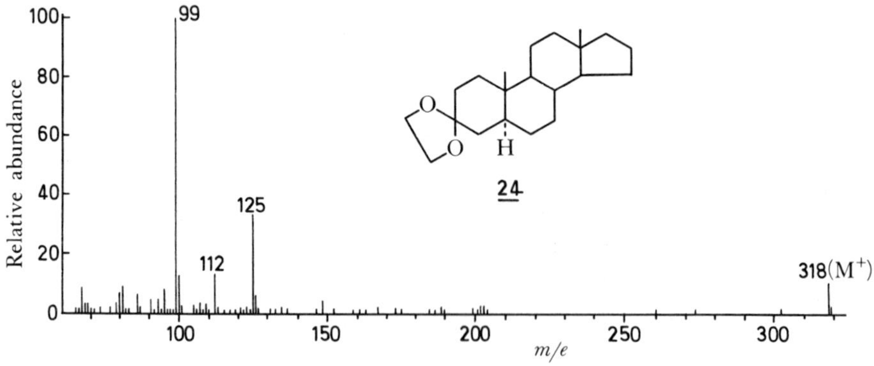

Figure 6-20

one-step preparation of the ethylene ketal from a ketone must be the preferred derivatization. Although the spectrum of 5α-androstan-3-one is very complex, containing large numbers of fragment ions of comparable abundance, the mass spectrum (Fig. 6-20) of the derived ethylene ketal (24) is very simple. The origins of the most abundant fragment ions (*m/e* 99 and 125) can be described in the following terms, as established by replacing the hydrogens thought to participate in hydrogen transfer reactions by deuterium (section 7-1A).

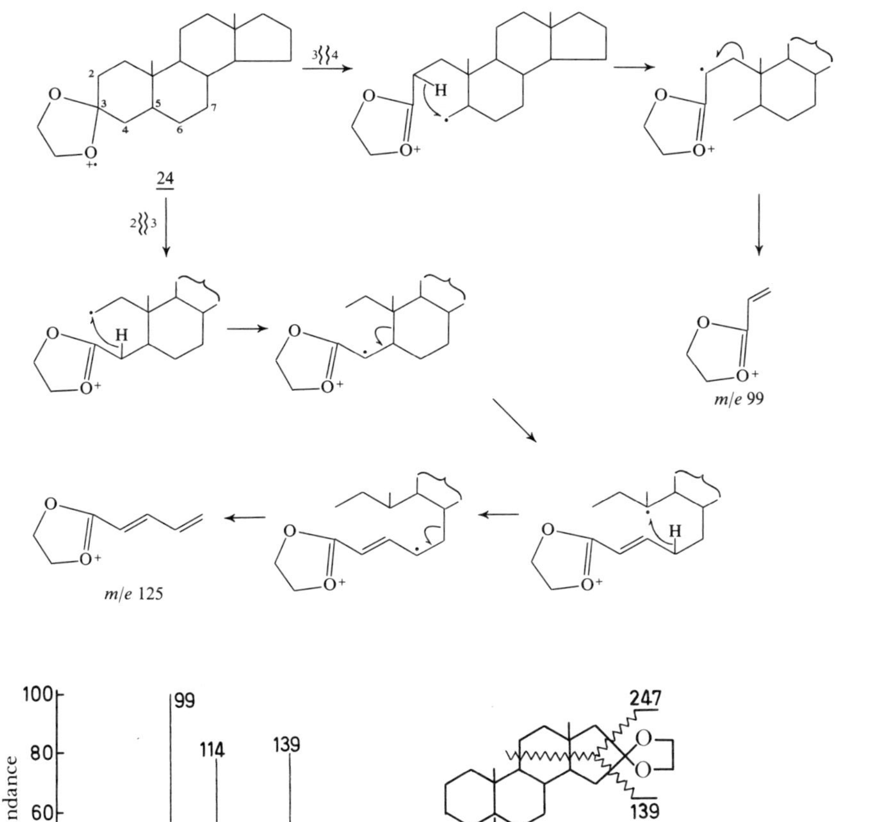

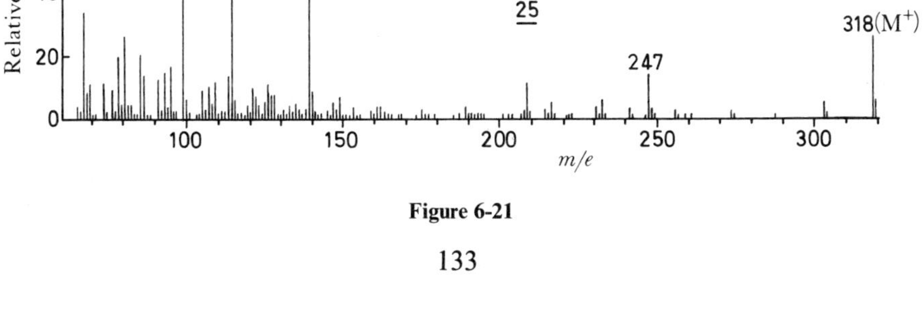

Figure 6-21

Such reaction sequences are so energetically favourable that analogous processes are found to occur, irrespective of the site of functionalization of the tetracyclic nucleus.[13] For example, the following fragmentations may be anticipated to occur in the mass spectrum (Fig. 6-21) of 5α-androstan-16-one ethylene ketal (25), and indeed abundant fragment ions are observed at *m*/*e* 139 and 247.

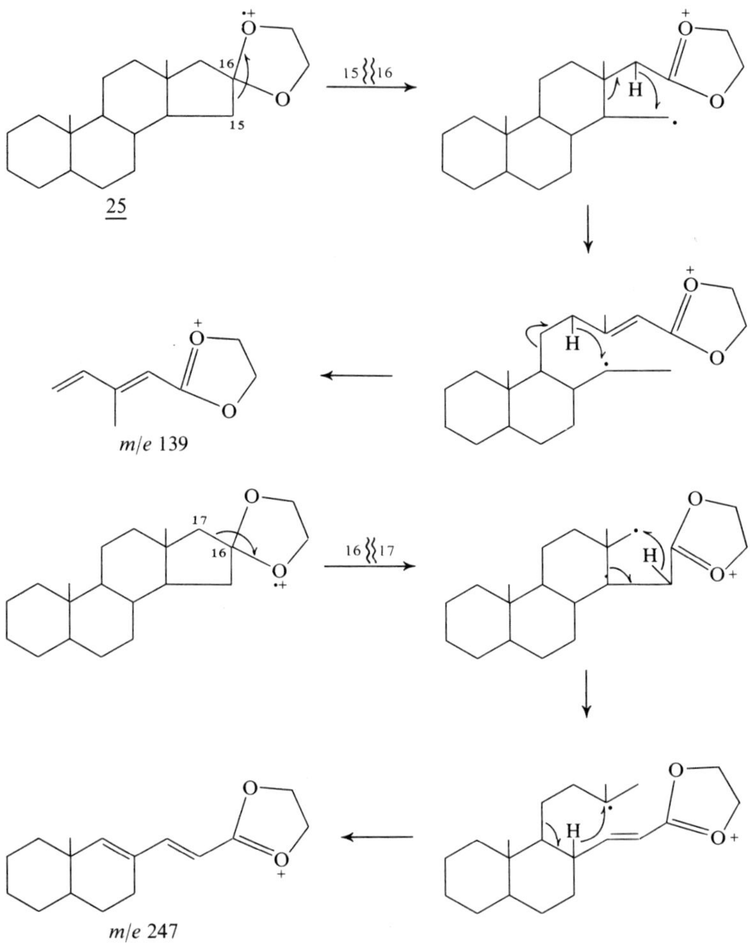

Shifts in the *m*/*e*-values of these conjugated oxonium ions will of course give information as to the nature and position of substituents which have been added to the basic skeleton. An application is to be found in the identification of the substance which controls sexual reproduction in the aquatic fungus *Achlya bisexualis*.[14] This substance, named antheridiol, is active at concentrations as low as 2×10^{-11} g/ml, and because of the scarcity of material a high-resolution mass spectrum was taken (using a computer and element map output—see section 12-3) on a few micrograms. The molecular formula was established as $C_{29}H_{42}O_5$. The ions of highest mass in the C—H, C—H—O and C—H—O_2 groups were $C_{19}H_{23}$, $C_{21}H_{30}O$ and $C_{22}H_{30}O_2$, respectively, and the species $C_{19}H_{25}O$ and $C_{19}H_{27}O_2$ were particularly abundant. Such a distribution is strongly suggestive of a steroid containing two oxygen atoms in the tetracyclic system. Assuming the abundant $C_{19}H_{27}O_2$ ion to arise by simple cleavage of the C-17, C-20 bond (see 26 for numbering of the steroid skeleton), the steroid nucleus would have to contain two double bonds. At this point, a steroid-like structure was assumed as a working hypothesis with the remaining ten carbons and three oxygens present as a side chain at C-17. The position of a third oxygen atom was revealed since the most abundant ion corresponded to $C_{22}H_{32}O_3$, suggesting C-22 as the point of attachment. The presence of two hydroxyl groups in the hormone was established by a high-resolution mass spectrum of an acetylated product (M^+ of diacetate $C_{33}H_{46}O_7$), and the presence of two double bonds via similar measurements on a hydrogenation product (M^+:$C_{29}H_{46}O_5$). Further considerations of the mass spectra, and of infrared (carbonyl bands for α,β-unsaturated ketone and α,β-unsaturated γ-lactone groups) and ^{1}H n.m.r. spectra, led to the proposal 27 as the most likely structure for the hormone. In particular, the tetrahydro-derivative of the hormone afforded an ethylene ketal 28 which gave an abundant ion at *m/e* 141 ($C_7H_9O_3^+$) in accordance with the presence of a 3-hydroxy-7-keto steroid.

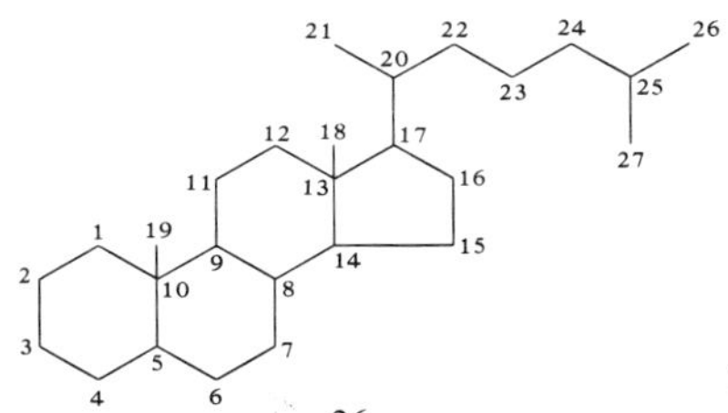

26

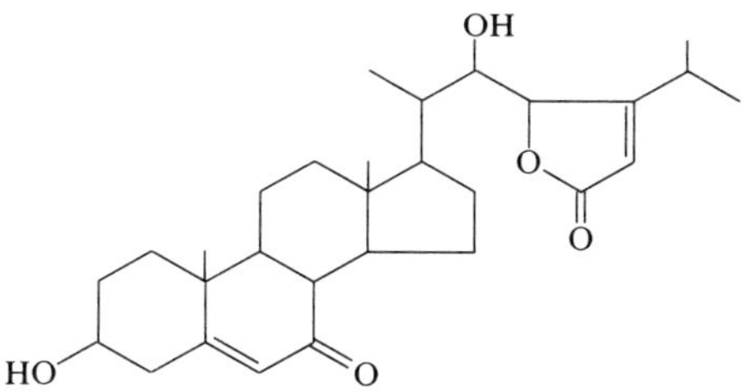

27

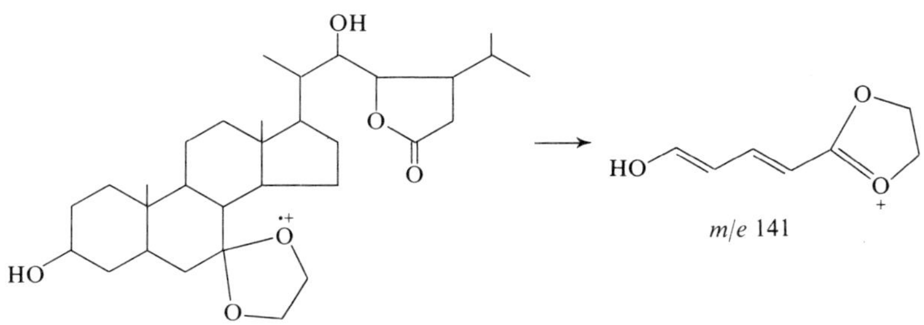

28

m/e 141

The structure 27 has subsequently been confirmed by synthesis.[15] The example is one of many appearing in the recent literature which attest to the power of high-resolution mass spectrometry. The method is particularly valuable when working with sub-microgram quantities. If larger amounts of an unknown are available, then the information provided by the mass spectrometric method is frequently an ideal complement to that provided by u.v., i.r. and n.m.r. spectroscopy.

D. Analysis of biopolymers. During the last five years, there have been considerable advances in analyzing the three main types of biopolymers (peptides, nucleotides and saccharides) by mass spectrometry. A common problem is the preparation of suitable derivatives so that the volatility of substances under investigation is optimized. In addition, only microgram quantities are often available, so mass spectrometry is the analytical method of choice. Much of the recent work carried out in this general area has been reviewed by Das and Lederer,[16] and Cooks and Johnson.[17]

(a) *Peptides.* Much pioneering work was carried out in this area by Biemann,[1] who showed over a decade ago that sequence information for small peptides could be obtained by mass spectrometry after reduction of the peptide to a polyamino alcohol with lithium aluminium hydride. After much further effort, an important advance was the observation by Lederer and co-workers[18] in 1965 that fortuitine, an acyl nonapeptide methyl ester (29, mol. wt. 1359) could be completely sequenced by mass spectrometry.

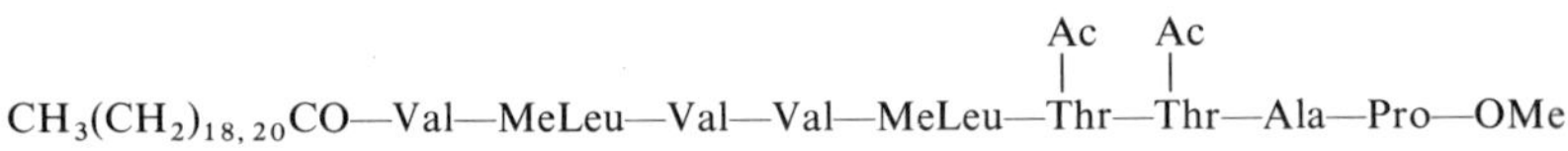

29

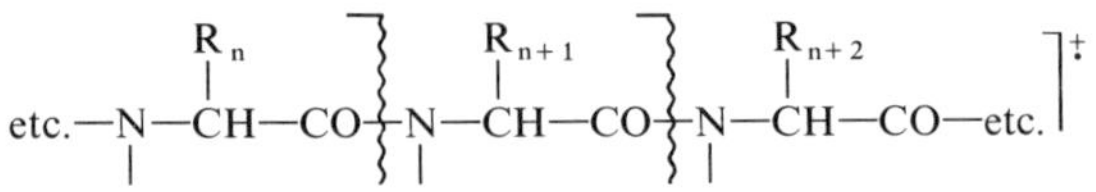

30

The sequence analysis was possible because a relatively large fraction of the total ion current was carried by ions formed by cleavage of each peptide bond (see 30) with charge retention in each case by the left-hand portion of the molecule, i.e., by the portion containing the N-terminus. Ions formed by such cleavages are known as sequence ions, and it will be obvious that the mass difference between the consecutive nth and $(n + 1)$th sequence ions will be characteristic of the amino acid in the $(n + 1)$th position from the N-terminus. Extraction of the sequence information was facilitated by the fact that all the useful sequencing ions (which necessarily contain the terminal N-acyl group) could be recognized because they occurred as 'doublets' situated 28 mass units apart, since the acyl group contained 18 and 20 methylene groups to similar extents, i.e., the peptide was a mixture of two components.

The most useful subsequent findings largely constitute improvements to the same approach. An important milestone, again from Lederer's laboratory,[19] was the suggestion that peptides should be permethylated prior to examination in the mass spectrometer. Since the —CO—NH— group is converted to a —CO—N(CH_3)— group by this procedure, intermolecular hydrogen bonding between amide linkages of adjacent chains is eliminated and volatility improved. Moreover, cleavage at the peptide bonds is now enhanced.

If the mass-spectrometric method is to compete with classical methods of protein sequencing, it must be possible to take the oligopeptides produced by enzymic cleavage of proteins, and derivatize these in high yield using quantities of only 0·05 to 1·0 micromoles (0·05 to 1·0 mg for a mol. wt. = 1000). The best derivatization sequence appears to be acetylation followed by permethylation. The acetylation[20] is carried out by means of acetic anhydride in methanol (1 : 4) at room temperature for 3 hr, under which conditions basic free amino groups in the peptide are acetylated (e.g., the N-terminus of the peptide and the basic amino group in the side chain of lysine); hydroxyl groups (e.g., of tyrosine, serine or threonine) are not acetylated under these conditions. Two methods of permethylation have been employed: (a) the N-acyl peptide derivative in dimethylformamide is treated with an excess of methyl iodide in the presence of silver oxide, or (b) the N-acyl peptide derivative is treated with an excess of dimethyl sulphoxide anion ($^-CH_2SOCH_3$) in DMSO, followed by addition of excess methyl iodide; the base $^-CH_2SOCH_3$ is generated by reaction of sodium hydride with DMSO. Both procedures work quite well, but the latter is to be preferred when working with small quantities since adsorption of material on the surface of the silver oxide is potentially a problem; in addition, the silver oxide procedure can lead to extensive C-methylation of glycine and aspartic acid.[21] The permethylation procedure results in the following conversions.

$$\text{—CONH—} \longrightarrow \text{—CON(CH}_3\text{)—}$$

$$\text{—CONH}_2 \longrightarrow \text{—CON(CH}_3\text{)}_2$$

$$\text{—}\overset{|}{\text{C}}\text{H(OH)} \longrightarrow \text{—}\overset{|}{\text{C}}\text{H(OCH}_3\text{)}$$

$$\text{—COOH} \longrightarrow \text{—COOCH}_3$$

The structures of the derivatized amino acids, following acetylation and permethylation, are given in Table 6-8, together with relevant mass numbers and associated ions. The sulphur-containing amino acids (cysteine, cystine and methionine) and arginine have been omitted from the table, since they require modification prior to methylation.

Table 6-8
Integral mass numbers corresponding to derivatized amino acids

Amino acid	Structure of derivative	A N-terminal mass	B Mass of residue	C C-terminal mass	Associated ions† (commonly observed)
Glycine (Gly)	$-N(CH_3)-CH_2CO-$	114	71	102	
Alanine (Ala)	$-N(CH_3)-CH(CH_3)CO-$	128	85	116	Loss of —CO from A or B
Valine (Val)	$-N(CH_3)-CH(CH(CH_3)_2)CO-$	156	113	144	Loss of —CO from A or B
Leucine‡ (Leu)	$-N(CH_3)-CH(CH_2CH(CH_3)_2)-CO-$	170	127	158	Loss of —CO from A or B plus further loss of ketene from A (*m/e* 100)
Serine (Ser)	$-N(CH_3)-CH(CH_2OCH_3)CO-$	158	115	146	Loss of CH_3OH from A or B (increases with increasing temperature)
Threonine (Thr)	$-N(CH_3)-CH(CH(OCH_3)CH_3)CO-$	172	129	160	Loss of CH_3OH from A or B (increases with increasing temperature); occasional loss of side-chain minus H (58 m.u.) from B or C
Aspartic Acid (Asp)	$-N(CH_3)-CH(CH_2COOCH_3)CO-$	186	143	174	Loss of —CO from A or or B
Glutamic Acid (Glu)	$-N(CH_3)-CH(CH_2CH_2COOCH_3)CO-$	200	157	188	Loss of —CO from A or B
Asparagine (Asn)	$-N(CH_3)-CH(CH_2CON(CH_3)_2)CO-$	199	156	187	Loss of —CO from A or B
Glutamine (Gln)	$-N(CH_3)-CH(CH_2CH_2CON(CH_3)_2)CO-$	213	170	201	Loss of —CO from A or B

Table 6-8 (*cont.*)

Amino acid	Structure of derivative	A N-terminal mass	B Mass of residue	C C-terminal mass	Associated ions† (commonly observed)
Phenylalanine (Phe)	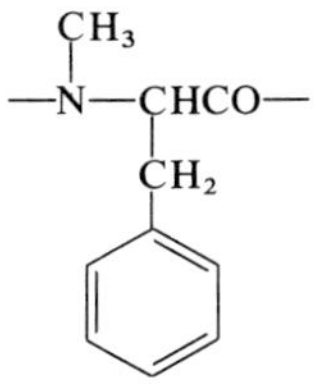	204	161	192	Loss of —CO from A; *m/e* 91, 162
Tyrosine (Tyr)	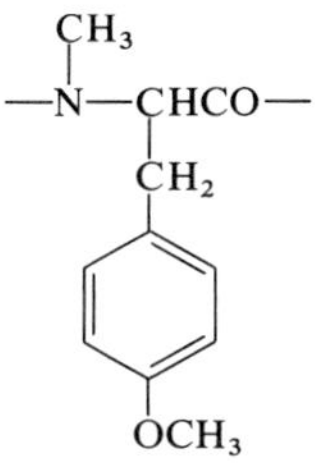	234	191	222	Loss of —CO from A; *m/e* 121, 161, 192
Tryptophan (Trp)	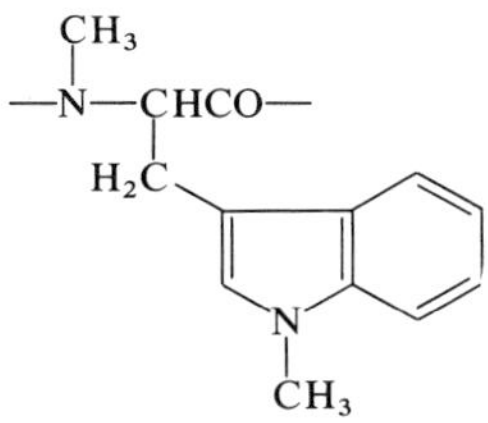	257	214	245	Loss of —CO from A; *m/e* 144, 215
Lysine (Lys)	CH_3 over —N—CHCO— with $(CH_2)_4N(CH_3)COCH_3$	241	198	229	Loss of —CO from A; *m/e* 171
Histidine§ (His)	CH_3 over —N—CH—CO— with CH_2-Imidazole (CH_3)	208	165	196	*m/e* 95; ± 14 m.u. satellites of B.
Ornithine¶ (Orn)	CH_3 over —N—CHCO— with $(CH_2)_3N(CH_3)COCH_3$	227	184	215	Loss of —CO from A
Proline (Pro)	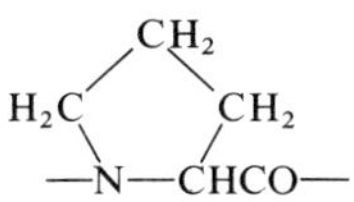	140	97	128	

† Associated ions which are normally abundant indicated by underlining. ‡ Not differentiated from isoleucine. § Permethylation of this residue not always successful. ¶ Result of hydrazinolysis of arginine.

The sequencing of a hexapeptide, produced by enzymic hydrolysis of a protein,[22] may be illustrated by reference to Fig. 6-22 and Table 6-8. The abundant ions at *m/e* 170 and 142 indicate N-terminal leucine (although it should be noted that mass spectrometrically this cannot be differentiated from *iso*-leucine). Following *m/e* 170, the next most abundant ion in the spectrum is at *m/e* 327 and this corresponds to a Leu.Glu. sequence. Further sequence ions of decreasing abundance follow at *m/e* 497, 610 and 707, giving the mass differences 170, 113 and 97, and so the sequence can be further extended to Leu.Glu.Gln.Val.Pro. The relatively weak satellite ions lying 14 mass units to the high mass side of the sequence ions *m/e* 327, 497, 610 and 707 correspond to a minor amount of C-methylation. The high relative abundance of the sequence ions is noteworthy; it is emphasized that even if non-sequence ions were of comparable abundance to sequence ions in a given region of the spectrum, the interpretation of the spectrum would not necessarily be complicated because only rarely would extraneous processes lead to mass differences corresponding to an amino acid which was not actually present.

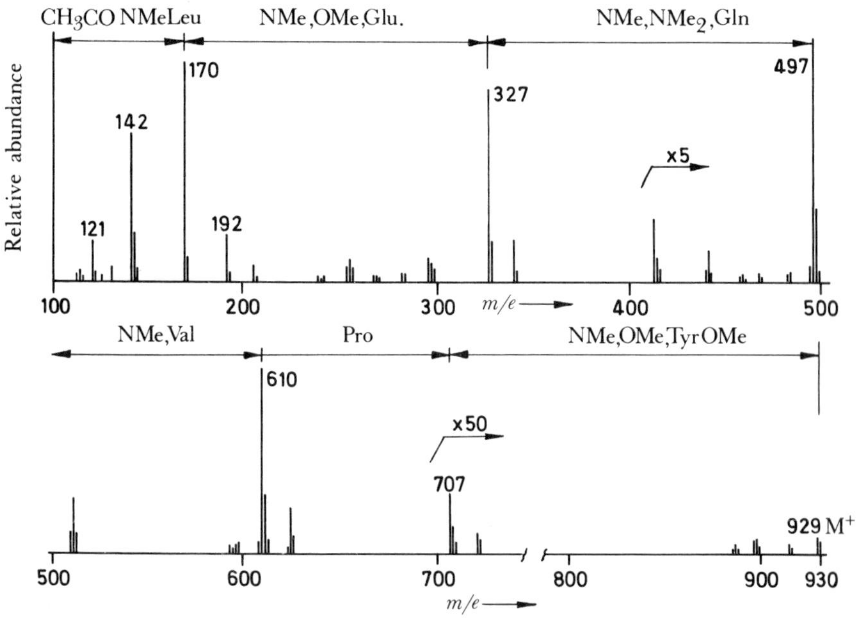

Figure 6-22

The molecular ion in the spectrum can be identified (with its associated ^{13}C isotope peak) as the highest mass ion in the spectrum, given also the identification aid that normally peaks will appear at 15 and 31 m.u. less (loss of CH_3 and OCH_3 from M^+). In the present instance (Fig. 6-22), the C-terminal amino acid is given by the difference of 222 (929–707), thus defining the complete sequence as Leu . Glu . Gln . Val . Pro . Tyr. The presence of tyrosine is further confirmed by the occurrence of an abundant ion at *m/e* 121 (r), and its C-terminal position also indicated by the ion at *m/e* 192 (s).

r, *m/e* 121

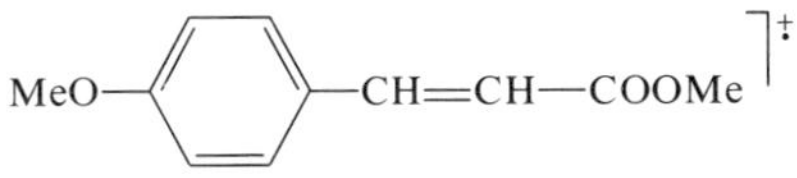

s, *m/e* 192

It is of course true that if the peptides being examined are products of enzymic digests, then those peptides which are produced by 'in-chain' cleavage of the original protein will, in many cases, carry only specific amino acids at the C-terminus. Hence identification of M^+ is aided in tryptic digests, since the enzyme trypsin cleaves only after lysine and arginine† (i.e., at the C-terminal side of Lys and Arg), while the enzyme chymotrypsin cleaves only after hydrophobic residues (e.g., at the C-terminal side of phenylalanine, tyrosine and tryptophan; cf. Fig. 6-22).

In obtaining spectra such as that shown in Fig. 6-22, it is advantageous to move the tip of the direct insertion probe into the ionization chamber by fractional movements. In this way, a temperature gradient may be obtained at the probe tip, and traces of non-peptide impurities (which are almost inevitably present in material isolated in small amounts after the use of column chromatography, paper chromatography, etc.) preferentially volatilize first. The progress of this fractionation may be followed by displaying fast scans of the spectrum on an oscilloscope. Since such impurities are usually of a molecular weight considerably less than penta- to decapeptides, it is frequently easier to obtain 'clean' spectra of such peptides than of di- to tetrapeptides.

Fractionation of mixed peptides on the probe tip[22, 23] can lead to sequence information being obtained from enzymic digests of proteins without the tedious isolation of pure peptides. The procedure[22] is to pass the enzymic digest through a Sephadex column which gives a separation according to molecular weight of peptides present. For mass spectrometric purposes, fractions containing mixtures of peptides of molecular weight less than 1000 are derivatized. For example, a derivatized Sephadex fraction of a tryptic digest of a protein yielded spectra which changed continuously from 150°C to 250°C, finally giving a spectrum which appeared to be that of a single component peptide at 260°C. Consecutive scans obtained about the middle of the temperature gradient are reproduced in Figs. 6-23(a) and 6–23(b),‡ the former at a source temperature of 200°C and the latter at a temperature of 220°C. Interpretation is effected by comparing consecutive scans and

† As mentioned previously, permethylation of arginine residues can cause complications; usually, the sequence only as far as the amino acid preceding arginine is obtained, and hence the usefulness of the above generalization is at present somewhat limited.

‡ In order to simplify the diagrams, the normal amino acid abbreviations are used to represent the derivatized amino acids.

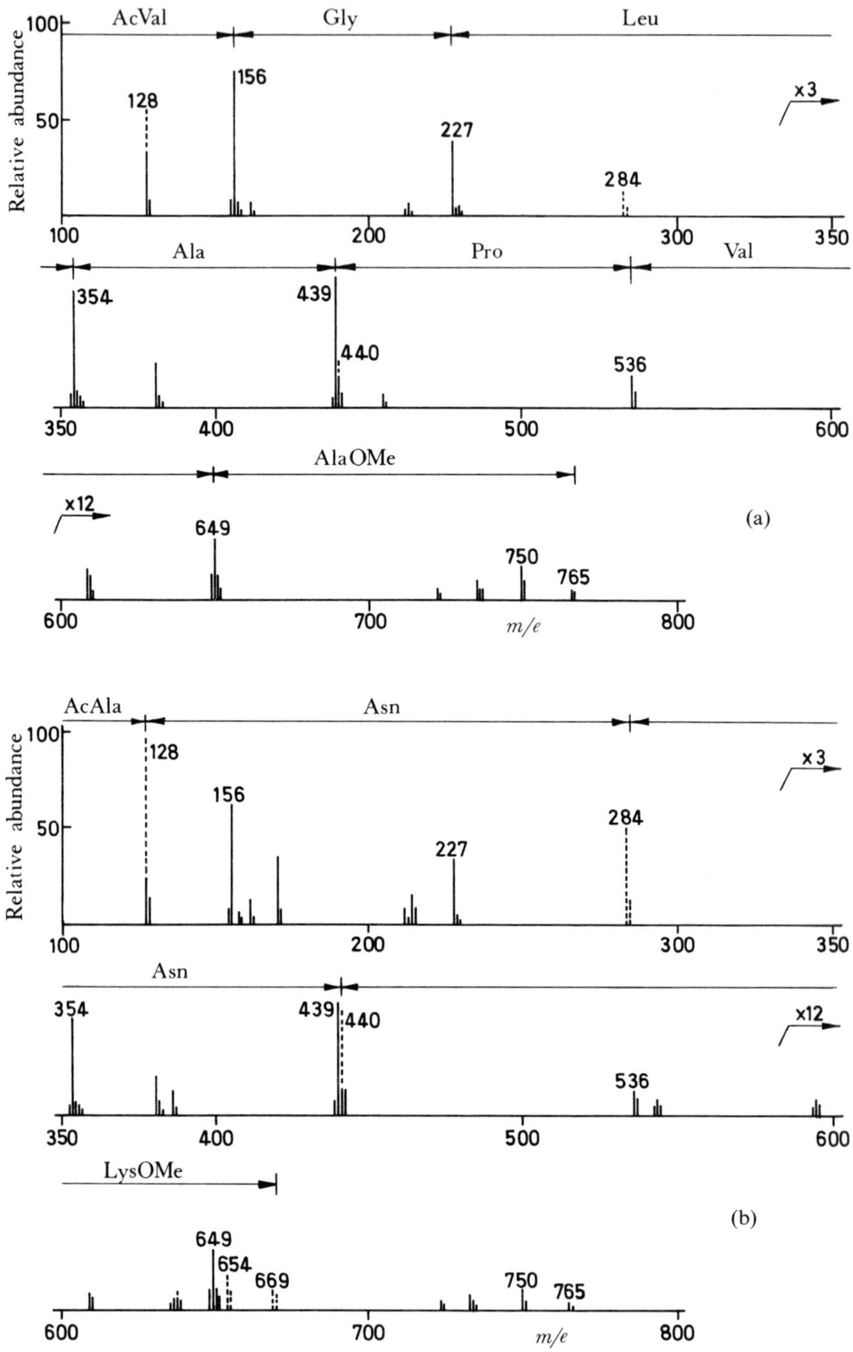

AcVal
Gly
Leu
Relative abundance
100
50
128
156
227
284
x3
100
200
300
350
Ala
Pro
Val
354
439
440
536
350
400
500
600
AlaOMe
x12
649
750
765
(a)
600
700
m/e
800
AcAla
Asn
Relative abundance
100
50
128
156
227
284
x3
100
200
300
350
Asn
354
439
440
x12
536
350
400
500
600
LysOMe
(b)
649
654
669
750
765
600
700
m/e
800

Figure 6-23

looking for changes in peak intensity as one component becomes more or less intense relative to the other. A component which is more abundant in the scan at 200°C (Fig. 6-23(a)) has the sequence Val. Gly. Leu. Ala. Pro. Val. Ala.; the second component, which gives rise to ions of increased abundance at 220°C (Fig. 6-23(b)), is a tetrapeptide of sequence Ala. Asn. Asn. Lys. (see dotted lines in the figures). A third component, volatilizing at a temperature of 260°C, was shown to have the sequence Gln. Tyr. Tyr. Thr. Val. Phe., although the Phe residue was not C-terminal; additional residues beyond this point were present but not identified.

Further enzymic digests of the initially produced peptides which are too large to handle by mass spectrometry, coupled with the use of different enzymes for the initial digest (to create overlaps in sequence), provide a strategy for the sequencing of an enzyme by mass spectrometry.

On occasions it is advantageous to obtain high-resolution measurements on ions in peptide sequencing. For example, the elucidation of the sequence of scotophobin, a pentadecapeptide which induces a specific behaviour in untrained mice, has been reported with the aid of high-resolution mass spectrometry.[24] This peptide, which was extracted from rats trained (by electrical shock treatment) to fear the dark, appeared to induce the same behaviour in untrained mice.

About 300 μg of scotophobin were extracted and after an amino acid analysis only 100 μg remained, which was surprisingly considered too small a quantity for 'wet' chemical methods. After esterification and trifluoroacetylation, the mass spectrum was obtained, to accuracies of four millimass units. Although the peptide pyrolysed on the probe, it was possible, by elemental composition determinations of the ions, to identify di- and tripeptide fragments that overlapped sufficiently to permit sequencing of the peptide (see also Table 11-2).

The foregoing examples illustrate the principles of peptide sequencing by mass spectrometry. Review articles of the work carried out up to 1968 are provided by Lederer[25a] and Shemyakin[25b] who, with their co-workers, represent two of the main groups contributing to the field. Shemyakin *et al.*[26] review later material. Attention is drawn to the possibility of desuphurizing sulphur-containing peptides by the use of Raney nickel.[20, 27] Although adsorption of the peptide onto the Raney nickel is a problem, especially when working with small quantities of peptide, good results can apparently be obtained if the desulphurization is carried out in dimethylformamide solution at 20°C for 2 days in the presence of a tenfold amount by weight of the catalyst.[27]

(b) *Carbohydrates*. Much of the early work in this field has been reviewed by Kochetkov and Chizhov,[28] and Heyns *et al.*;[29] Cooks and Johnson[17] cover the literature from 1968–1970.

Carbohydrates are invariably derivatized prior to analysis by electron-impact mass spectrometry, and most work has been carried out on peracetates, pertrimethylsilyl ethers or permethyl ethers. In the case of monosaccharides, structural information may, if necessary, be obtained by g.c./m.s. following exhaustive trimethylsilylation. In oligosaccharides one would ideally wish to determine the number of sugars involved, their sequence, nature of linking (e.g., 1 → 4, 1 → 6, etc.), and stereochemical details. The present situation is that molecular ions have been obtained for derivatized mono- to pentasaccharides, and that prominent sequence ions frequently allow the ordering of monosaccharide units; using sugars with known modes of linking (e.g., 1 → 4, 1 → 6, etc.), criteria for distinguishing between the various possibilities have been developed in some cases, but it is not yet clear how widely such criteria might be applied with confidence to sugars of unknown structure. In general, mass spectrometry has not proved sensitive to stereoisomeric differences between carbohydrates. However, a useful approach[30] is based on the fact that cyclic acetals and boronic esters of stereoisomers are frequently structural isomers and may give very different mass spectra. For example, the cases of D-galactose (31) and D-glucose (32) may be cited. The former has two pairs of *cis*-1,2-diol groupings in the α-pyranoid form and therefore forms mainly a pyranoid 1,2:3,4-di-O-isopropylidene derivative (33). D-glucopyranose on the other hand has only one *cis*-diol group in the α-form 32 (none in the β-form), but in the otherwise less favourable furanoid α-isomer 34 is capable of reacting with two molecules of acetone to form 1,2:5,6-di-O-isopropylidene-D-glucofuranose (35).

The partial mass spectra (above *m/e* 90) of the isomers 35 and 33 are reproduced in Figs. 6-24(a) and (b), respectively, and are clearly quite different. Note, for example, that only the 1,2:3,4-di-O-isopropylidene derivative loses acetic acid from the $M^+ - 15$ ion, and that the 1,2:5,6-di-O-isopropylidene derivative produces a more abundant ion at *m/e* 159 due to the relatively facile formation of t.

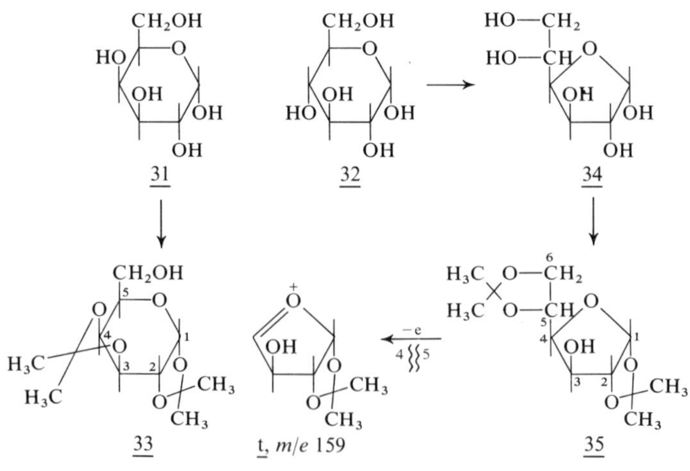

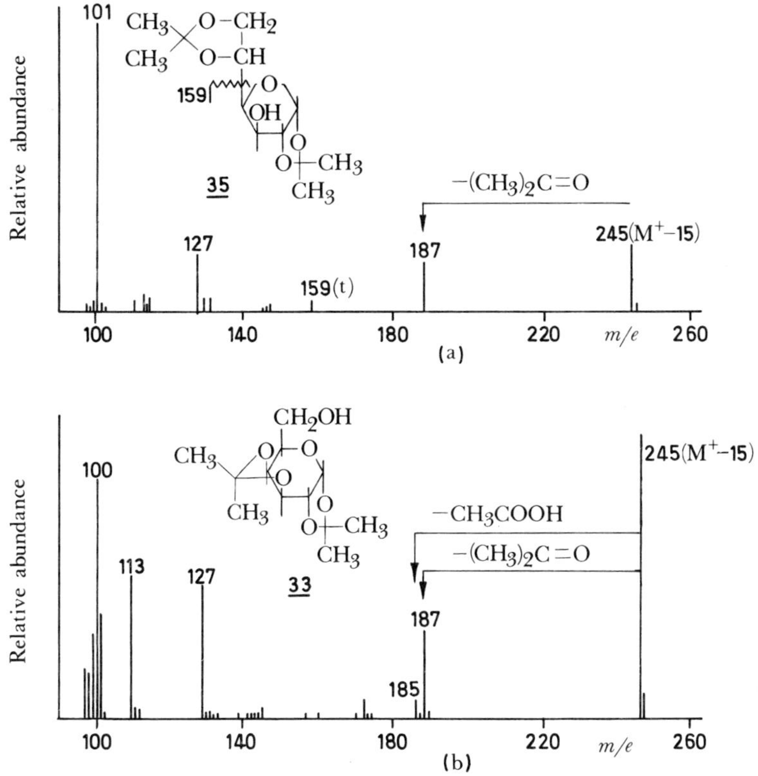

Figure 6-24

Having illustrated how stereochemical information may be derived it remains only for us to illustrate the principles via which the numbers, sequence and nature of linkage of monomer units may be obtained. Further to important work on di- and trisaccharides,[31] molecular ions of 1-phenylflavazole peracetates of di- to pentasaccharides have been obtained.[32] The mass spectrum of the 1-phenylflavazole peracetate (36) of maltopentose is reproduced in Fig. 6-25.

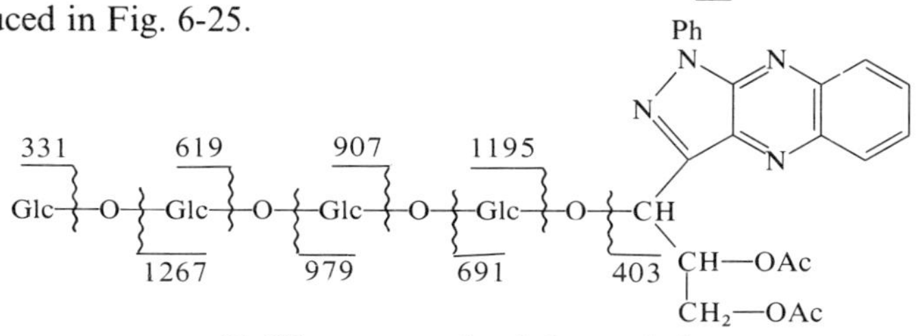

36 (Glc = peracetylated glucose ring)

The incorporation of the aromatic residue into the saccharide molecule represents an important application of earlier theoretical considerations (section 4-2). The aromatic moiety has a low ionization potential and therefore some molecular ions are produced in which the total internal excitation energy is insufficient to cause ionization in the sugar part of the derivative. Hence, these molecular ions do not have available to them the low-energy decomposition pathways of sugar peracetates which are ionized in the sugar portion, and M^+ ions survive long enough to be recorded at the collector. In addition, the derivatives, which are prepared using *o*-phenylenediamine and phenylhydrazine and purified using cellulose t.l.c.,[33] fluoresce; and this can be an advantage when experiments are carried out on $\leqslant$ 1 μmole of material.

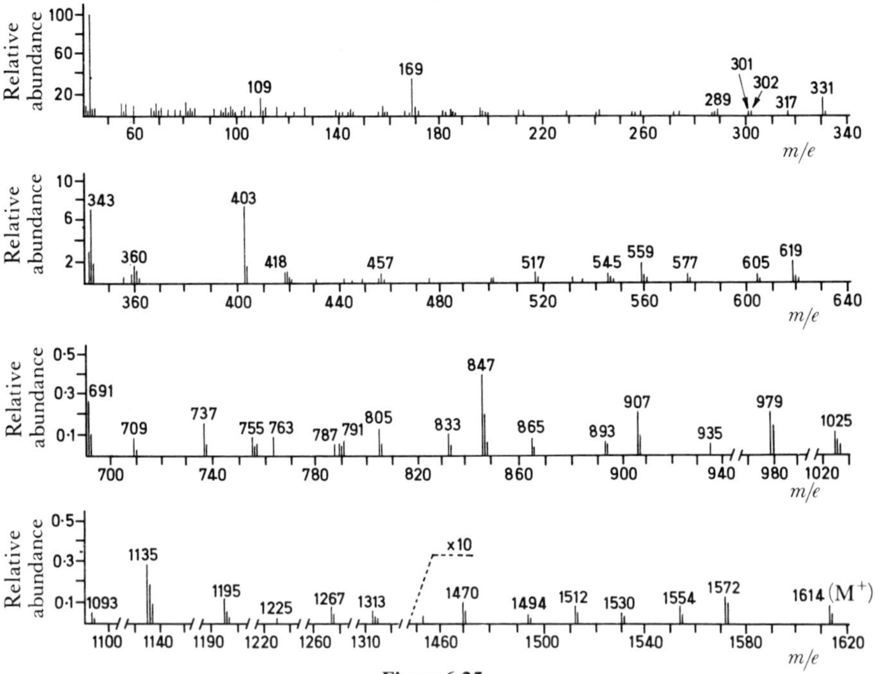

Figure 6-25

The major sequence ion cleavages of the maltopentose derivative <u>36</u> are indicated in the structure. In the case where each sugar corresponds to a peracetylated glucose ring, as in the present case, these sequence ions differ in mass by 288 m.u. These ions, together with their daughters due to acetic acid (60 m.u.) and/or ketene (42 m.u.) loss, constitute a large fraction of the abundant ions above *m/e* 350 in the spectrum.

In the spectra of the tri- and tetrasaccharide 1-phenylflavole peracetates, it is further observed that when the first glycosidic linkage (numbering from the derivatized end of the molecule) is $1 \rightarrow 6$, the $[M^+ - 42]/[M^+]$ abundance ratio is high (e.g., 9, 18, 30 in three representative samples) but low (e.g., 0·4, 0·3 and 0·4) in compounds having $1 \rightarrow 4$ linkages. The nature of the second and subsequent glycosidic linkages is correlated with the ease with which acetic acid is lost by the fragment ions belonging to the sequence not retaining the flavazole (viz., *m*/*e* 619, 907, 1195 in 36). In all cases, these ions lose acetic acid readily, if, and only if, the lowest-numbered glycosidic linkage (adjacent to that cleaved in forming the ion in question) is 1 → 4. [Sequence ion − AcOH]/[sequence ion] ratios of 1·0–2·5 are observed for 1 → 4 linkages, but the ratios for 1 → 6 linkages are 0·04–0·10.

The above method is of course restricted to oligosaccharides which can form 1-phenylflavazoles, but it illustrates well the general principles involved, and it should prove possible to modify the approach using other hetero-aromatic groups[34] and other hydroxyl-protecting groups (e.g., OMe replacing OAc).

(c) *Nucleotides, nucleosides and bases*. The first study of the mass spectra of nucleosides was carried out by Biemann and McCloskey.[35] They were able to obtain molecular ions in numerous cases without derivatization of the parent nucleoside by vapourizing the samples (at the sub-microgram level if necessary) about 1 cm from the ionizing electron beam. However, not all nucleosides afford molecular ions under these conditions, and therefore field ionization is of some potential in this area (section 9-2). Yet at the time of writing, field-ionization sources are not widely available nor, when available, easy to operate routinely with continuing high sensitivity. Detailed studies have therefore been undertaken of the mass spectra of pertrimethylsilyl derivatives of mononucleosides, mononucleotides and bases,[36] and of pertrimethylsilyl derivatives of dinucleotides.[37] Base sequence information can be derived from the spectra of derivatized dinucleotides.[37, 38]

However, since elegant and very sensitive 'wet' chemical methods are available for the sequencing of nucleotides, it seems at the present that the main advantage of mass spectrometry in this area lies in the identification of unusual bases, frequently those isolated from transfer RNAs. For example, it was possible to isolate from an enzymic hydrolysis of yeast tRNA very small amounts of an unknown nucleoside, containing impurities which stemmed from the isolation procedure.[39] In this case, it was advantageous to record a complete high-resolution spectrum; peaks from the nucleoside will contain relatively large numbers of oxygen and/or nitrogen atoms, which allow their differentiation from impurity peaks. Normally, mass spectra of nucleosides contain abundant ions due to B + H (and frequently also B + 2H) where B represents the base of the nucleoside (see also section 9-2, and Fig. 9-2). Indeed, the unknown afforded a pair of intense peaks of *m*/*e* 200 and 201, which surprisingly were shown from the exact mass measurements to contain sulphur ($C_7H_8N_2O_3S$ and $C_7H_9N_2O_3S$, respectively). Hence, it is possible to infer that the structure of the new nucleoside is represented by 37; this formulation is supported by the presence of a $C_5H_9O_4$ ion corresponding to the pentoside unit.

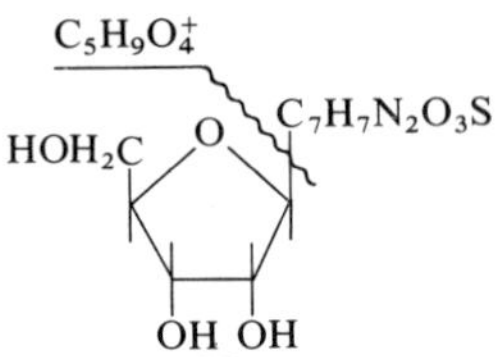

37

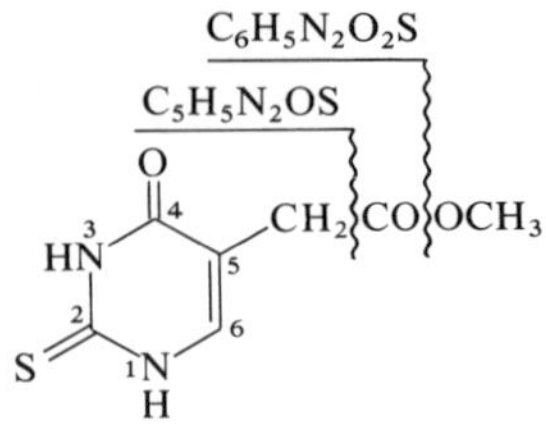

38

The anticipated molecular ion species was not evident in the mass spectrum, but ions of the compositions $C_{11}H_{13}N_2O_6S$ and $C_{12}H_{14}N_2O_7$ (corresponding to $M^+ - CH_2OH$ and $M^+ - H_2S$) indirectly confirm the postulated molecular formula. The nitrogen content indicates a pyrimidine rather than a purine ring. In addition, ions with the lowest number of oxygen atoms correspond to $C_5H_5N_2OS$ and $C_5H_6N_2OS$, and these are compatible with a substituted thiouracil (e.g., 38), rather than with a substituted uracil. The minimum number of carbon atoms in the $C_nH_mN_2O_2S$ group is 6, and in the $C_nH_mN_2O_3S$ group is 7, and such data are most compatible with 38. The placing of the side chain at C-5 (as opposed to C-6) is arbitrary on the basis of the mass spectrometry data, but is supported by ultraviolet spectra. The position of sulphur at C-2 (rather than C-4) is based upon H_2S loss upon electron impact being anticipated to be facile if it can proceed with formation of a C-2,C-5′ ether bridge (39).

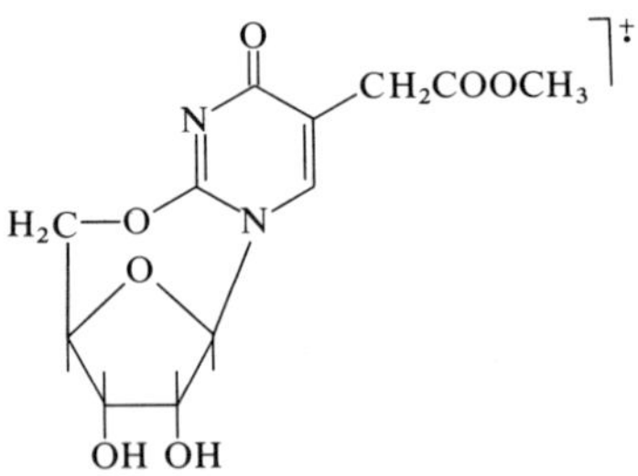

39

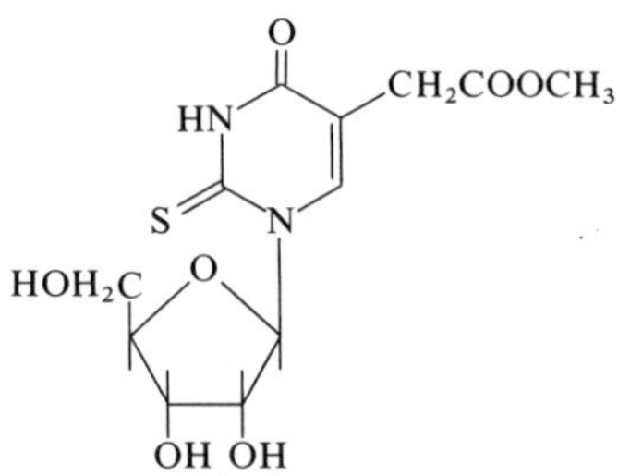

40

The complete structure 40 would of course be very tentative on the basis of mass-spectrometric evidence alone, but is supported by ultraviolet spectra and chemical transformations (e.g., demonstration of the presence of a saponifiable group).

6-6. Organometallic compounds

A. Isotope patterns. Interpretation of the mass spectra of organometallic compounds is frequently complicated by the occurrence of a number of metals as a mixture of naturally occurring isotopes. This is especially true if more than one metal atom is present, but conversely a particular isotopic pattern may be useful in identifying one or more metal atoms. An example of the 70 eV spectrum of a multi-isotopic species, tetramethyltin, is shown in Fig. 6-26.[40] The characteristic isotopic pattern of tin, which has ten naturally occurring stable isotopes, is evident. Successive losses of methyl groups from the molecular ion are observed and some metal hydride species are found.

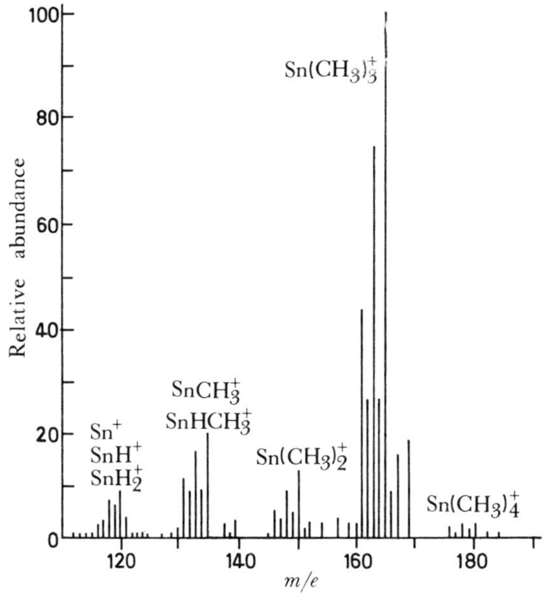

Figure 6-26

It can be envisaged from Fig. 6-26 that spectra will become very complicated if two or more multi-isotopic atoms are present, and will be complicated still further if two groups of ions of different composition are sufficiently close to overlap (e.g., the groups of ions having different compositions are almost overlapping in Fig. 6-26).

As a simple example of calculation of isotopic abundances in organometallic compounds, consider the molecular ion of $Re(CO)_5Cl$. Rhenium has two naturally occurring isotopes, ^{185}Re and ^{187}Re, of respective abundances 37·07 and 62·93 per cent. The two naturally occurring isotopes of chlorine, ^{35}Cl and ^{37}Cl, have respective abundances of 75·53 and 24·47 per cent. The ion $ReCl^{+\cdot}$ in the mass spectrum of $Re(CO)_5Cl$ would therefore be found at three different nominal *m*/*e*-values of 220, 222 and 224. The abundance at each of the respective *m*/*e*-values is obtained by multiplying together the abundances of the constituent atoms, i.e., $^{185}Re^{35}Cl^{+\cdot}$ at *m*/*e* 220 will have an abundance (relative to the total abundance of $ReCl^{+\cdot}$) of $0{\cdot}3707 \times 0{\cdot}7553$ ($=0{\cdot}2800$). The abundances at *m*/*e* 222 and 224 are similarly obtained (*m/e* 222 has two components). The calculated abundances are: *m/e* 220, 28·0 per cent; *m/e* 222, 56·6 per cent; *m/e* 224, 15·4 per cent. The calculation of peak distribution in the molecular ion, $Re(CO)_5Cl^{+\cdot}$, is further complicated by the natural abundance (1·112 per cent) of ^{13}C.

Computer programs may be written to facilitate the interpretation of the spectra of molecules containing more than one multi-isotopic atom. For example, the calculated peak distribution for the molecular ion of π-$C_5H_5(CO)_3MoGeEt_3$ is shown in Fig. 6-27.[41]

Nominal mass	Relative abundance
398	18·4132
399	2·8661
400	35·4639
401	30·6747
402	70·4708
403	49·9924
404	93·0164
405	68·2517
406	100·0000
407	51·7477
408	78·3215
409	20·0667
410	32·7513
411	4·8487
412	4·7385
413	0·6562
414	0·0730

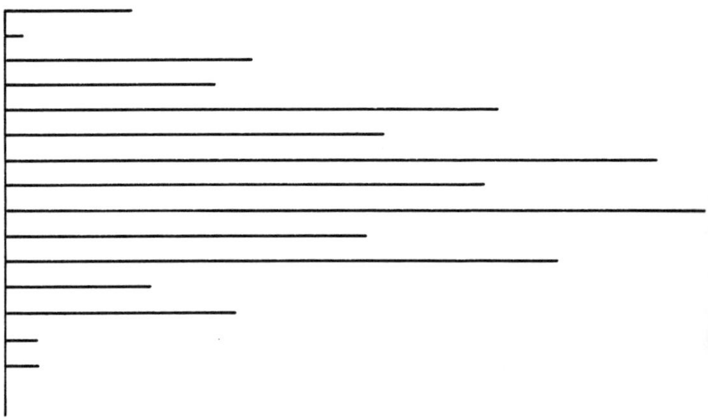

Figure 6-27

B. Thermal decomposition. Another complication which may influence organometallic spectra is thermal decomposition in the mass spectrometer. It appears that many organometallic compounds are susceptible to such effects, and in these cases analytical and energetic data can be seriously affected by spurious features in the mass spectra. For example, at a heated inlet temperature of 150°C, $Fe(CO)_5$ decomposes thermally to an extent of 20–30 per cent to form mainly $Fe(CO)_4$ and CO.[42] As the temperature is increased, the abundances of $Fe(CO)_4^+$ and CO^+ increase, whereas those of the other ions in the spectra decrease. Under conditions where decomposition occurs an appearance potential of 8·48 eV was measured for $Fe(CO)_4^+$, whereas at low temperatures this was increased to 9·1 eV. It seems likely that the ionization potential of $Fe(CO)_4$ is the quantity measured in the former case.

As a general rule, thermal decomposition should be suspected in organometallic mass spectrometry when the ligand ions (e.g., CO^+, NO^+, $C_5H_5^+$) give excessively dominant peaks. If the decomposition results from the deposition of catalytic metal films, a small amount of air will poison the catalyst and reduce the decomposition.

C. Ionization potentials. We have seen (section 2-1) that the ionization process at threshold consists of removal of an electron from the highest occupied molecular orbital. If this molecular orbital in organometallic compounds has a significant contribution from the metal orbitals, then we would anticipate that the ionization potentials of organometallics would show some correlation with those of the free metals.

The ionization potentials of metal carbonyls (Table 6-9) certainly lend support to the theory that electron removal takes place from a molecular orbital constituted principally from the metal *d*-orbitals in these cases. For example, the ionization potentials of the hexacarbonyls in the series Cr, Mo, W show a slight increase as the corresponding ionization potential of the free metal increases. In general, the ionization potential of a given carbonyl is not greatly different ($<1{\cdot}6$ eV greater) from that of the corresponding metal, and in all cases is much less than the I.P. of carbon monoxide (14·0 eV).

Table 6-9

Ionization potentials of metal carbonyls and corresponding free metals

	Ionization potential (eV)		
Compound	Electron impact†	Photon impact†	Metal
$V(CO)_6$	7·53		6·74
$Cr(CO)_6$	8·35	8·03	6·76
$Mn(CO)_5$	8·44		7·43
$Fe(CO)_5$	8·29	7·90	7·87
$Co(CO)_4$	8·30		7·86
$Ni(CO)_4$	8·52	8·25	7·63
$Mo(CO)_6$	8·36	8·12	7·10
$W(CO)_6$	8·49	8·18	7·98

† Average values from Refs. [43] and [44].

D. Fragmentation processes. It is not proposed to attempt here any cataloguing of the reactions of organometallic compounds, because of the enormous volume of relevant publications.[44] Instead, a few important and typical fragmentations have been selected to illustrate general principles.

The highly thermally stable metallocenes [$M(C_5H_5)_2$, where M is a metal] have low ionization potentials and yield abundant molecular ions which fragment by successive loss of C_5H_5 units.[45] The stability of the molecular ions of metallocenes is reflected in the stability of the same ions in aqueous solution and in the solid phase. For example, cobaltocene, $Co(C_5H_5)_2$, having a nineteen-electron configuration, is oxidized with air in aqueous solution to form the cobalticinium ion, $Co\,(C_5H_5)_2^+$, which may be isolated as its PF_6^- salt.

A large number of substituted metallocenes have been shown to undergo some unusual rearrangements, involving the organic side chain, in their mass spectra. For example, ejection of C_5H_4CO has been observed from the molecular ion of the substituted ferrocenes 41, followed by further elimination of C_5H_5 (see 41 → 42 → 43).[46]

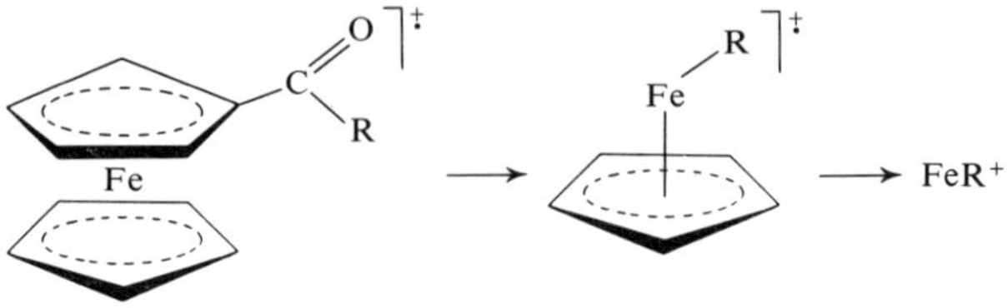

41 42 43

The importance of this rearrangement in the mass spectrum† of $C_5H_5—Fe—C_5H_4CO_2H$ is evident in Fig. 6-28, where the base peak corresponds to ion 42.

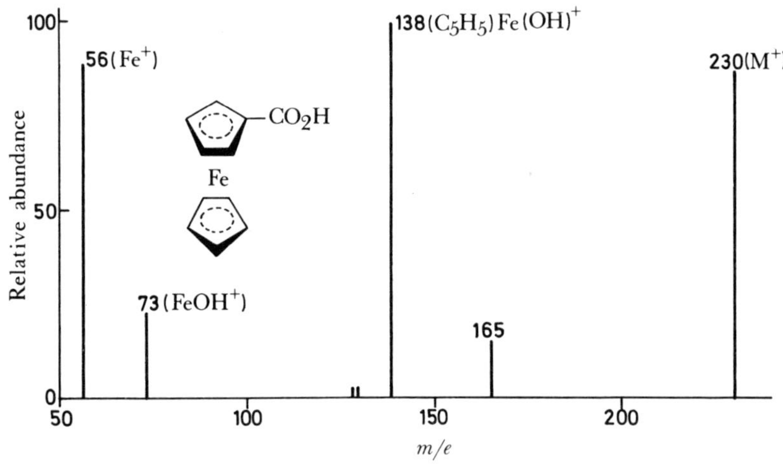

Figure 6-28

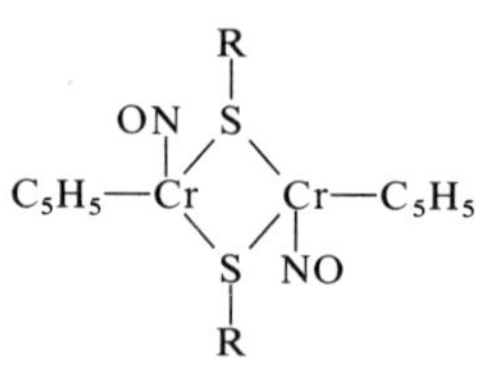

44

Migrations of ligands are also relatively common in the breakdown of organometallic ions. For example, the spectra of the binuclear metal nitrosyls with sulphur bridges (44, R = Me, Ph) exhibit an ion at *m/e* 182 corresponding to the composition $Cr(C_5H_5)_2^{+\cdot}$.[47] Migration of CO is known in the breakdown of binuclear carbonyl molecular ions, and caution should be

† Normalized for ^{56}Fe and ^{12}C.

exercised in structural interpretation that more metal—CO bonds than are actually present are not attributed to one of the metal atoms.

The mass spectrum of iron pentacarbonyl (Fig. 6-29) illustrates the stepwise loss of a ligand from a molecular ion.[48] Ions are observed for all the species $Fe(CO)_n^+$ (n = 0–5) and metastable peaks are observed for all the relevant transitions.[49]

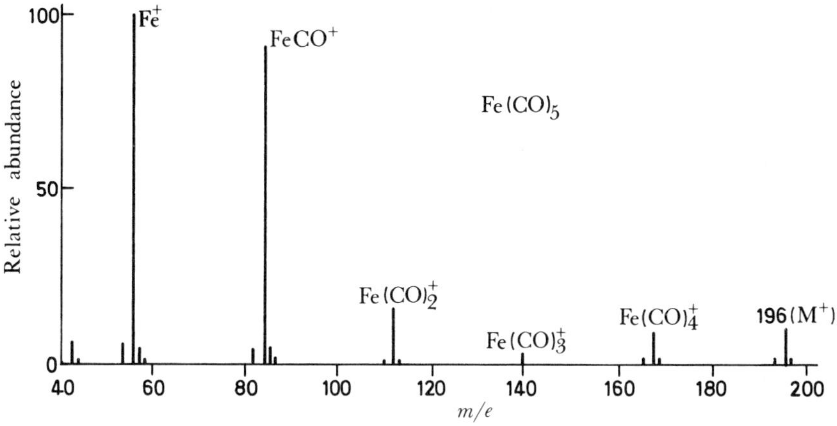

Figure 6-29

The final product of the sequence of decompositions is the singly charged iron atom Fe^+. Since the heats of formation of $Fe(CO)_5$ and of CO are known, it is a simple matter to calculate the heat of formation of Fe^+ from $Fe(CO)_5$ by measuring the appearance potential of Fe^+ (see section 5-2). Other transition-metal carbonyls may be similarly treated, and it is found in general that ΔH_f for metal ions formed by the above process is considerably greater than the true value found by other means (see Table 6-10). Although formation of the metal ions in an excited state from their carbonyls is a possibility, the discrepancy is probably due to loss of internal energy to the several CO molecules eliminated. The minimum excess energy of $X(CO)_n^+$ which is theoretically needed to produce X^+ will not in practice produce a measurable

Table 6-10
Directly and indirectly determined heats of formation of some metal ions

Ion	Parent molecule	ΔH_f (ion)(kcal/mole)[50]† from carbonyl	spectroscopic
Cr^+	$Cr(CO)_6$	257 (1075)	251 (1050)
Fe^+	$Fe(CO)_5$	296 (1238)	282 (1180)
Ni^+	$Ni(CO)_4$	331 (1385)	279 (1167)
Mo^+	$Mo(CO)_6$	363 (1518)	321 (1343)
W^+	$W(CO)_6$	425 (1777)	387 (1602)

† Figures in parentheses are in units of kJ/mole.

concentration of X^+ ions, since the probability of imparting no internal energy to the *n*CO groups as they depart is essentially negligible.

It is apparent from the large amount of data concerning the mass spectra of complexes containing mixed ligands that there is usually a greater tendency for some ligands to be eliminated in preference to others. For example, from studies of nitrosyl carbonyl complexes, it is evident that CO groups are somewhat more readily lost than NO groups in ionic decompositions [see the primary decompositions in the mass spectrum (Fig. 6-30)† of $Fe(CO)_2(NO)_2$].[51] This is due to a greater electron-donating ability of NO compared with the CO ligand. When the strong electron-donating substituent $P(OEt)_3$ is introduced into nitrosyl carbonyl complexes, a stronger metal–ligand bond is formed and the metal preferentially retains the $P(OEt)_3$ ligand.

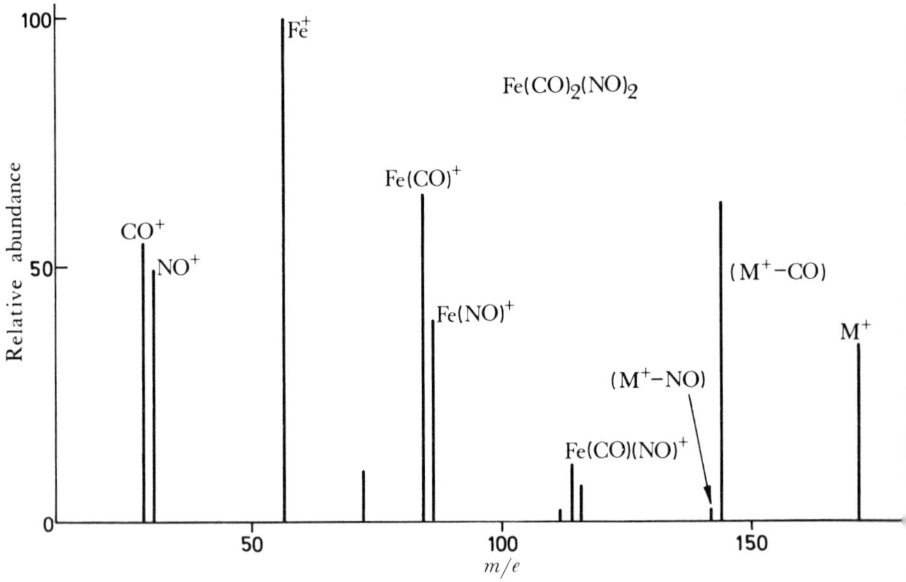

Figure 6-30

E. Detection of polymeric species. Mass spectrometry is a useful technique for detecting the presence of dimeric or polymeric species in organometallic compounds. For example, a number of instances are known of association (through bridging bonds) in solution of electron-deficient organometallics, e.g., the aluminium alkyls. The mass spectra of Me_3Al, Me_2AlH, Et_2AlH and Et_2AlOEt have all revealed evidence for dimers and trimers, particularly at low source temperatures. For example, the highest mass ion in the spectrum of trimethylaluminium corresponds to $Me_5Al_2^+$ and carries 3 per cent of the total ion current at a source temperature of 50°C.[52] This ion is

† Normalized for ^{56}Fe and ^{12}C.

presumably generated by elimination of a methyl radical from the ionized dimeric species 45.

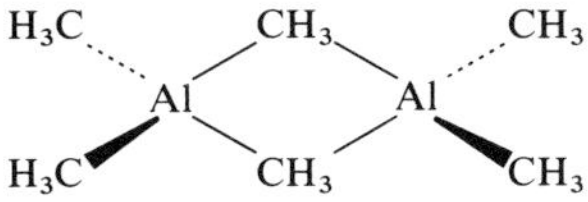

45

6-7. Skeletal rearrangement

We have seen in the previous section that group migrations are quite common in the mass spectra of organometallic compounds. Such rearrangements (generally known as *skeletal rearrangements*) are less common in the spectra of ordinary organic compounds, but do occur with moderate frequency. A large number of examples have been documented in two review articles.[53, 54]

Frequently group migration seems to be initiated by the generation of an electron-deficient site upon ionization or fragmentation. An isomerized molecular ion, or fragment ion, is then generated. For example:

$$\overset{+}{A}\text{—B—C} \longrightarrow \text{C—}\overset{+}{A}\text{—B} \longrightarrow \text{C—}\overset{+}{A} \quad + \text{B}$$

To exemplify this kind of behaviour, an example is taken from the numerous cases of migrations to carbonium ion centres. In the mass spectrum (Fig. 6-31) of methyl α-bromophenylacetate (46), the $M^+ - Br$ ion undergoes loss of CO in a one-step process with associated methoxyl migration, presumably to the carbonium ion centre.[55] In the following representation of the rearrangement, the structures of the ions are assumed and not proven.

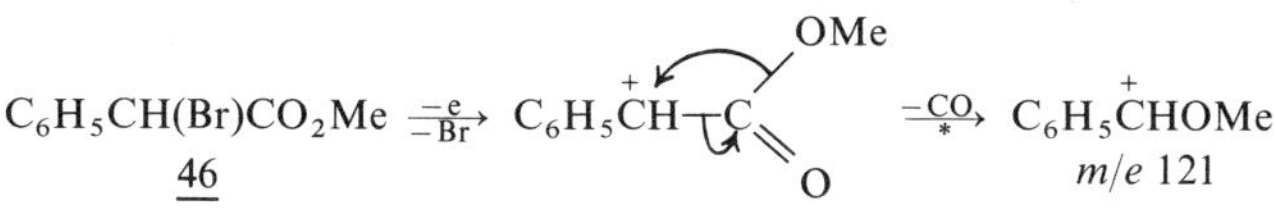

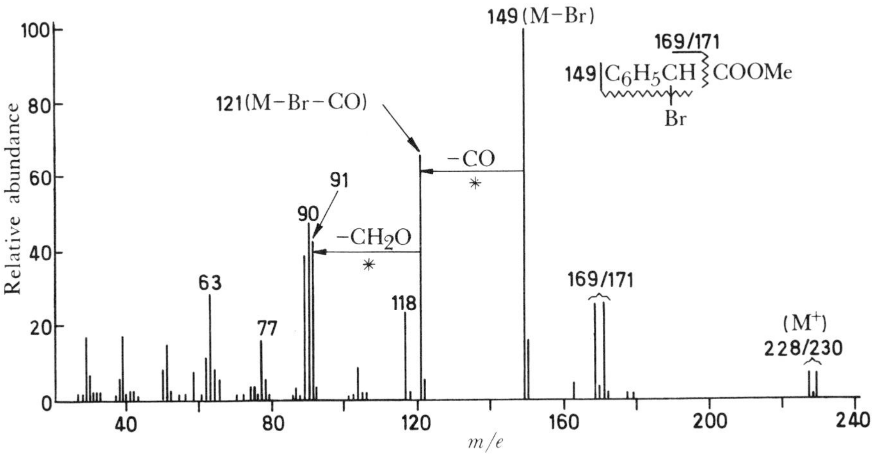

Figure 6-31

Such rearrangements do of course generate units which are not present in the original molecule, and may on occasions lead to incorrect structural conclusions being made from a mass spectrum.

6-8. Temperature and ion abundances

Mass spectra can of course be greatly simplified on reduction of the electron-beam energy from 70 eV to a nominal 12–15 eV. At low beam energies, only the primary reactions of low activation energy prevail, and consecutive decompositions are minimized (Fig. 4-4). However, it remained for Spiteller and co-workers to show that in the case of large molecules, the effect of source temperature upon relative ion abundances is also very considerable.[56, 57] The simplest mass spectra are obtained by using a 'cold' ion source ($\sim$70°C) at low voltage (say, 12 eV). Particularly noteworthy is the fact that a temperature decrease of 250°C may, in certain classes of compounds, lead to an increase in the intensity of the parent ion peaks by a factor of 200–300 compared to those recorded at the higher temperature. In general, the smaller the amount of energy required for the cleavage of the parent ion, the greater is the influence of temperature upon the M^+ abundance. For example, the abundance of the molecular ion peak in the mass spectrum of the saturated hydrocarbon triacontane ($C_{30}H_{62}$) is of the order of 1 per cent of the *m/e* 57 ($C_4H_9^+$) abundance at 70 eV and 340°C, but is 86 per cent of the *m/e* 57 abundance at 70 eV and 70°C. Although such large variations are somewhat unusual, and might seem surprising at first glance, the effects are more readily understandable if a calculation of the thermal energy of $C_{30}H_{62}$ at 340°C is made; one obtains a value of about 3 eV (1 eV = 23·06 kcal/mole = 96·49 kJ/mole). In a large molecule, the thermal energy is significant relative to the average internal energy gained upon electron impact with 70 eV electrons. It is instructive to enquire why the positive ions generated in the mass spectrometer decompose while the comparably excited neutral molecules often do not. The main factor is that some bonds are generally considerably weaker in the ion than in the neutral molecule. When there is not a large difference in the bond strengths in the neutral and cation radical (e.g., an unsubstituted aromatic hydrocarbon), it is noteworthy that most of the cation radicals do not have the energy to decompose (see section 4-2); only those ions which are (with low probability) highly excited by the electron beam do decompose.

References

[1] K. Biemann, *Mass Spectrometry*, McGraw-Hill, 1962.
[2] H. Budzikiewicz, C. Djerassi and D. H. Williams, *Structure Elucidation of Natural Products by Mass Spectrometry*, Vols. 1 and 2, Holden-Day, San Francisco, 1964.

[3] F. W. McLafferty, *Interpretation of Mass Spectra*, W. A. Benjamin, New York, 1966.
[4] H. Budzikiewicz, C. Djerassi and D. H. Williams, *Mass Spectrometry of Organic Compounds*, Holden-Day, San Francisco, 1967.
[5] J. H. Beynon, R. A. Saunders and A. E. Williams, *The Mass Spectra of Organic Molecules*, Elsevier, Amsterdam, 1968.
[6] (a) J. H. Beynon, *Mass Spectrometry and its Applications to Organic Chemistry*, Elsevier, Amsterdam, 1960, pp. 262–263; (b) G. Spiteller, *Massenspectrometrische Structuranalyse Organische Verbindungen*, Verlag Chemie, Weinheim, 1966, pp. 88, 97–98.
[7] M. Kraft and G. Spiteller, *Org. Mass Spec.* **2**, 541 (1969).
[8] F. W. McLafferty, *Anal. Chem.* **34**, 2 (1962).
[9] G. Remberg, E. Remberg, M. Spiteller-Friedmann and G. Spiteller, *Org. Mass Spec.* **1**, 87 (1968); G. Remberg and G. Spiteller, *Chem. Ber.* **103**, 3640 (1970).
[10] C. Djerassi, *Pure Appl. Chem.* **6**, 575 (1963).
[11] K. Biemann and G. Spiteller, *Tetrahedron Letters* 299 (1961); *J. Am. Chem. Soc.* **84**, 4578 (1962).
[12] B. Gilbert, J. A. Brissolese, J. M. Wilson, H. Budzikiewicz, L. J. Durham and C. Djerassi, *Chem. and Ind.* 1949 (1962).
[13] H. Audier, A. Diara, M. de J. Durazo, M. Fetizon, P. Foy and W. Vetter, *Bull. Soc. chim. Fr.* 2827 (1963); Z. Pelah, D. H. Williams, H. Budzikiewicz and C. Djerassi, *J. Am. Chem. Soc.* **86**, 3722 (1964).
[14] G. P. Arsenault, K. Biemann, A. W. Barksdale and T. C. McMorris, *J. Am. Chem. Soc.* **90**, 5635 (1968).
[15] J. A. Edwards, J. S. Mills, J. Sundeen and J. H. Fried, *J. Am. Chem. Soc.* **91**, 1248 (1969).
[16] B. C. Das and E. Lederer in *Topics in Organic Mass Spectrometry*, A. L. Burlingame, ed., Wiley Interscience, New York, 1970, pp. 255–325.
[17] R. G. Cooks and G. S. Johnson in *Specialist Periodical Reports: Mass Spectrometry*, Vol. 1, The Chemical Society, London, 1971.
[18] M. Barber, P. Jollès, E. Vilkas and E. Lederer, *Biochem. Biophys. Res. Comm.* **18**, 469 (1965).
[19] B. C. Das, S. D. Gero and E. Lederer, *Biochem. Biophys. Res. Comm.* **29**, 211 (1967).
[20] D. W. Thomas, B. C. Das, S. D. Gero and E. Lederer, *Biochem. Biophys. Res. Comm.* **33**, 519 (1968).
[21] D. W. Thomas, *F.E.B.S. Letters* **5**, 53 (1969).
[22] H. R. Morris, D. H. Williams and R. Ambler, *Biochem. J.* **125**, 189 (1971).
[23] F. W. McLafferty, R. Venkataraghavan and P. Irving, *Biochem. Biophys. Res. Comm.* **39**, 274 (1970).
[24] D. M. Desiderio, G. Ungar and P. A. White, *Chem. Comm.* 432 (1971).
[25] (a) E. Lederer, *Pure Appl. Chem.* **17**, 489 (1968); (b) M. M. Shemyakin, *Pure Appl. Chem.* **17**, 313 (1968).
[26] M. M. Shemyakin, Yu. A. Ovchinnikov, E. I. Vinogradova, A. A. Kirushkin, M. Yu. Feigina, N. A. Aldonova, Yu. B. Alakhov, V. M. Lipkin, A. I. Morishnikov, B. V. Rosinov and S. A. Kazaryan, *F.E.B.S. Letters* **7**, 8 (1970).
[27] A. A. Kiryushkin, V. A. Gorlenko, B. V. Rosinov, Yu. A. Ovchinnikov and M. M. Shemyakin, *Experientia* **25**, 913 (1969).
[28] N. K. Kochetkov and O. S. Chizhov, *Adv. Carbohydrate Chem.* **21**, 39 (1966).
[29] K. Heyns, H. F. Grützmacher, H. Scharmann and D. Müller, *Fortschr. Chem. Forsch.* **5**, 448 (1966).
[30] D. C. De Jongh and K. Biemann, *J. Am. Chem. Soc.* **86**, 67 (1964).

[31] N. K. Kochetkov, O. S. Chizhov and N. V. Molodtsov, *Tetrahedron* **24**, 5587 (1968).
[32] G. S. Johnson, W. S. Ruliffson and R. G. Cooks, *Chem. Comm.* 587 (1970).
[33] *Methods Carbohydrate Chem.* **2**, 136 (1963).
[34] N. K. Kochetkov, O. S. Chizhov, N. N. Malysheva and A. I. Shiyonok, *Org. Mass Spec.* **5**, 481 (1971).
[35] K. Biemann and J. McCloskey, *J. Am. Chem. Soc.* **84**, 2005 (1962); see also S. M. Hecht, A. S. Gupta and N. J. Leonard, *Analyt. Biochem.* **30**, 249 (1969).
[36] A. M. Lawson, R. N. Stillwell, M. M. Tacker, K. Tsuboyama and J. A. McCloskey, *J. Am. Chem. Soc.* **93**, 1014 (1971).
[37] D. F. Hunt, C. E. Hignite and K. Biemann, *Biochem. Biophys. Res. Comm.* **33**, 378 (1968).
[38] J. J. Dolhun and J. L. Wiebers, *J. Am. Chem. Soc.* **91**, 7755 (1969).
[39] L. Baczynskyj, K. Biemann and R. H. Hall, *Science* **159**, 1481 (1968).
[40] R. W. Kiser in *Characterisation of Organometallic Compounds*, Part I, M. Tsutsui, ed., Wiley-Interscience, New York, 1969, p. 147.
[41] A. Carrick and F. Glockling, *J. Chem. Soc.* (**A**), 40 (1967).
[42] S. Pignataro and F. P. Lossing, *J. Organometallic Chem.* **11**, 571 (1968).
[43] Ref. [40], p. 171.
[44] *Specialist Periodical Reports: Mass Spectrometry*, Vol. 1, The Chemical Society, London, 1971, chapter 5.
[45] L. Friedman, A. P. Irsa and G. Wilkinson, *J. Am. Chem. Soc.* **77**, 3689 (1955).
[46] A. Mandelbaum and M. Cais, *Tetrahedron Letters* **51**, 3847 (1964).
[47] F. I. Preston and R. I. Reed, *Chem. Comm.* 51 (1966).
[48] R. E. Winters and R. W. Kiser, *Inorg. Chem.* **3**, 699 (1964).
[49] R. E. Winters and J. H. Collins, *J. Phys. Chem.* **70**, 2057 (1966).
[50] J. L. Franklin *et al.*, *Ionization Potentials, Appearance Potentials and Heats of Formation of Gaseous Positive Ions*, NSRDS–NBS 26, U.S. Govt. Printing Office, 1969.
[51] A. Foffani, S. Pignataro, G. Distefano and G. Innorta, *J. Organometallic Chem.* **7**, 473 (1967).
[52] D. B. Chambers, G. E. Coates, F. Glockling and M. Weston, *J. Chem. Soc.* (**A**), 1712 (1969).
[53] P. Brown and C. Djerassi, *Angew. Chem. Internat. Edn.* **6**, 477 (1967).
[54] R. G. Cooks, *Org. Mass Spec.* **2**, 481 (1969).
[55] R. G. Cooks, J. Ronayne and D. H. Williams, *J. Chem. Soc.* (**C**), 2601 (1967).
[56] M. Spiteller-Friedmann, S. Eggers and G. Spiteller, *Monatsh.* **95**, 1740 (1964).
[57] M. Spiteller-Friedmann and G. Spiteller, *Chem. Ber.* **100**, 79 (1967).

7. Isotopic labelling

7-1. Synthesis of isotopically labelled organic compounds

This topic will be dealt with only very briefly, since the development of the subject up to 1964 in relation to mass spectrometry has previously been reviewed.[1] Much of the work published up to 1958 is covered by Murray and Williams.[2]

The most useful isotopes from the mass spectrometric viewpoint are ^{2}H, ^{13}C, ^{15}N and ^{18}O.

A. Incorporation of deuterium. Deuterium is by far the most common isotope used in mass spectrometric studies, since it is easy to incorporate, and since essentially isotopically pure (99·7 per cent) deuterium oxide is cheaply available.

The active hydrogens of —OH, —NH and —SH groups are conveniently replaced by deuterium in the mass spectrometer itself, by repeated introduction of D_2O through the heated inlet system prior to running the sample in question. The following reaction sequences summarize some of the other more important methods of introducing deuterium:

1. $$RCH_2COCH_2R \xrightarrow[\text{base}]{D_2O} RCD_2COCD_2R$$

2. $$RCD_2COCD_2R \xrightarrow[\text{Villiger}]{\text{Baeyer}} RCD_2OH + RCD_2COOH$$

3. $$RCH_2COOH \xrightarrow[\text{sealed tube at } 200^\circ]{D_2O/DCl} RCD_2COOH(D)$$

4. $$-CH{=}CH- \xrightarrow[D_2]{Pd/C} -CHD{-}CHD-$$

5. $$-\overset{\overset{\displaystyle O}{\|}}{C}- \xrightarrow[(NaBD_4)]{LiAlD_4} -\overset{\overset{\displaystyle OH}{|}}{C}D-$$

6.

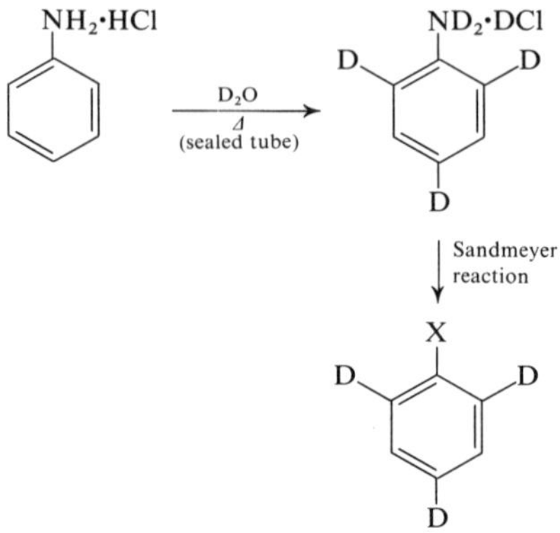

Numerous syntheses of deuteriated organic compounds are facilitated through the ready availability of d_6-benzene, d_6-acetone, d_4-acetic acid and d_3-methyl iodide.

The relative ease with which specifically deuteriated analogues of organic compounds may be synthesized is illustrated by reference to the preparation of numerous deuteriated analogues of 8-methylnonan-4-one.[3]

7.

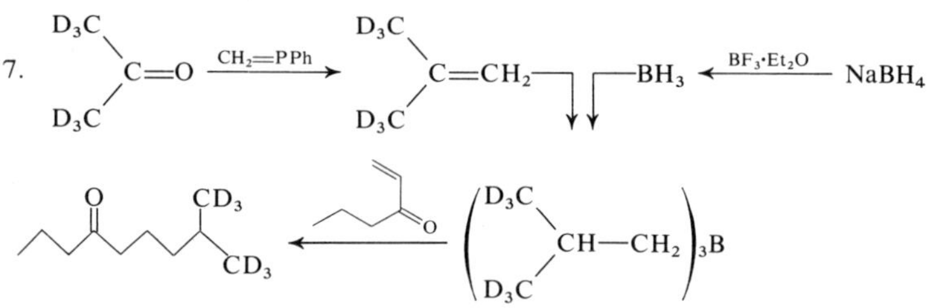

8.

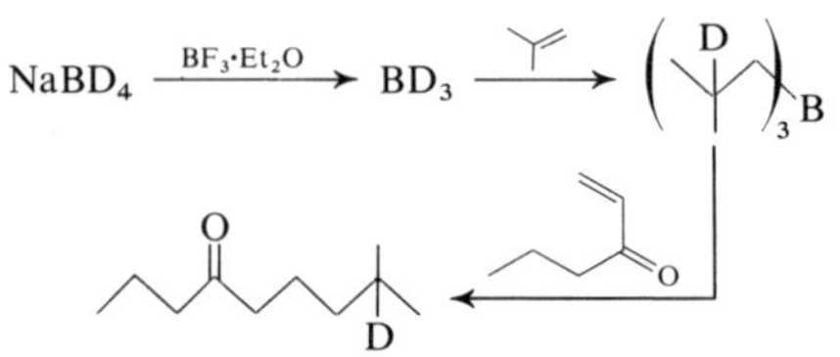

9.

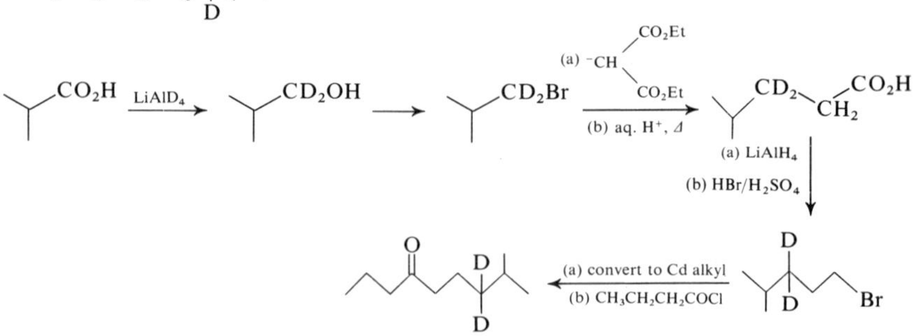

10. 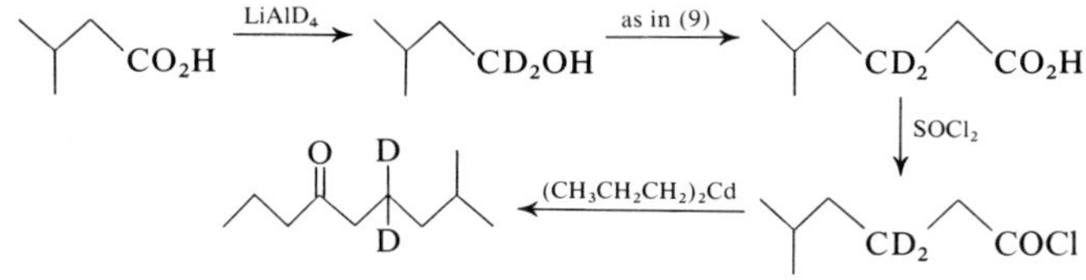

11. 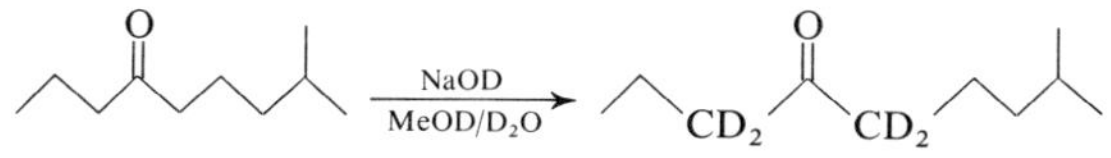

12. 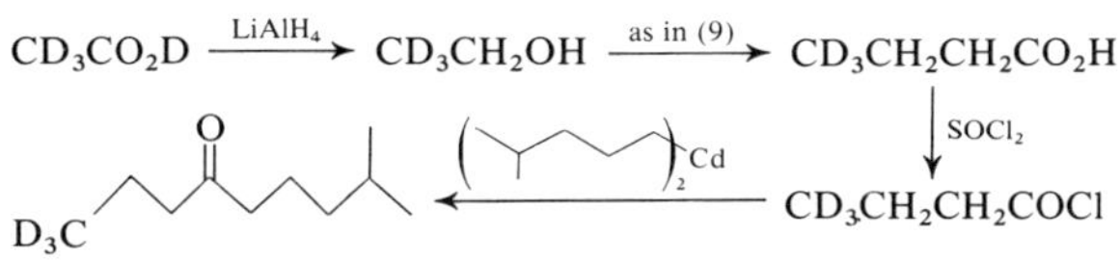

B. Incorporation of ^{13}C. At the time of writing, the cheapest source of ^{13}C for organic syntheses is $Ba^{13}CO_3$ (~£6 per gram, enriched with 65 per cent ^{13}C). $^{13}CH_3I$ is moderately priced at less than £50 per gram for 65 per cent isotopic incorporation.

Organic compounds specifically labelled with ^{13}C have been used increasingly in recent years to investigate mass-spectrometric reactions[4] and particularly for the elucidation of carbon-scrambling mechanisms in organic ions (see also section 7-2).

$^{13}CO_2$, prepared by reaction of acid with $Ba^{13}CO_3$, provides a convenient source of ^{13}C for synthesis of organic compounds:

13. $$Ba^{13}CO_3 \xrightarrow[\text{(b) RMgBr}]{\text{(a) H}^+} R^{13}CO_2H$$

Specific examples of incorporation of ^{13}C into organic molecules are outlined by the following schemes:

14.[4d] 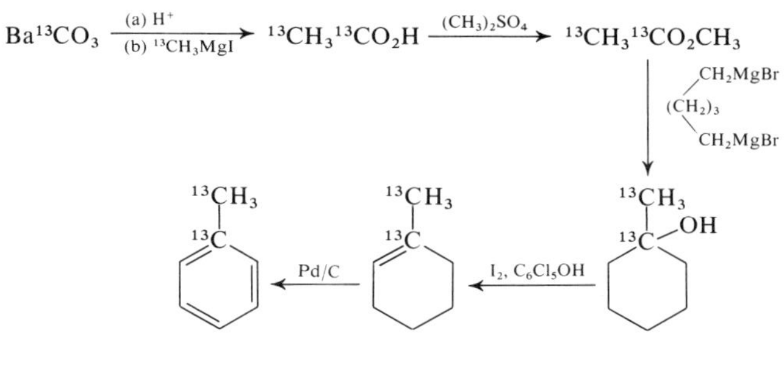

15.[4b] $$Ph^{13}CHO \xrightarrow{Ph\cdot CH_2MgBr} Ph\cdot{}^{13}CH{=}CH\cdot Ph$$

16.[4c]

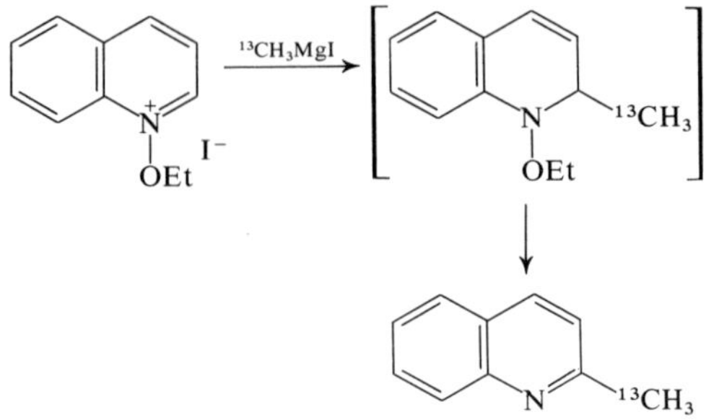

C. Incorporation of ^{18}O. Although most mass-spectrometric investigations of labelled organic molecules have involved ^{2}H and/or ^{13}C, fragmentation mechanisms have also been investigated employing ^{18}O as a label.[5] $H_2{}^{18}O$ is a common commercial source of ^{18}O and incorporation of the label into the common oxygen-containing organic functional groups can be readily effected:

Acids and derivatives

17. $RCO_2H \xrightarrow[HCl]{H_2{}^{18}O} RC^{18}O_2H$

18.

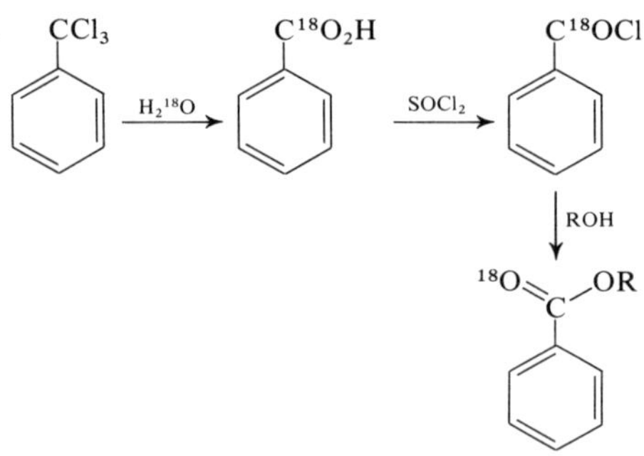

Aldehydes, ketones

19. $R{-}CO{-}R' \xrightarrow[CF_3CO_2H]{H_2{}^{18}O/\text{dioxan}} R{-}C^{18}O{-}R'$

($R' = $ alkyl, H)

Phenols

20.

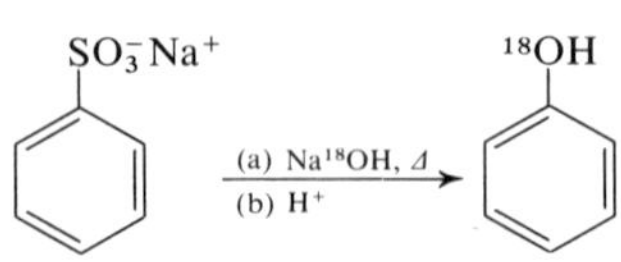

D. Incorporation of ^{15}N. Although ^{15}N has been used as a label for the elucidation of mass spectral decomposition pathways,[6] the commonest use of ^{15}N in mass spectrometry is the determination of isotopic incorporation in natural products. Metabolic pathways can be deduced by introducing ^{15}N-labelled compounds (usually amino acids) *in vivo* and investigating the extent of isotopic incorporation into other metabolites.[7]

Common sources of ^{15}N are $^{15}NH_3$, $^{15}NH_4Cl$ and $K^{15}NO_3$, and amino acids labelled with ^{15}N are commercially available. A few syntheses of ^{15}N-labelled compounds are shown below:

Amides

21. $RCOCl \xrightarrow{^{15}NH_3} RCO^{15}NH_2$

Primary amines

22. $RCO^{15}NH_2 \xrightarrow{LiAlH_4} RCH_2{}^{15}NH_2$

$RCO^{15}NH_2 \xrightarrow{NaOBr} R^{15}NH_2$

Pyrimidines
(e.g., cytosine)

23.

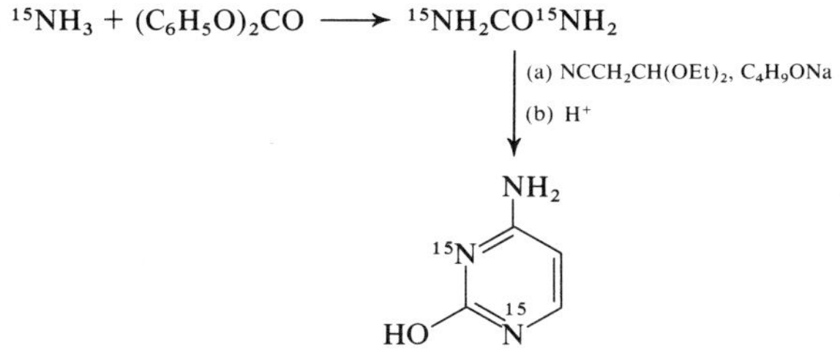

7-2. Unimolecular reactions of isotopically labelled organic ions

Deuterium labelling has proved to be a very useful technique in organic mass spectrometry in uncovering site-specific hydrogen-transfer reactions. A classic example of this is found in the γ-hydrogen (McLafferty) rearrangement of carbonyl compounds. The 70 eV mass spectrum of ethyl butyrate (1) contains a prominent ion at m/e 88, whereas in the deuterium-labelled compound 2, this peak is shifted to m/e 89.[8] This confirms that the transfer of a γ-hydrogen occurs (see 2 → 3), but reveals nothing about the detailed mechanism of the reaction. We have depicted 2 → 3 as a concerted process, but molecular orbital calculations indicate that transfer of the γ-hydrogen may occur before loss of the neutral olefin.[9]

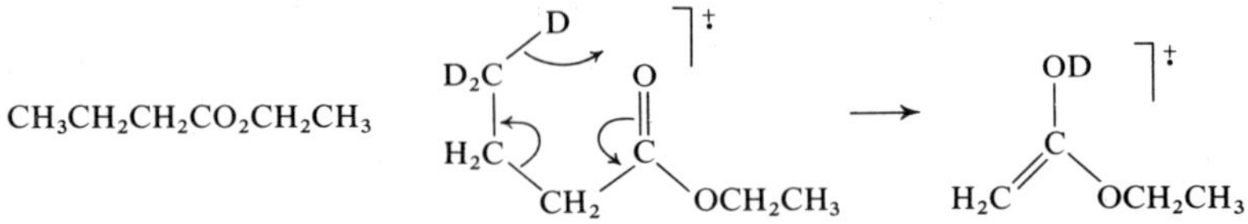

1 2 3, *m/e* 89

However, in many known mass-spectral reactions it is impossible to determine the specificity of the processes involved. We have seen (section 4-5) that the McLafferty rearrangement becomes apparently non-specific as ions of progressively longer lifetimes (e.g., metastables) are sampled. The reason for this is internal hydrogen scrambling (requiring a low activation energy) within the parent ion, and experience has shown that this phenomenon is common, even for reactions in the source in many cases.

For example, scrambling of hydrogen atoms has been shown to be commonplace in aromatic and substituted aromatic molecular ions. Thus, complete hydrogen equivalence exists in the molecular ion of 1,3,5-$[^2H]_3$-benzene (4) prior to acetylene loss in a metastable transition.[10]

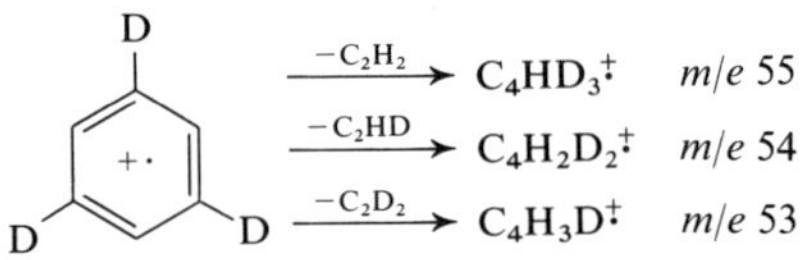

4, *m/e* 81

If no scrambling occurs, then only loss of a C_2HD fragment would be expected from 4. In fact, loss of C_2H_2, C_2HD and C_2D_2 occur in the approximate ratio of 1:3:1 (see Fig. 7-1), which is the expected ratio if complete H/D scrambling and no isotope effect are assumed. One arrives at this

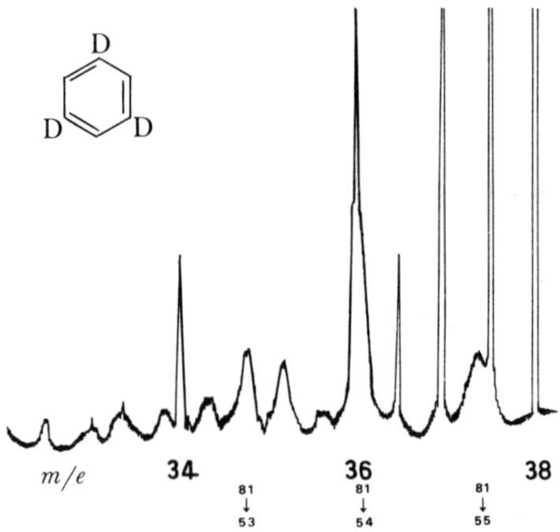

Figure 7-1

figure by determining the number of possible ways of selecting $(C_2)H_2$, $(C_2)HD$ and $(C_2)D_2$ from a scrambled ion $(C_6)H_3D_3^{+\cdot}$. It is evident that the number of ways of selecting H_2 from H_3 (and D_2 from D_3) is 3. The number of ways of selecting HD from H_3D_3 is 9. Hence the ratio of 3:9:3 is obtained.

In the above case, the determination of the statistical ratio of products is a simple matter, but in more complicated examples, formula (7-1) is a useful asset, and gives the number of ways of choosing n atoms from p atoms. Thus the number of ways of choosing H_2 from H_6 is 15, which provides a useful check that the calculation for the above acetylene eliminations from 4 are correct (3 + 9 + 3).

$$\frac{p!}{n!(p-n)!} \tag{7-1}$$

The foregoing experiment carried out on deuteriated benzene does not of course distinguish between carbon and hydrogen scrambling, or indeed throw light on the possibility that both processes might be occurring.

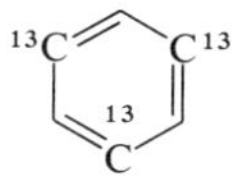

5

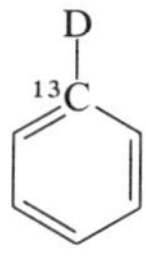

6

If no carbon scrambling occurs, then acetylene should be lost from 5 exclusively as $^{13}CCH_2$; the observed ratio of 1:3:1 for loss of C_2H_2, $^{13}CCH_2$ and $^{13}C_2H_2$ in metastable transitions is consistent with complete carbon scrambling.[11] Metastable transitions from 6 have further established that C—H bonds are broken and reformed during the scrambling process.[12, 13] Therefore two independent scrambling mechanisms are operative and the structure of the $C_6H_6^{+\cdot}$ species which loses acetylene remains uncertain. The occurrence of a scrambling process prior to decomposition of the benzene molecular ion is not surprising when it is considered that the lowest activation energy decomposition requires about 4 eV. Since a scrambling process involves simultaneous making and breaking of bonds, its activation energy might be considerably lower than this.

In the case of nitrobenzene, however, it has been demonstrated[14] that the nitro-substituent does not shift around the ring prior to elimination of NO. The spectrum of 1-^{13}C-nitrobenzene (7) shows that the $M^+ -$ NO ion eliminates a CO molecule containing all the ^{13}C label.

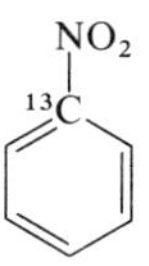

7

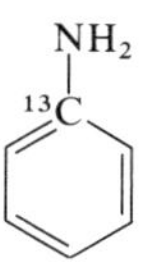

8

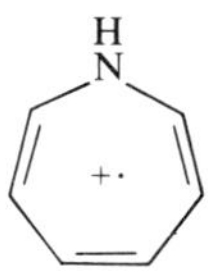

9

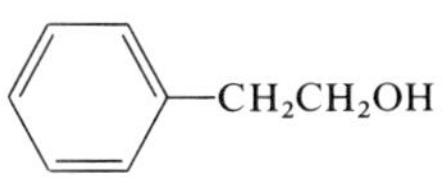

10

The molecular ion of 1-^{13}C-aniline (8) also demonstrates almost complete retention of positional identity.[15] More than 90 per cent of the HCN lost is eliminated as $H^{13}CN$. This rules out the possibility that all the molecular ions rearrange to the azepinium isomer (9) before elimination of HCN.

Examples are known where scrambling occurs over part of an organic ion. For example, the metastable peaks for loss of 'C_2H_4O' from a number of specifically deuteriated derivatives of 2-phenylethanol (10) have established that exchange of the hydroxyl and aromatic (but not aliphatic) hydrogens occurs prior to this reaction.[16]

Deuterium labelling of bicyclic aromatic molecules has revealed evidence for complete or partial hydrogen randomization prior to decomposition of the molecular ion. For example, the molecular ion of 1-cyano-[2,4-$^2H_2$]-naphthalene (11) eliminates HCN and DCN in the ratio 5:2 in metastable transitions (see *m/e* 105·8 and *m/e* 104·0, respectively in Fig. 7-2) and the distribution of normal peaks in the $M^{+\cdot}$ – HCN region is close to that calculated on the basis of complete hydrogen scrambling over both rings.[17]

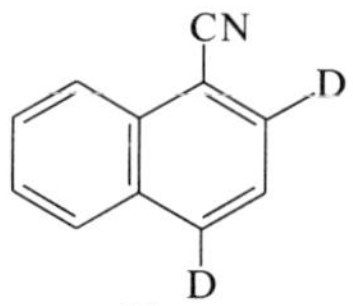

11

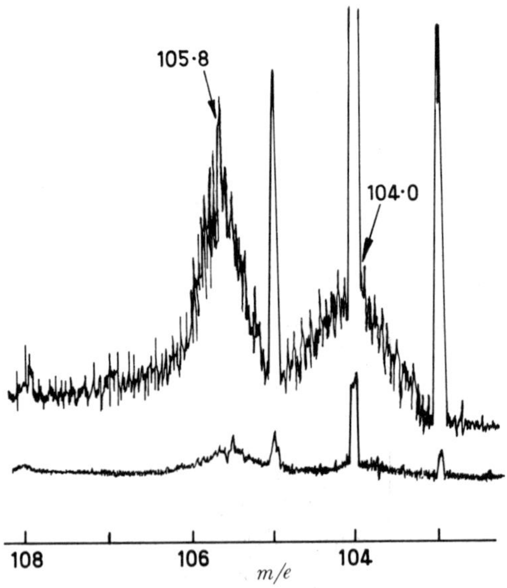

Figure 7-2

In aliphatic ions, too, there are many known cases of hydrogen scrambling prior to decomposition. For example, butyl ions derived from the series of compounds of structure 12 undergo complete H/D scrambling before methane loss in metastable transitions.[18] Similarly, the *t*-butyl compounds 13 yield butyl ions which undergo complete H/D scrambling.[19]

$CH_3CH_2CD_2CH_2X$

12

X = Cl, Br, I, CO_2Me, CO_2Et, CO_2Ph

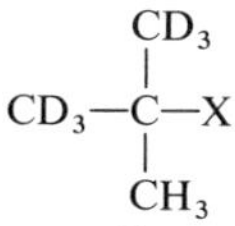

13

X = Br, I

The very similar abundances of competing metastable transitions (see section 4-4) of $C_4H_9^+$ ions generated from *n*-, *iso*-, *sec*- and *tert*-butyl structures indicate isomerization of $C_4H_9^+$ ions to a common structure, or mixture of structures, and such a reversible interconversion of structures provides a mechanism for H/D scrambling.

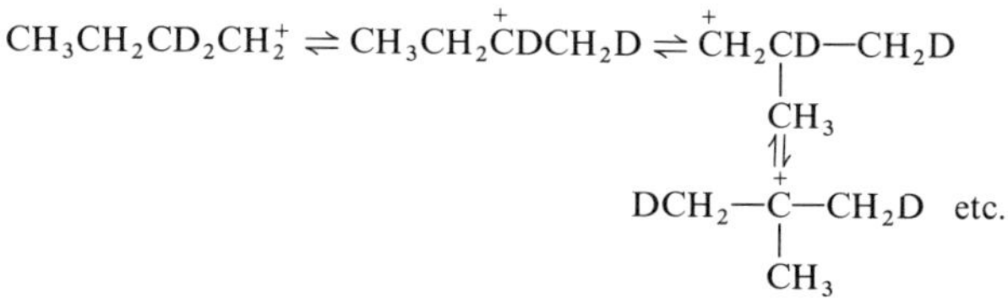

Studies of the ion kinetic-energy spectrum (section 3-2) of the ^{2}H- and ^{18}O-labelled ethyl benzoate (14) have helped to elucidate the mechanism of breakdown of the $M^+ - C_2D_4$ ion, which corresponds formally to the ionized benzoic acid molecular ion (15).[20]

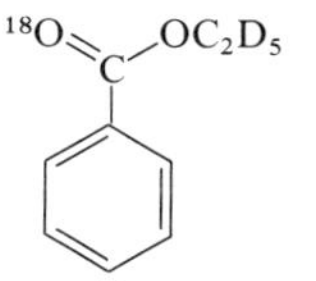

14

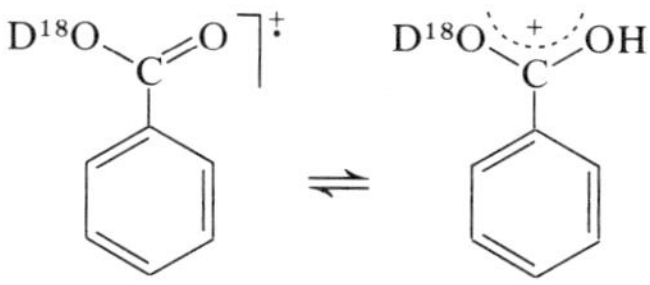

15 16

The $M^+ - C_2H_4$ ion from unlabelled ethyl benzoate eliminates OH in a metastable transition, whereas the $M^+ - C_2D_4$ ion from 14 eliminates ^{18}OD, ^{18}OH, OD and OH. The partial IKE spectrum of 14 provides a rapid method of determining the relative abundances of these metastable transitions in the first drift region (see Fig. 7-3).

The first point to emerge from consideration of Fig. 7-3 is that hydroxyl radicals containing H are lost in almost twice the abundance of those containing D. This is consistent with an equilibration between *ortho*- and carboxyl hydrogens prior to metastable decomposition (an equilibrium independently established to occur in the molecular ion of benzoic acid itself), and is confirmed by the IKE spectrum of $C_6H_5CO_2C_2D_5$. Second, the hydroxyl radicals containing ^{18}O are lost with almost equal probability compared with those containing ^{16}O, as confirmed by the spectrum of $C_6H_5C^{18}OOC_2H_5$. One possibility is that structures of the type 16 exist in equilibrium with structure 15.

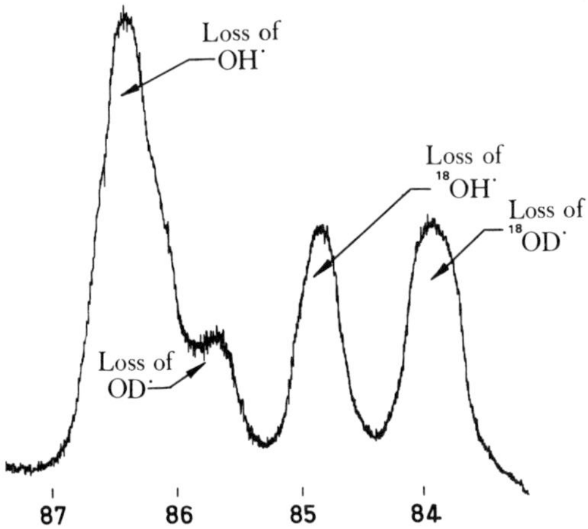

Figure 7-3

Finally, Fig. 7-3 shows that the ratio of the probabilities of losing OH and ^{18}OH from 15 is almost 2:1, whilst the ratio of the probabilities of losing OD and ^{18}OD is 1:2. The two oxygen atoms, therefore, have not become completely equivalent with respect to the available H and D atoms in the doubly labelled $M^+ - C_2D_4$ ion. There is a greater probability of the deuterium atom being attached to the ^{18}O atom and of the *ortho*-hydrogen atoms being attached to the ^{16}O atom.

The overall conclusions to be reached from these results are that equilibration of *ortho* and carboxyl-hydrogen atoms occurs via reactions such as 15 $\rightleftharpoons$ 16 and not via transfer of hydrogen between the two oxygen atoms. Complete equivalence of the oxygen atoms with respect to the available H and D atoms in 15 cannot be reached until several half-revolutions about the Ar—C bond and interconversions of 15 and 16 have taken place. It is calculated[20] that about eight half-revolutions about the Ar—C bond must have taken place to explain the results in Fig. 7-3.

Several important points should be gathered from the foregoing examples of internal rearrangements prior to fragmentation:

(a) If the scrambling is uncovered in metastable transitions (as it can be most conveniently and unambiguously), then some form of internal rearrangement has a lower activation energy than the fragmentation process which uncovered the scrambling.

(b) It follows from the first point that scrambling is usually more likely to be observed the higher the activation energy required for unimolecular decomposition of the species being examined.

(c) Randomization of H and C within hydrocarbon ions is a fairly common occurrence. In contrast, the incorporation of heteroatoms (e.g., O, N) into ions often increases the energy barrier to randomization processes relative to fragmentation processes. In general, carbon–heteroatom bonds are more likely to retain their integrity upon electron impact than are carbon–hydrogen bonds.

(d) The occurrence of complete H (or C) scrambling within a given species does not in general give any information about the nature of the decomposing structure.

The use of mass spectra of isotopically labelled species in elucidating the mechanisms of biosynthetic pathways is elegantly illustrated by the following example.[21]

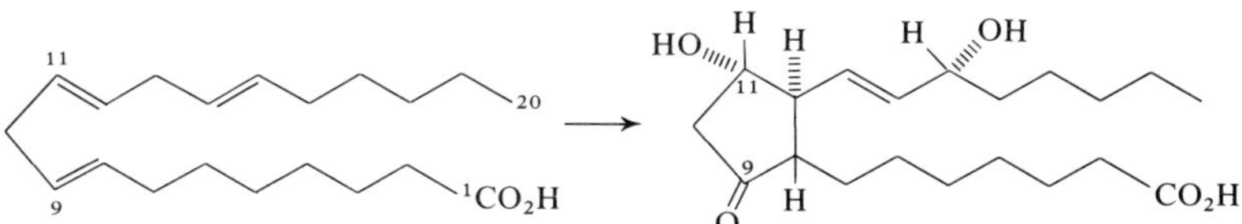

17 18

The biosynthesis of prostaglandin E_1 (18) from 8,11,14-eicosatrienoic acid (17) in the presence of atmospheric oxygen involves incorporation of 3 atoms of oxygen. To investigate the origin of the oxygens at C-9 and C-11, [2-^{14}C]8,11,14-eicosatrienoic acid was incubated with a preparation of vesicular gland for 10 min in an atmosphere containing $^{18}O_2$. The trihydroxy derivative of 18 was then prepared by adding ice-cold ethanolic $NaBH_4$ to the incubation mixture. After esterification (ethyl ester) and methylation of the hydroxy groups, followed by permanganate-periodate oxidation and further esterification, the diester 19 was formed.

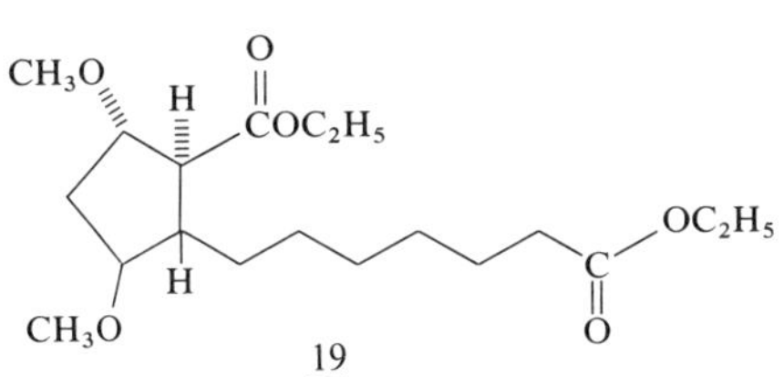

19

The mass spectra of a reference sample of 19, and of a sample of 19 formed in an atmosphere containing $^{18}O_2$, were then compared. The spectra were obtained using a directly linked glc/mass-spectrometer system (chapter

11), about 5 μg of material being injected into the column. Three fragment ions which retained both oxygen atoms in the ring were evident: at *m/e* 343 $(M - CH_3)^+$, 313 $(M - OC_2H_5)^+$ and 297 $(M - (CH_3 + C_2H_5OH))^+$. In the spectrum of 19 derived from prostaglandin E_1 formed in an atmosphere containing $^{16}O_2$ (43 per cent), $^{16}O^{18}O$ (1 per cent) and $^{18}O_2$ (56 per cent), peaks were also seen at *m/e* 347, 317 and 301. Furthermore, these peaks (due to fragments containing two ^{18}O atoms) were 12–17 times more intense than peaks at *m/e* 345, 315 and 299, clearly demonstrating that the oxygen atoms of the hydroxyl group at C-11 and of the keto group at C-9 originate from the same molecule of oxygen.

7-3. Deuterium isotope effects

A factor that must be considered in some decompositions of deuterium-labelled organic ions is the primary deuterium isotope effect. Such an effect is observed when an X—H bond is broken in the rate-determining step of a decomposition. (Most cases reported have involved C—H bond rupture, but cases are also known for X = O, N.) In view of the widespread use of deuterium labelling to determine scrambling processes and specific hydrogen transfers it is important to ascertain the magnitude of the deuterium isotope effect and the conditions under which it may be expected to occur.

The origin of the primary deuterium isotope effect lies in the differences in zero-point energy between X—H and X—D bonds. Because of the higher vibrational frequency of the X—H bond, its zero-point energy is higher and less energy is required to reach the transition state for reaction. For the organic ions studied, the effect of substituting a deuterium for a hydrogen atom is to increase the activation energy (for a C—H bond rupture) by between 0·05 and 0·14 eV. The variation of the kinetic isotope effect k_H/k_D with internal energy may be understood qualitatively by reference to the simplified rate equation:

$$k = \nu\left(\frac{E - E_0}{E}\right)^{s-1} \tag{7-2}$$

We may therefore write eq. (7-3) for the isotope effect k_H/k_D in a partially deuteriated organic ion:

$$\frac{k_H}{k_D} = \frac{\nu_H}{\nu_D}\left(\frac{E - E_0^H}{E - E_0^D}\right)^{s-1} \tag{7-3}$$

It is emphasized that eq. (7-2) has severe limitations close to threshold, but eq. (7-3) can nevertheless give valuable qualitative information about the variation of k_H/k_D with internal energy. It is evident that (a) k_H should always be greater than k_D for the simple bond rupture, (b) k_H/k_D should decrease with increasing energy, and (c) k_H/k_D should be infinite at the activation energy for C—H rupture and should approach 1·4 at high energies ($\nu_H/\nu_D = \sqrt{2} \simeq 1{\cdot}4$).

Calculations of the variation of isotope effect with internal energy have been made[22] using the improved equation of the QET and Fig. 7-4(a) shows a calculated curve for the loss of H and D from deuteriated toluenes, corrected for the number of H and D atoms in the scrambled molecular ion.[23] This variation of isotope effect with internal energy in isolated organic ions is analogous to the variation of isotope effect with temperature for thermally equilibrated organic neutrals. A calculated thermal curve for a ^{12}C—H vs. ^{12}C—D bond fission is shown in Fig. 7-4(b).[24]

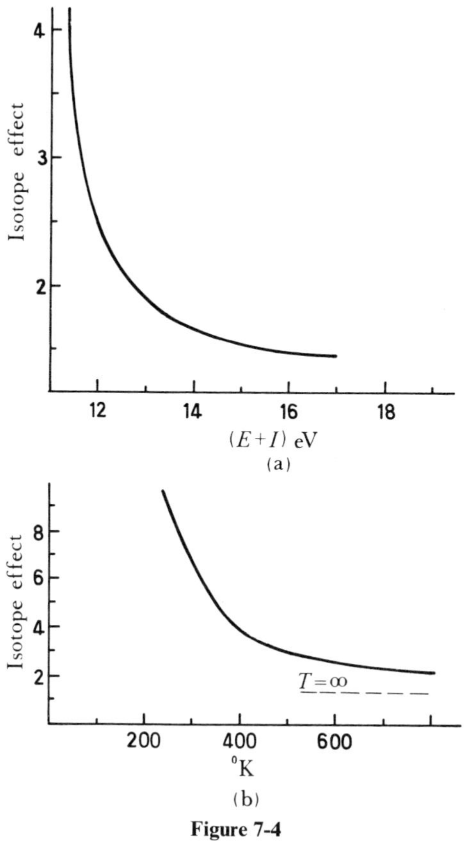

Figure 7-4

Theory therefore predicts that high-isotope effects should be observed at ion energies just above the threshold for decomposition. This condition is satisfied for metastable decompositions, and in practice high deuterium isotope effects are often found for metastable transitions involving C—H bond cleavage. For example, the molecular ion of CH_3CD_3 eliminates H in preference to D in a ratio of 600:1.[25] Very large isotope effects are expected for the metastable transitions from small ions, because the energy range giving metastables for such ions is often very narrow due to the rapid rise of k with E. If partial deuteriation introduces a new reaction (loss of D) whose activation energy is 0·10 eV higher than that for loss of H, then the higher-energy reaction might almost fall outside the metastable-energy window.

Partial H/D scrambling is a considerable problem in the evaluation of deuterium isotope effects and necessitates studies on species deuteriated in different parts of the molecule. If reactions are specific, with zero or complete scrambling, then the situation is somewhat simplified.

The mass spectra[23] of the deuteriated toluenes 20–22 provide evidence for the fall in isotope effect with increasing internal energy predicted by eq. (7-3).

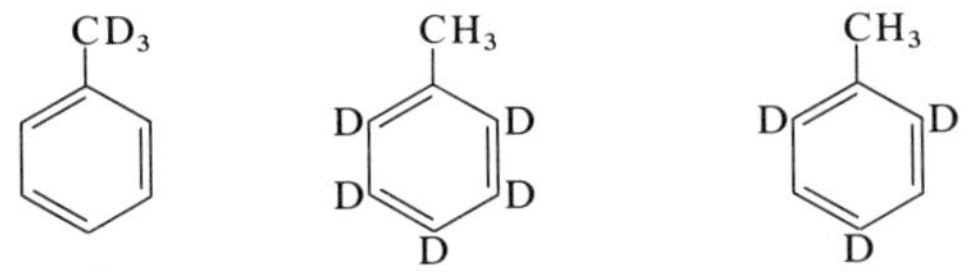

20 21 22

The ratio of metastable ion abundances (first drift region) for loss of H and D respectively from the molecular ions of 20–22 are shown in Table 7-1. These ratios are then corrected by division by the 'statistical' ratio, which is the ratio expected on the basis of no isotope effect and complete scrambling (i.e., 5/3 for 20 and 22, 3/5 for 21). No assumption need be made about the structures of the decomposing ion.

Table 7-1

$[M^+ - H]/[M^+ - D]$ *for unimolecular metastable ions from deuteriated toluenes at 70 eV*

	20	21	22
$[m^*_H]/[m^*_D]$ (observed)	4·68 ± 0·09	1·68 ± 0·99	4·77 ± 0·15
Observed/statistical	2·81 ± 0·06	2·80 ± 0·15	2·86 ± 0·09

The observed/statistical ratios are the same within experimental error and this is consistent with (a) an isotope effect of 2·8 and (b) complete H/D scrambling in these low-energy, metastable molecular ions. The assumption that k_H/k_D is equal to the corrected metastable ratio is valid because no further decompositions of the '$C_7H_7^+$' ions are occurring for the energy range measured. It is notable that the isotope effect is not very large, in contrast to the case for ethane (see above). This is because the metastable-energy window is wider for toluene compared with ethane, due to the larger number of degrees of freedom.

The relative abundances of $M^+ - H$ and $M^+ - D$ ions formed in the source at 70 eV from compounds 20–22 are shown in Table 7-2. The observed/statistical ratios show that (a) there is a preference for loss of α-hydrogens, and the partial spectra (see Fig. 7-5) of isomers 20 and 22 demonstrate this, and (b) the isotope effect is less than that for the metastable transitions.

Table 7-2

$[M^+ - H]/[M^+ - D]$ *for daughter ions from deuteriated toluenes at 70 eV*

	20	21	22
$[M^+{-}H]/[M^+{-}D]$	1·76 ± 0·06	1·14 ± 0·02	2·90 ± 0·04
Observed/ statistical	1·06 ± 0·04	1·90 ± 0·03	1·74 ± 0·02

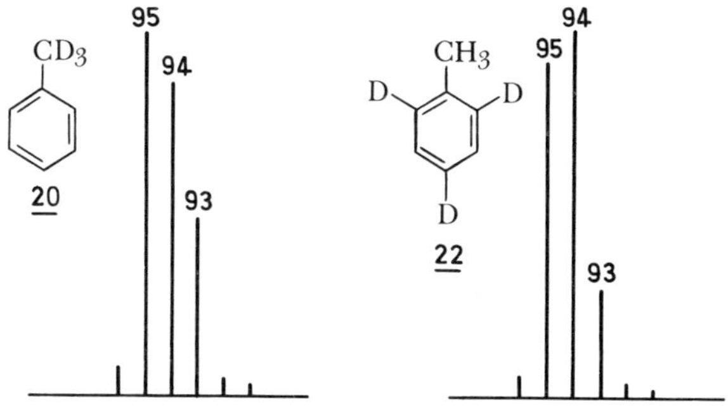

Figure 7-5

In certain cases it is possible to utilize the primary deuterium-isotope effect in mass spectrometry to yield mechanistic information, as in neutral chemistry. Isotope effects upon reactions not involving elimination of a hydrogen-containing neutral may be investigated by use of a competing reference reaction from the organic ion under study. For example, an isotope effect upon loss of CO from the molecular ion of *p*-bromophenol has been detected[26] by measurement of competing metastable ion abundances for CO and Br loss from compounds 23 and 24 (see Table 7-3). An isotope effect upon CO loss is reflected in the substantial decrease in the metastable abundance relative to that for bromine loss upon deuteriation.

Table 7-3

Competing metastable abundance ratios for reactions from the molecular ion of p-bromophenol, the M^+*-ethylene ion of p-bromophenetole and their deuteriated derivatives*

Number	Compound	Decomposing ion (A^+)	$m^*[A^+{-}Br]/m^*[A^+{-}CO]$
23	*p*-BrC_6H_4OH	M^+	20 ± 1
24	*p*-BrC_6H_4OD	M^+	62 ± 4
25	*p*-$BrC_6H_4OC_2H_5$	$M^+ - C_2H_4$	19 ± 1
26	*p*-$BrC_6H_4OC_2D_5$	$M^+ - C_2D_4$	50 ± 3

This isotope effect may be used to yield mechanistic information on the elimination of ethylene from the molecular ion of phenetoles. First, the similarity in metastable abundance ratios (see section 4-4) between the molecular ions of the phenols (compounds 23 and 24) and their M^+ – ethylene counterparts from the phenetoles (25 and 26 respectively) confirms that the decomposing metastable ions have the same structure and energy, whether formed directly from *p*-bromophenol or indirectly from *p*-bromophenetole. Second, the isotope effect in the case of compound 26 demonstrates that the reaction 25 → 23 is occurring (4-membered transition state), since the direct reaction 25 → 27 (via a 6-membered transition state) would not present an opportunity for an isotope effect on the further loss of CO.

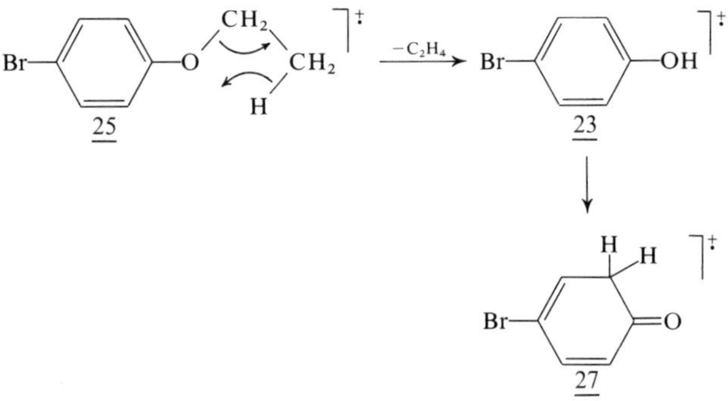

References

[1] H. Budzikiewicz, C. Djerassi and D. H. Williams, *Structure Elucidation of Natural Products by Mass Spectrometer*, Vol. 1, Holden-Day, San Francisco, 1964; for a recent account of deuterium labelling, see A. F. Thomas, 'Deuterium Labelling in Organic Chemistry', Appleton-Century-Crofts, New York, 1971.

[2] A. Murray and D. L. Williams, *Organic Syntheses with Isotopes*, Interscience, New York, 1958.

[3] G. Eadon and C. Djerassi, *J. Am. Chem. Soc.* **92**, 3084 (1970).

[4] See, for example: (a) R. G. Cooks and S. L. Bernasek, *J. Am. Chem. Soc.* **92**, 2129 (1970); (b) P. F. Donaghue, P. Y. White, J. H. Bowie, B. D. Roney and H. J. Rodda, *Org. Mass Spec.* **2**, 1061 (1969); (c) P. M. Draper and D. B. Maclean, *Canad. J. Chem.* **48**, 746 (1970); (d) K. L. Rinehart, A. C. Buchholz, G. E. Van Lear and H. L. Cantrill, *J. Am. Chem. Soc.* **90**, 2983 (1968).

[5] See, for example: (a) R. H. Shapiro and K. B. Tomer, *Org. Mass Spec.* **3**, 333 (1970); (b) J. H. Bowie and P. Y. White, *J. Chem. Soc.* (**B**), 89 (1969); (c) T. H. Kinstle, O. L. Chapman and M. Sung, *J. Am. Chem. Soc.* **90**, 1227 (1968).

[6] See, for example: (a) R. T. M. Fraser and N. C. Paul, *J. Chem. Soc.* (**B**), 1407 (1968); (b) P. N. Rylander, S. Meyerson, E. L. Eliel and J. D. McCollum, *J. Am. Chem. Soc.* **85**, 2723 (1963).

[7] See, for example: (a) A. K. Bose, K. G. Das, P. T. Funke, I. Kugajevsky, O. P. Shukla, K. S. Khanchandani and R. J. Suhadolnik, *J. Am. Chem. Soc.* **90**, 1038 (1968); (b) P. D. Shaw and J. A. McCloskey, *Biochemistry* **6**, 2247 (1967).

[8] K. Biemann, *Mass Spectrometry: Organic Chemical Applications*, McGraw-Hill, 1962, p. 121.
[9] F. P. Boer, T. W. Shannon and F. W. McLafferty, *J. Am. Chem. Soc.* **90**, 7239 (1968).
[10] K. R. Jennings, *Zeitschrift für Naturforschg.* **22a**, 454 (1967).
[11] I. Horman, A. N. H. Yeo and D. H. Williams, *J. Am. Chem. Soc.* **92**, 2131 (1970).
[12] R. Dickinson and D. H. Williams, *J. Chem. Soc.* (**B**), 249 (1971).
[13] W. O. Perry, J. H. Beynon, W. E. Baitinger, J. W. Amy, R. M. Caprioli, R. N. Renaud, L. C. Leitch and S. Meyerson, *J. Am. Chem. Soc.* **92**, 7238 (1970).
[14] F. Benoit and J. L. Holmes, *Chem. Comm.* 1031 (1970).
[15] (a) K. L. Rinehart, A. C. Bucholz and G. E. Van Lear, *J. Am. Chem. Soc.* **90**, 1073 (1968); (b) A. V. Robertson and C. Djerassi, *J. Am. Chem. Soc.* **90**, 6992 (1968).
[16] N. M. M. Nibbering and Th. J. de Boer, *Org. Mass Spec.* **1**, 365 (1968).
[17] R. G. Cooks, I. Howe, S. W. Tam and D. H. Williams, *J. Am. Chem. Soc.* **90**, 4064 (1968).
[18] B. Davis, D. H. Williams and A. N. H. Yeo, *J. Chem. Soc.* (**B**), 81 (1970).
[19] A. N. H. Yeo and D. H. Williams, *Chem. Comm.* 737 (1970).
[20] R. H. Shapiro, K. B. Tomer, R. M. Caprioli and J. H. Beynon, *Org. Mass Spec.* **3**, 1333 (1970).
[21] B. Samuelsson, *J. Am. Chem. Soc.* **87**, 3011 (1965).
[22] M. Vestal and G. Lerner, *Fundamental Studies Relating to the Radiation Chemistry of Small Organic Molecules*, Aerospace Research Laboratory Report 67–0114, U.S. Defence Documentation Center, Alexandria, Va., 1967.
[23] I. Howe and F. W. McLafferty, *J. Am. Chem. Soc.* **93**, 99 (1971).
[24] H. Eyring and F. W. Cagle, *J. Phys. Chem.* **56**, 889 (1952).
[25] U. Löhle and Ch. Ottinger, *J. Chem. Phys.* **51**, 3097 (1969).
[26] I. Howe and D. H. Williams, *Chem. Comm.* 1195 (1971).

8. Collision processes

8-1. Chemical ionization mass spectrometry and ion–molecule reactions

In chemical ionization mass spectrometry,[1] ionization of the substance under investigation is effected by reactions between the molecules of the substance and a set of ions which serve as ionizing reactants.[2] To achieve this, the usual experimental conditions are a pressure of 1 torr of a reactant gas (usually a hydrocarbon) and 10^{-4} torr of sample. Under these conditions it is much more probable that a sample molecule will react with an ion than be directly ionized by an electron.

Most studies to date have used methane as the source of reactant ions. Electron impact on the methane molecules produces largely $CH_4^{+\cdot}$ and CH_3^+ ions which undergo ion–molecule reactions with neutral CH_4 to produce CH_5^+ (47 per cent of the total ionization) and $C_2H_5^+$ (41 per cent of the total ionization).

$$CH_4^{+\cdot} + CH_4 \longrightarrow CH_5^+ + CH_3\cdot$$
$$CH_3^+ + CH_4 \longrightarrow C_2H_5^+ + H_2$$

Other ion–molecule reactions involving neutral methane occur to a small extent but the two ions, CH_5^+ and $C_2H_5^+$, are largely responsible for ionization of the sample molecules. The ions which are produced as a result of collision between the sample molecules and CH_5^+ or $C_2H_5^+$, along with those formed by subsequent breakdown, constitute the chemical ionization spectrum. Ion–molecule reactions between sample ions and sample molecules are avoided by keeping the concentration of sample appreciably lower than 0·6 per cent. Unlike electron-impact and photo-ionization, the ionization process is not controlled by Franck–Condon considerations, but involves slow, equilibrium adjustments of electronic states and atomic positions:[2] see also the 'collision complex', section 8-4. Hence the appearance of chemical ionization spectra is different from E.I. and P.I. spectra.

Although several classes of compounds have been studied by chemical ionization mass spectrometry, the spectra of hydrocarbons, esters and some natural products will serve here to illustrate the potential of the method. The chemical ionization spectrum of *n*-octadecane (*n*-$C_{18}H_{38}$) is given in Fig. 8-1(b) and compared with the corresponding electron-impact spectrum

(Fig. 8-1(a)). The methane C.I. spectra of saturated hydrocarbons are normally dominated by an ($M^+ - 1$) peak and *n*-octadecane is no exception. In this instance the advantage of C.I. over E.I. mass spectrometry for molecular-weight determination is evident (see Figs. 8-1(a) and (b), from which ^{13}C isotope peaks are omitted).

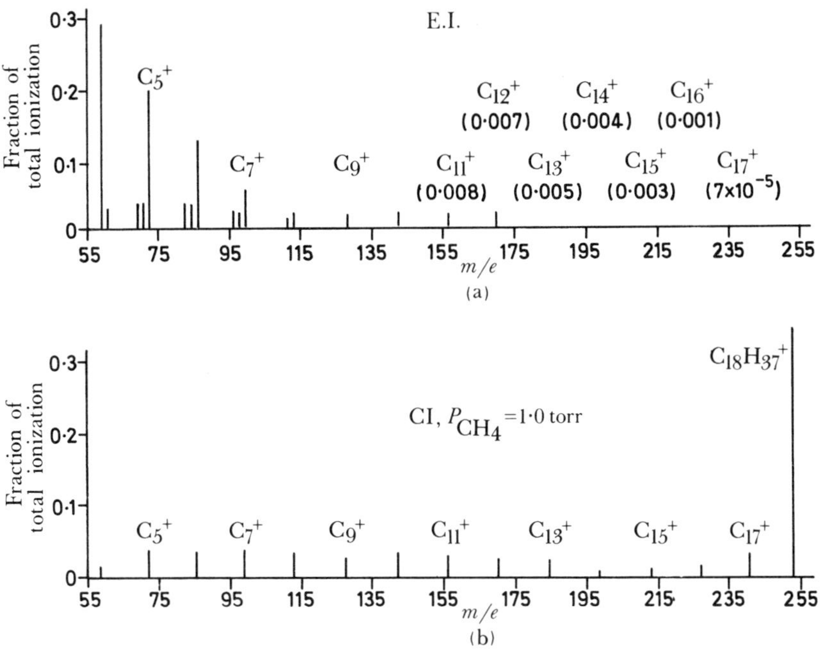

Figure 8-1

The generation of the $M^+ - 1$ species may be rationalized by considering CH_5^+ as a Brønsted acid and $C_2H_5^+$ as a Lewis acid, participating in the following exothermic reactions:[2]

$$C_{18}H_{38} + CH_5^+ \longrightarrow C_{18}H_{37}^+ + CH_4 + H_2, \qquad \Delta H = -27 \text{ kcal (113 kJ)/mole}$$
$$C_{18}H_{38} + C_2H_5^+ \longrightarrow C_{18}H_{37}^+ + C_2H_6, \qquad \Delta H = -25 \text{ kcal (105 kJ)/mole.}$$

The other ions in the C.I. spectrum of *n*-octadecane are all $C_nH_{2n+1}^+$ ions, whose production may be rationalized as follows, for example,

$$C_{18}H_{38} + CH_5^+ \longrightarrow C_8H_{17}^+ + C_{10}H_{22} + CH_4$$

It has been suggested[2] that the chemical ionization process in normal paraffins may be regarded as a random attack of the reactant ions along the length of the alkane chain, since (a) the $M^+ - 1$ intensity shows no dependence on molecular size, and (b) the alkyl ions of $m/e < (M^+ - 1)$ all have approximately the same intensity (Fig. 8-1(b)). Attack can involve either C—C or C—H electrons of the alkane, but C—C bond attack produces fragment alkyl ions of different m/e-values whereas C—H bond attack always produces $M^+ - 1$. Hence the latter ion will be the most abundant in the spectrum (given that the relative probabilities of C—C and C—H bond attack are not grossly different). In fact the observed intensities suggest that C—C attack is approximately four times more probable than C—H attack.

In branched paraffins, the intensity of the $M^+ - 1$ peak depends on the degree of branching, decreasing as the degree of branching increases. Enhanced intensities of alkyl ions are observed due to cleavage at the branch points. Thus pristane (1) exhibits a large $M^+ - 1$ peak but also enhanced ion abundances at $M^+ - 15$, $M^+ - 85$ and $M^+ - 155$.[2]

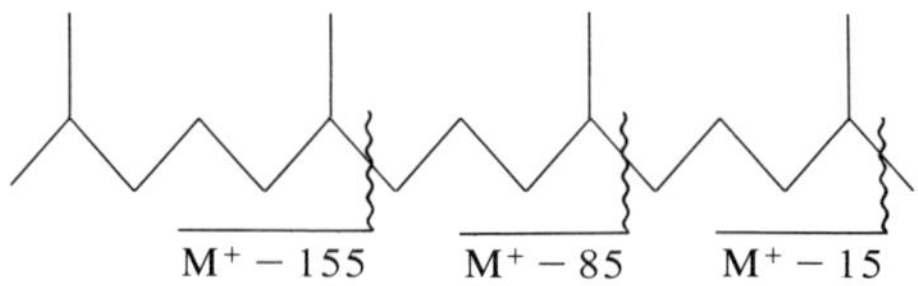

1

In saturated acyclic hydrocarbons, the occurrence of $M^+ + 1$ ions is not prevalent because the protonated molecular ions are very unstable and readily decompose either by H_2 loss or C—C bond fission. However, in hydrocarbons containing either rings or double bonds, $M^+ + 1$ ions (sometimes called *quasi-molecular ions*, QM^+) are observed. Cycloalkanes[3] and olefins[4] exhibit $M^+ + 1$ ions although $M^+ - 1$ ions still predominate. However, in the C.I. mass spectra of alkylbenzenes[5] very abundant $M^+ + 1$ ions are usually found, accompanied by $M^+ + 29$ and $M^+ + 41$ ions. The latter two species arise from collision between the alkylbenzene and $C_2H_5^+$ or $C_3H_5^+$ ions which exist in the methane plasma. It is evident that ions occurring at higher m/e value than the molecular ion are a potential hazard in structural interpretations by chemical ionization mass spectrometry.

We have seen that, in the case of alkanes, random attack by the reactant ion apparently occurs. This is not the case for molecules possessing a permanent dipole, such as esters.[6] It has been shown that in the presence of a permanent dipole, the cross-section for ion–molecule reactions increases greatly[7] (see section 8-4) and hence the attacking ions are probably attracted to the polar carbalkoxy group of esters. In a series of propionate esters,[6] $M^+ - 1$ abundances are very low but $M^+ + 1$ ions become relatively abundant (Fig. 8-2(b)). The $M^+ + 1$ ions can arise via proton transfer from CH_5^+ (and possibly $C_2H_5^+$). Elimination of the alcohol chain as an olefin can then furnish protonated propionic acid (m/e 75), which is usually an abundant ion in the chemical ionization spectra of the propionates so far examined.

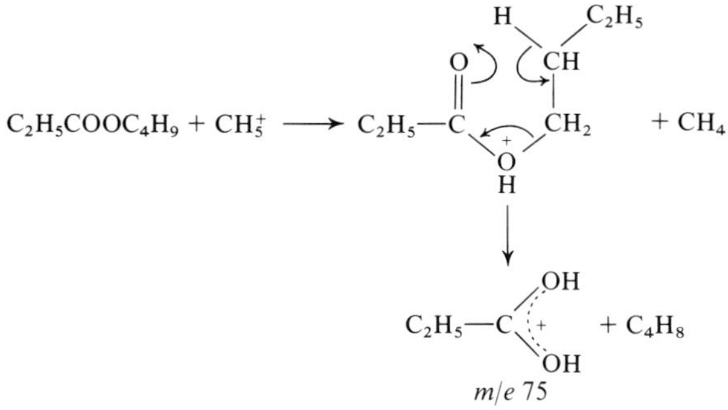

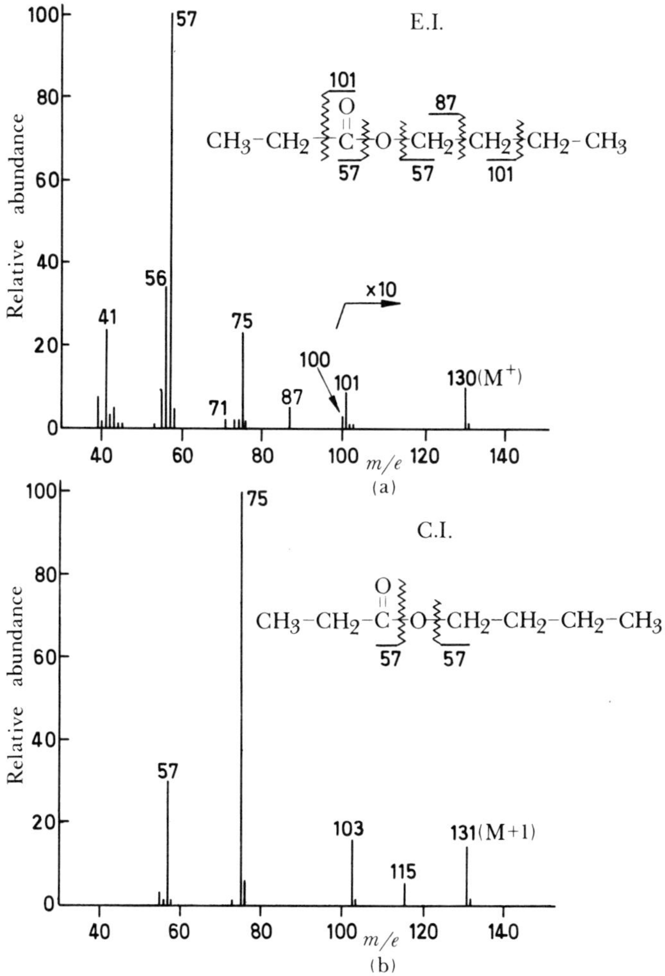

Figure 8-2

It is probable that the ions at m/e 103 and 115 in Fig. 8-2(b) arise due to the following reactions of the ester with $C_2H_5^+$ and $C_3H_5^+$ ions:

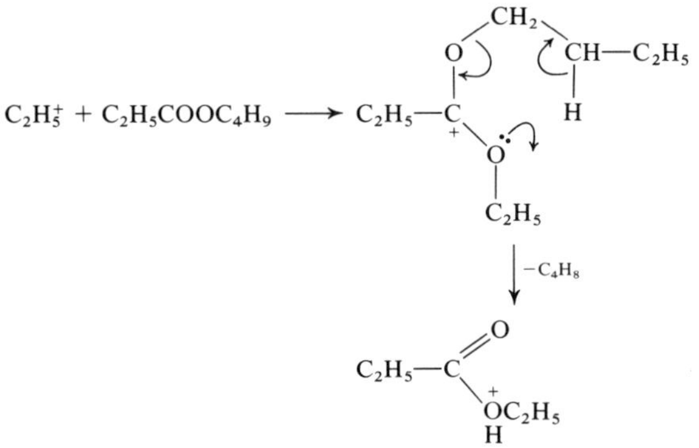

m/e 103

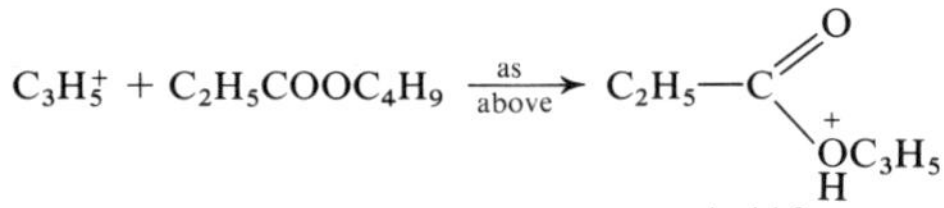

m/e 115

As in the case of n-alkanes, the simplicity of the chemical ionization spectrum relative to the electron-impact spectrum (Fig. 8-2(a)) suggests possible analytical advantages. However, much work remains to be done on other functional types before the true analytical potential of the technique can be evaluated.

Some of the initial work on natural products has been encouraging. For example, the molecular ion of amino acids is of low abundance in electron-impact spectra, but chemical ionization yields an abundant quasi-molecular ion, $M^+ + 1$[8] (contrast the E.I. and C.I. spectra of proline, Figs. 8-3(a) and (b) respectively).

The abundant QM^+ ions of amino acids are attributable to the facile protonation on the amino group. In fact the high proton affinity of the nitrogen atom is responsible for the abundant quasi-molecular ions in a number of nitrogen-containing natural products, notably alkaloids.[9] Amino compounds often have unstable molecular ions in E.I. mass spectra because of low-energy pathways to stable product immonium ions (see section 6-3C). Decompositions from the protonated (quaternary nitrogen) molecular ion are not so prevalent and a stable $M^+ + 1$ ion results.

Chemical ionization spectra have also been determined using isobutane ($CHMe_3$) as the ionizing reagent.[10] At a source pressure of 1 torr the predominant ion is the t-butyl ion ($C_4H_9^+$) formed by hydride abstraction from the isobutane molecule, for example,

$$C_4H_{10}^{+\cdot} \longrightarrow C_3H_7^+ + CH_3\cdot$$
$$C_4H_{10} + C_3H_7^+ \longrightarrow C_4H_9^+ + C_3H_8.$$

Except for the 3 per cent of $C_3H_3^+$ ions present at this pressure, the spectrum is virtually monoionic and the $C_4H_9^+$ ion behaves as a mild Brønsted acid, protonating the sample molecules. The tendency to protonate is not as high for $C_4H_9^+$ as for the CH_5^+ ion, since the respective proton affinities of C_4H_8 and CH_4 are 187 and 118 kcal/mole (782 and 493 kJ/mole).

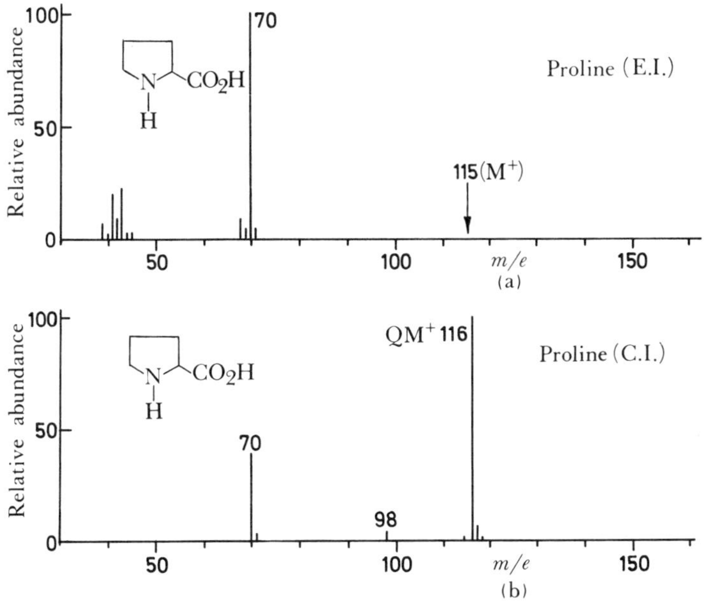

Figure 8-3

We have seen (in chapter 4) that the absence of collisions in normal low-pressure electron-impact mass spectrometry precludes the application of thermal-equilibrium kinetics. Indeed, many of the phenomena of E.I. organic mass spectrometry (e.g., substituent effects) have been explained in terms of the quasi-equilibrium theory for a system of individually thermally isolated ions. However, if ion–molecule reactions are permitted, as in chemical ionization mass spectrometry, thermal equilibrium may be attained and the Arrhenius equation is applicable.

The kinetics of decomposition of organic ions in the chemical ionization source may therefore be investigated[10, 11] and it is apparent that the technique offers a useful means of determining mechanisms and thermodynamical parameters, free from the influence of solvent interactions. For example, the rate of the simple cleavage of protonated benzyl acetates (2) to form substituted benzyl ions (3) has been investigated by i-C_4H_{10} chemical ionization. Appropriate intensity measurements and reasonable estimations of source residence times enable a rate constant to be evaluated for the reaction for each substituent X; and a Hammett correlation yields a ρ-value of $-3{\cdot}9$, in accord with condensed phase reactions in which a carbonium ion is formed in the rate-determining step.

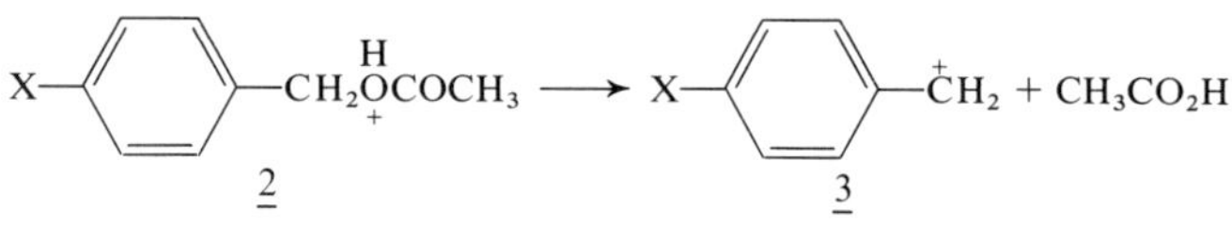

In summary, chemical ionization mass spectrometry has an advantage over conventional electron-impact mass spectrometry in the increased ion abundance in the molecular ion region (either M^+, $M^+ - 1$ or $M^+ + 1$). The spectra, too, are frequently simpler for C.I. However, ions of *m/e*-value corresponding to addition of hydrocarbon fragments to the molecular ion can be a drawback. It appears that C.I. holds considerable potential for investigation of reaction kinetics of organic ions. Finally, it has been reported[12] that a C.I. source is particularly useful when linked to a gas chromatograph, since the carrier gas for gas chromatography becomes the reactant gas for chemical ionization and hence no splitter or separator is required. GC/MS techniques will be discussed at length in Chapter 11.

The chemical ionization technique described above is one instance of a system for investigating ion–molecule reactions. The special feature of C.I. is the predominance of one species (the ionizing gas) in the C.I. source over other species. In the general case, the partial pressures of the participating species may be varied over a wide range.

The ion–molecule reactions undergone after ionization of pure species in a high-pressure source have been investigated for a number of organic compounds. A large number of hitherto unknown chemical reactions have been uncovered in this way and kinetic and/or thermodynamical information has been forthcoming. For example, the spectrum of ions produced from benzene has been measured at various pressures[13] and the results are shown in Fig. 8-4.

It is apparent from Fig. 8-4 that ions due to collision reactions (e.g., $C_{12}H_{11}^+$) increase relative to those due to unimolecular fragmentation (e.g., $C_6H_5^+$) as the benzene pressure is increased—in accord with kinetic principles.

As an example of the thermodynamical information obtainable from such studies it has been established[14] that an equilibrium exists between monomer and dimer ions:

$$C_6H_6^{+\cdot} + C_6H_6 \rightleftharpoons (C_6H_6)_2^{+\cdot}$$

Temperature studies on yields of these ions, followed by the usual kinetic plots, indicate that the dimerization is exothermic [$\Delta H^\circ = -15$ kcal (63 kJ)/mole], in agreement with condensed phase investigations. The high negative standard entropy change ($\Delta S^\circ = -96$ J deg^{-1} mole^{-1}) is in accord with the loss of at least three degrees of translational freedom in dimer formation.

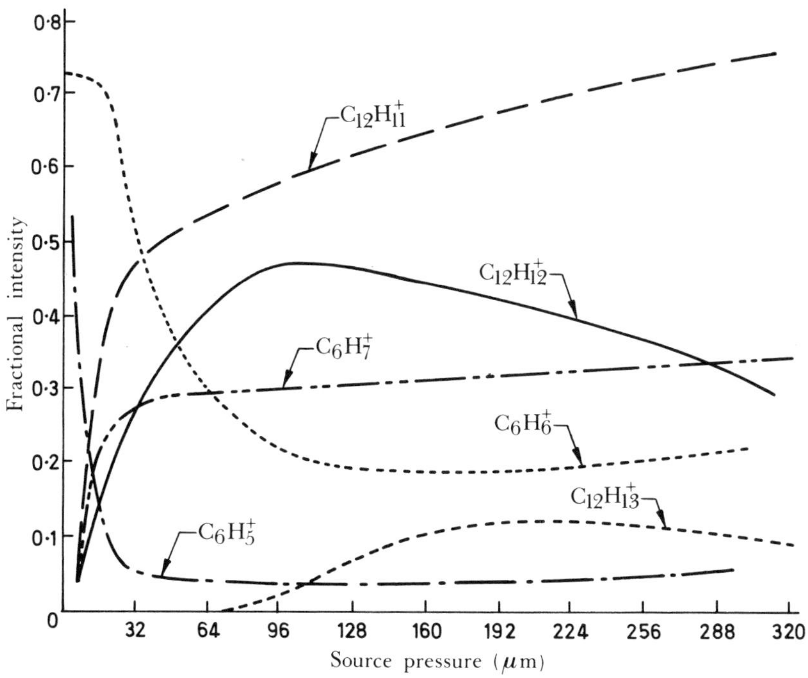

Figure 8-4

8-2. Ion-cyclotron resonance spectroscopy

The study of ion–molecule reactions and, in particular, their application to ion-structure elucidation, has been further broadened in recent years by the introduction of Ion-Cyclotron Resonance (ICR) mass spectrometry.[15, 16] In contrast to the high-pressure mass spectrometry discussed previously in this chapter, ICR owes its suitability for observation of collision processes to the long ion lifetimes involved (up to milliseconds) and not merely to high source pressures.

A schematic view of a typical ICR cell is shown in Fig. 8-5. Ions are generated by electron impact in a uniform magnetic field B (perpendicular to the plane of the paper) and are restricted to a circular path perpendicular to the direction of the magnetic field. The angular frequency, ω_c, of this motion is independent of the velocity of the ions and is given by

$$\omega_c = eB/m, \tag{8-1}$$

where m/e, as usual, is the mass-to-charge ratio for a given ion. The angular frequency ω_c therefore depends on the m/e-value and is known as the cyclotron resonance frequency. If an alternating electric field of radio-frequency ω_1 is applied normal to B, an ion will absorb energy if $\omega_c = \omega_1$; and an ICR mass spectrum can be obtained by fixing ω_1 and scanning the magnetic field B so that ions of different m/e satisfy eq. (8-1). Absorption of energy by the ions (at *resonance*) is measured using an oscillator–detector system similar to those used in nmr and esr spectroscopy.

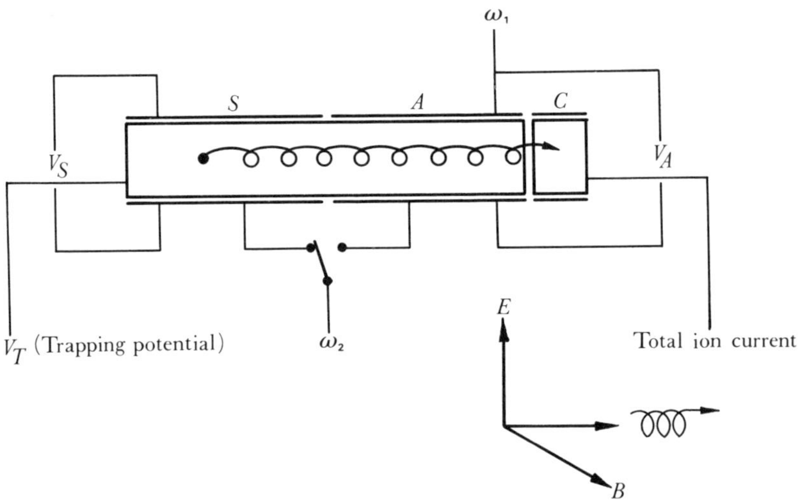

Figure 8-5

After formation in the source S, ions drift through the analyser A to the collector C by application of a small static electric field E, perpendicular to B. The velocity of the ions, in a direction perpendicular to both B and E (see Fig. 8-5) is given by

$$v = E/B \tag{8-2}$$

In the final region of the ICR cell, the total ion current is measured at the collector.

Some typical parameters for an ICR mass spectrometer are shown below:[16]

$$\text{Operating pressure} = 10^{-7} \text{ to } 10^{-4} \text{ torr}$$
$$\text{Length of source region} = \text{length of analyser region} = 6{\cdot}35 \text{ cm}$$
$$\text{For } Ar^+,\ m/e\ 40, \text{ and for } B = 8000 \text{ gauss and } E = 0{\cdot}25 \text{ V/cm:}$$
$$\omega_c = 307 \text{ kHz}$$
$$v = 3 \times 10^3 \text{ cm/s}$$
$$\text{Time spent in analyser region} = 2 \times 10^{-3} \text{ s.}$$

Since the residence times are of the order of milliseconds, there is a high probability of observing ion–molecule reactions at low pressures and the availability of various double resonance and pulsing techniques facilitate the elucidation of the mechanisms of such reactions. The *double-resonance* technique[17] is very useful in determining the precursor of a given daughter ion. Consider the general ion–molecule reaction $A^+ + B \rightarrow C^+ + D$. The magnetic field is first adjusted so that the C^+ ions absorb energy at an oscillator frequency ω_1. A second radiofrequency field of frequency ω_2 is then applied to the source of the ICR cell and ω_2 is scanned until the cyclotron resonance frequency of A^+ is attained. At this frequency the A^+ ions will absorb energy and their translational energy will consequently increase. Since the rate constants of ion–molecule reactions involving A^+ depend on the translational energy, the yield of C^+ will change if A^+ is indeed a precursor. At a fixed magnetic field,

$$\frac{\omega_1}{\omega_2} = \frac{m_A}{m_C} \tag{8-3}$$

so different precursors of C^+ can be identified by scanning ω_2.

The application of this powerful technique is illustrated simply by an experiment to determine the relative acidities of aliphatic alcohols ROH in the gas phase free from solvent interactions. It is possible, by the use of ion-cyclotron pulsed double resonance to determine in which direction the following equilibrium lies:[18]

$$ROH + R'O^- \rightleftharpoons RO^- + R'OH$$

A 1:1 mixture of ROH and R′OH (with a comparable amount of water to facilitate the production of RO^- and $R'O^-$) are mixed in the ICR cell under conditions for negative ion spectra and at pressures sufficiently high ($\sim 10^{-5}$ torr) for ion–molecule reactions to occur. The $R'O^-$ ion is irradiated and any change in the RO^- intensity is observed. If a decrease in the RO^- intensity occurs this is consistent with an exothermic reaction (in the direction left $\rightarrow$ right) whose specific rate constant decreases with increasing kinetic energy. The conclusion is verified by investigating the transfer of a proton from R′OH to RO^-, by irradiating the RO^- ion and observing any change in the intensity of $R'O^-$. If no change occurs then the equilibrium lies largely to the right and $R'O^-$ is being produced from a source other than the above reaction.

Suppose this experiment shows that ΔG° is negative for the reaction $L \rightarrow R$. If ΔS° is assumed to be zero, then ΔH° is negative and this provides a measure of the relative acidities of the two alcohols (ROH more acidic).

By this means the following order of acidities has been found: neopentyl alcohol > *t*-butyl > isopropyl > ethyl > methyl > water; and *t*-butyl ≈ *n*-pentyl ≈ *n*-butyl > *n*-propyl > ethyl.

The order obtained here for gas-phase acidities is the reverse of that found for the acidities of alcohols in solution. Thus, *t*-butoxide anion is less basic in the gas phase than methoxide, in contrast to solution behaviour.

This important result may be analysed in terms of differences in polarizability of the alkyl groups. In the gas phase *t*-butyl is more capable of stabilizing the negative charge in the $(CH_3)_3CO^-$ ion by induced polarization than is methyl in the CH_3O^- ion. The situation in solution is reversed because solvent molecules can readily envelop and stabilize the CH_3O^- ion by hydrogen bonding etc., whereas the bulky *t*-butyl group partially shields the negative charge from interaction with solvent molecules in the $(CH_3)_3CO^-$ ion. Thus an inductive mechanism of electron donation by methyl groups, often traditionally used to rationalize the high basicity of $(CH_3)_3C{-}O^-$ in solution, is clearly invalid since such an effect would also have to operate in the gas phase.

A particularly important application of ICR spectroscopy has been found in the problem of ion-structure elucidation. For example, ion–molecule reactions in the ICR spectrometer have been devised[19] to distinguish between the structures of the $C_3H_6O^{+\cdot}$ ions generated from acetone, from 2-hexanone (4) via a McLafferty rearrangement, from 1-methylcyclobutanol (5) via loss of ethylene, and from 5-nonanone (6) via two consecutive McLafferty rearrangements (section 6-3B).

$CH_3CO(CH_2)_3CH_3$

4

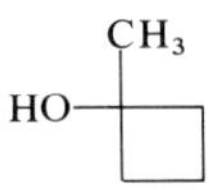

5

$CH_3(CH_2)_3CO(CH_2)_3CH_3$

6

A comparison of two $C_3H_6O^{+\cdot}$ ions generated from different compounds is achieved by 'mixing' the different $C_3H_6O^{+\cdot}$ ions and observing their respective reactions with neutral species present in each mixture. Deuterium labelling is, of course, necessary to distinguish between the two ions of differing origins.

For example, in a mixture of acetone and hexadeuterioacetone, the following reactions occur:

$$[CD_3COCD_3]^{+\cdot} + CH_3COCH_3 \longrightarrow CD_3{-}\overset{\overset{\displaystyle CD_3}{|}}{C}{=}\overset{+}{O}{-}COCH_3 + CH_3\cdot \qquad m/e\ 107$$

$$\longrightarrow CH_3{-}\overset{\overset{\displaystyle CH_3}{|}}{C}{=}\overset{+}{O}{-}COCD_3 + CD_3\cdot \qquad m/e\ 104$$

However, when 2-hexanone and CD_3COCD_3 are mixed, no ions are observed at *m/e* 104 or 107, so the $C_3H_6O^{+\cdot}$ ion from 2-hexanone must have a different structure from that obtained by ionization of acetone. The most feasible structure for the McLafferty rearrangement ion is the enol ion 7.

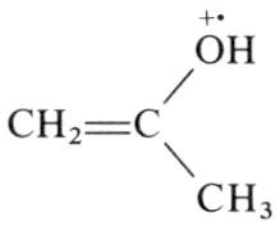
7

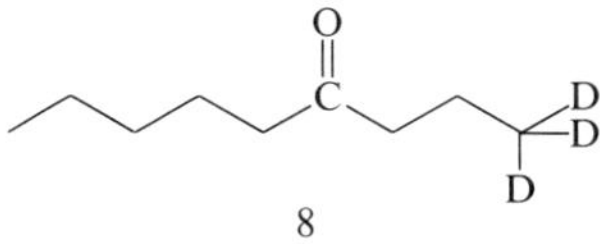
8

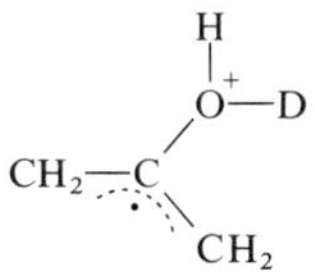
9

It has also been observed that, in ions produced by single and consecutive McLafferty rearrangements, only the hydrogens bonded to oxygen appear to be transferred to neutral ketones in ion–molecule reactions. For example, double-resonance experiments establish that the $C_3H_5DO^{+\cdot}$ double McLafferty rearrangement ion from d_3-nonan-4-one (8) transfers only an H atom to neutral molecules. This result suggests an enol structure (7, CH_2D) and not the symmetrical oxonium ion (9) from which transfer of both H and D would be anticipated.

Ion-cyclotron resonance spectroscopy has established itself as a powerful means of investigating ion–molecule chemistry in the gas phase. However it should be noted that ion lifetimes in the ICR cell are significantly longer than lifetimes in a conventional mass spectrometer. Consequently the ions undergoing ion–molecule reactions will be of low energy and largely those ions with insufficient energy to undergo unimolecular reactions.

8-3. Collision-induced metastables

A considerable enhancement of internal energy may occur when organic ions of high kinetic energy (e.g., 8 keV) collide with neutral molecules. The extent of decomposition of the organic ions consequently increases and the fragmentations may be conveniently observed in the metastable drift regions of a conventional mass spectrometer, as *collision-induced metastables*.

Several instrumental techniques have been devised for observation of these higher-energy metastable decompositions. For example, the target gas for collision may be introduced into the electrostatic analyser by gentle baking or via a small leak. Collisional activation of the organic ions then results and the resulting metastable decompositions may be observed in the second drift region. Because of the higher average energy after collision compared with the normal energy in the second drift region, reactions are accessible via collisional excitation that are not normally observable as metastables. For example, collision-induced metastable decompositions from a number of molecular ions derived from aromatic compounds have been investigated in this way and several transitions, some requiring two or more steps, could be observed in each case.[20] The resulting *collision-induced mass spectrum* corresponds closely to the 70 eV mass spectrum.

These results illustrate that several electron volts may be added by collision and we may represent the situation as follows:

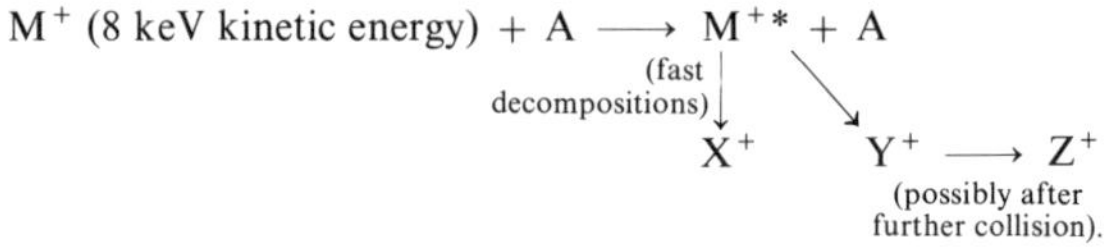

Since metastable decompositions in the first field-free region (between the source and electric sector) may be sensitively measured (see section 3-1B), it has been found that this region may be successfully employed as the collision chamber.[21] Neutral gases, such as argon and helium, have been used as the target, and under controlled pressure conditions it is possible to determine the optimum pressure for observation of collision-induced metastables. The effect of increasing pressure on the collision-induced spectrum parallels that of increasing the electron beam energy in the normal spectrum: there is an increase in the prevalence of decompositions requiring more energy and this is illustrated for nitrobenzene in Fig. 8-6.

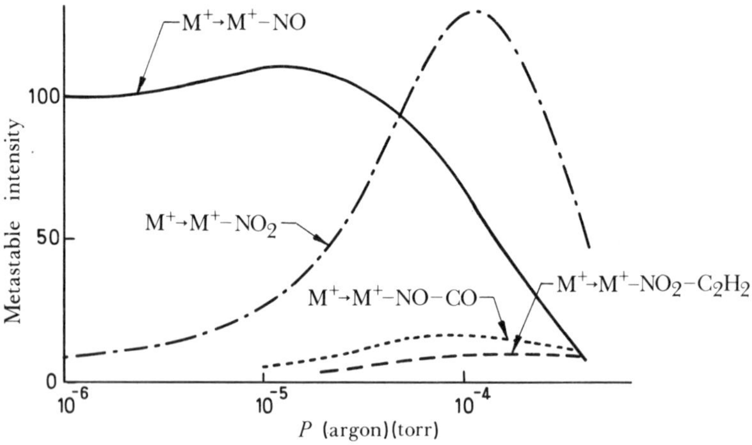

Figure 8-6

It has also been confirmed from measurements in the first drift region that high-pressure collision-induced spectra closely resemble the 70 eV normal spectra. This is illustrated in Table 8-1 for benzonitrile and anisole.[21]

Loss of total ion intensity occurs with increasing pressure due to collisional scattering. Hence an optimum pressure exists for the measurement of collision-induced metastables and has been found to be about 10^{-4} torr for a number of cases studied, although the optimum pressure varies with such factors as the target gas, the length of the drift tube and the kinetic energy of the incident ion.

Table 8-1
Collision-induced spectra for (a) benzonitrile and (b) anisole†

Ion	AP eV	1×10^{-4} torr Ar	5×10^{-4} torr	9×10^{-4} torr	70 ev spectrum
(a) 103	9·7	6460	654	210	163
76	13·3	64	53	44	49
75	15·2	12	12	10	15
51	15·7	12	14	13	13
50	15·0	12	20	33	23
(b) 108	8·6	5610	447	168	69
93	11·3	11	6	5	7
78	10·9	37	34	32	27
77	13·0	23	19	22	10
65	12·3	18	19	22	33
51	16·0	8	14	14	8
39	14·3	4	8	<5	15

† $\sum$ fragment ions = 100.

The basic use of collision-induced metastables is simply to provide a more detailed picture of fragmentation processes, not only from molecular ions but also from fragment ions. Further, since such decompositions occur from higher-energy precursors than normal metastable decompositions, collisional activation minimizes rearrangement reactions.[22] It was pointed out in section 4-1 that higher-energy molecular ions frequently decompose preferentially via direct cleavage than via rearrangement, due to the relative properties of the k vs. E curves (see Fig. 4-1).

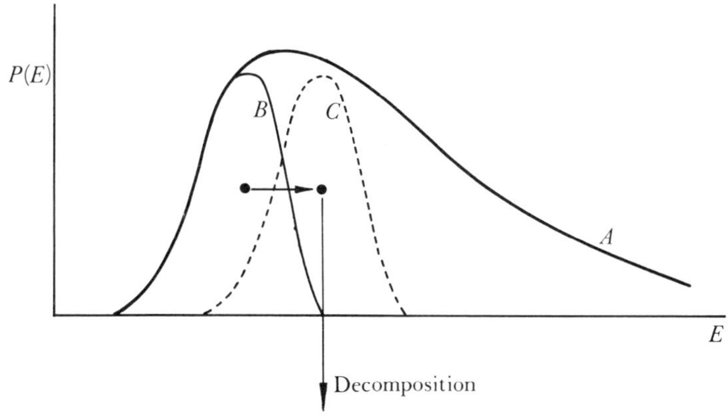

Figure 8-7

In certain instances, collision-induced metastables may provide suitable systems for kinetic investigations of decompositions from higher-energy states. Changes in isotope effects, for example, may be investigated between metastable decompositions with and without a collision gas. Thus the primary deuterium isotope effect for loss of hydrogen from the scrambled toluene molecular ion in the first drift region decreases from $2{\cdot}80 \pm 0{\cdot}06$ to $2{\cdot}33 \pm 0{\cdot}08$ on introduction of argon at a pressure of 2×10^{-4} torr.[23]

In summary, it appears that collision-induced metastable decompositions occur as a result of the conversion of a small fraction of the translational energy of organic ions after acceleration into internal energy by collision with neutral molecules (see also section 8-4). Electronic excitation probably occurs followed by internal conversion of the excess energy into vibrational energy, with resulting decomposition. A representation of the supposed situation for a molecular ion is shown in Fig. 8-7.

Curve *A* represents the hypothetical energy distribution of a molecular ion after electron impact and only the lower-energy ions represented by curve *B* remain after leaving the source. On impact with neutral molecules in the first drift region, the energy distribution of M^+ is then elevated to curve *C* and many of the molecular ions will then have sufficient energy to decompose in less than a microsecond. Further collisional excitation of the primary product ion(s) is then possible, with consequent further decomposition.

8-4. Mechanisms of ion–molecule reactions

We have discussed so far in this chapter a number of collisional processes between organic ions and neutral molecules and the way that these processes may be utilized in mass spectrometry. It is therefore pertinent that we should mention, if only briefly, some of the current views on the mechanisms of ion–molecule reactions. Complex mathematical theories have been developed which have served to explain in detail the mechanisms and rates of some reactions involving very small molecules and ions;[24] but here it is intended to present only the general principles governing the ion–molecule reactions involving organic ions.

It is an experimental fact that, in general, ion–molecule reactions are appreciably faster than reactions between neutral species. Bimolecular rate constants for a number of reactions are shown in Table 8-2; it is evident from the quoted figures that ion–molecule reactions in the gas phase have the highest rate constants.

First, it should be noted that most ion–molecule reactions proceed with no energy of activation. It is informative to relate the rate of ion–molecule reactions to the collision frequency. The simple theory of molecule collisions as applied to bimolecular reactions predicts the following rate constant:

$$k_2 = 2\sigma^2 \left(\frac{2\pi kT}{\mu}\right)^{1/2} e^{-E_a/kT}, \qquad (8\text{-}4)$$

where σ is the collision diameter, μ is the reduced mass of the reactants ($\mu = m_1 m_2/(m_1 + m_2)$) and E_a is the activation energy.

If the reaction possesses zero activation energy, then k_2 reduces to the pre-exponential term in eq. (8-4). However, the remaining expression is a maximum value, since the derivation assumes that every collision results in reaction: many bimolecular reactions require a specific orientation of the reactants, with the consequence that some, or many, collisions are unreactive. The value of k_2 may therefore be considerably less than the ideal predicted by eq. (8-4). Substituting reasonable values for the parameters in

Table 8-2
Examples of bimolecular rate constants

Reaction	k (cm^{-1} mol^{-1} s^{-1})	T (°C)
Gas-phase ion–molecule reactions:		
$Ar^{+\cdot} + H_2 \rightarrow ArH^+ + H\cdot$	$3{\cdot}5 \times 10^{-10}$	25
$H_2O^{+\cdot} + H_2O \rightarrow H_3O^+ + HO\cdot$	$1{\cdot}3 \times 10^{-9}$	25
$CH_4^{+\cdot} + CH_4 \rightarrow CH_5^+ + CH_3\cdot$	$1{\cdot}2 \times 10^{-9}$	100
$CH_2NH_2^+ + CH_3NH_2 \rightarrow CH_3NH_3^+ + CH_3N$	$2{\cdot}4 \times 10^{-9}$	100
$C_2H_2^{+\cdot} + C_2H_3F \rightarrow C_2H_3F^{+\cdot} + C_2H_2$	$1{\cdot}9 \times 10^{-9}$	100
$CH_3OCD_3^{+\cdot} + CH_3OCD_3 \rightarrow CH_3OCD_3H^+ + \cdot CH_2OCD_3$	$1{\cdot}15 \times 10^{-9}$	100
$C_4H_2^{+\cdot} + C_6H_6 \rightarrow C_{10}H_7^+ + H\cdot$	$4{\cdot}7 \times 10^{-10}$	—
Gas-phase reactions involving radicals:		
$C_2H_5\cdot + C_2H_5\cdot \rightarrow C_4H_{10}$	$2{\cdot}7 \times 10^{-11}$	25
$CH_3\cdot + C_2H_6 \rightarrow CH_4 + C_2H_5\cdot$	8×10^{-22}	25
$Br\cdot + H_2 \rightarrow HBr + H\cdot$	8×10^{-24}	25
Gas-phase reactions between molecules:		
$NO + O_3 \rightarrow NO_2 + O_2$	2×10^{-14}	25
$HI + C_2H_5I \rightarrow C_2H_6 + I_2$	$1{\cdot}6 \times 10^{-28}$	25
$NO_2 + NO_2 \rightarrow 2NO + O_2$	$1{\cdot}6 \times 10^{-31}$	25
Ions in aqueous solution:		
$CO_2 + OH^- \rightarrow HCO_3^-$	7×10^{-18}	25
$CH_3COOC_2H_5 + OH^- \rightarrow CH_3COO^- + C_2H_5OH$	$1{\cdot}6 \times 10^{-22}$	25

eq. (8-4) ($\sigma = 10^{-7}$ cm, $T = 373$°K, $m_1 = m_2 = 40/(6 \times 10^{23})$ g), we obtain a value for k_2 of 2×10^{-9} cm^3 mol^{-1} s^{-1}.

It therefore appears from Table 8-2 that many ion–molecule reactions in the gas phase exhibit this maximum rate constant, i.e., most collisions result in reaction (with the proviso that there are some limitations in the hard-sphere model used to derive eq. 8-4).

One reason for rapidity of ion–molecule reactions may be found in the treatment of Langevin.[25] Long-range attractive forces are produced

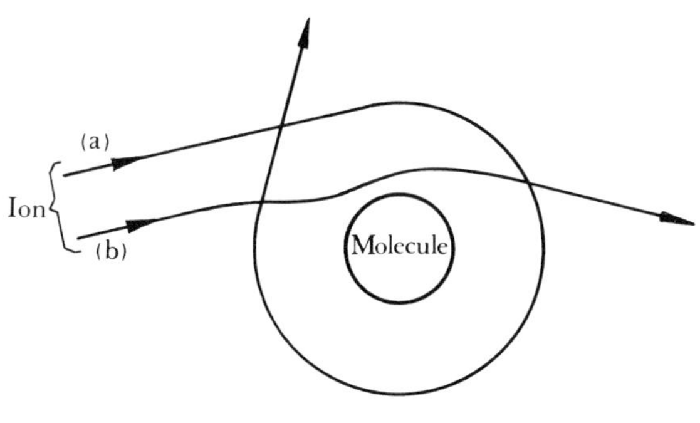

Figure 8-8

between the approaching ion and the molecule due to induction of polarization in the target molecule (or increased polarization if a permanent dipole is present already). Depending on the proximity and the relative velocity of the two species, these attractive forces may cause the distance of closest approach to be sufficiently small for reaction to occur. Furthermore, if the ion approaches the target molecule within a certain range of distances, the trajectory takes on a spiral or orbiting nature around the molecule. In such cases, the duration of the interaction is greatly increased (see Fig. 8-8(a)). There is, of course, a lower limit to the distance of approach due to short-range repulsive forces, and ions which approach on a close trajectory will not enter the metastable orbit (see Fig. 8-8(b)).

In summary, this simplified approach shows that the *cross-section* for an ion–molecule reaction is large due to (a) the long-range attractive forces between the two species and (b) the possible orbiting interaction between the two species (*collision complex*) which increases the duration of the interaction.

Evidence has been forthcoming that the collision complex may have a lifetime which is long compared with vibrational and rotational periods and extensive reorganization may occur prior to reaction. For example, isotopic labelling experiments have established that complete randomization of hydrogens occurs prior to dissociation of the $C_2H_7^+$ collision complex resulting from the reaction[26]

$$CH_3^+ + CH_4 \longrightarrow [C_2H_7^+] \longrightarrow C_2H_5^+ + H_2$$

Furthermore, the metastable $C_{12}H_{11}^+$ intermediate has been observed in the reaction[27]

$$C_6H_5^+ + C_6H_6 \longrightarrow [C_{12}H_{11}^+] \longrightarrow C_{10}H_9^+ + C_2H_2$$

On the basis of this and other evidence we may assume that the collision complex may potentially have a lifetime which is long compared with vibrational and rotational periods. In this case the complex loses 'knowledge' of its mode of formation and the internal energy is randomized among the internal degrees of freedom. The decomposition pathways are then characterized solely by the energy content, as in the QET (see chapter 4).

The conversion of translational energy into internal energy by collision (see collision-induced metastables—section 8-3) is an important occurrence. We have seen that collision between organic ions of high kinetic energy (up to 8 keV) and neutral molecules may result in a transfer of several eV to the internal modes of the organic ion. It is evident that at high relative velocity of the two species there is little or no opportunity for 'strong coupling' or orbiting and the interaction may be over in 10^{-15} s or less. The simple theory as outlined above breaks down and a hard-core collision model becomes appropriate. It appears that the conversion of kinetic energy of the ion may occur at impact either to vibrational energy or to electronic energy, depending not only on Franck–Condon factors involved, but also on the incident kinetic energy. It should be noted that even if as much as 5 eV is transferred to the internal modes of an ion having 5 keV of kinetic energy, this constitutes only 0·1 per cent conversion of kinetic into internal energy. Details of the mechanisms involved in such collisional excitations have been investigated largely for diatomic molecule-ions and the problem for organic ions is apparently complex.[28]

References

[1] M. S. B. Munson and F. H. Field, *J. Am. Chem. Soc.* **88**, 2621 (1966).
[2] F. H. Field, *Accts Chem. Res.* **1**, 42 (1968); see also J. M. Wilson in *Specialist Periodical Reports: Mass Spectrometry*, Vol. 1, D. H. Williams, ed., The Chemical Society, London, 1971.
[3] F. H. Field and M. S. B. Munson, *J. Am. Chem. Soc.* **89**, 4272 (1967).
[4] F. H. Field, *J. Am. Chem. Soc.* **90**, 5649 (1968).
[5] F. H. Field and M. S. B. Munson, *J. Am. Chem. Soc.* **89**, 1047 (1967).
[6] M. S. B. Munson and F. H. Field, *J. Am. Chem. Soc.* **88**, 4337 (1966).
[7] T. F. Moran and W. H. Hamill, *J. Chem. Phys.* **39**, 1413 (1963).
[8] G. W. A. Milne, T. Axenrod and H. M. Fales, *J. Am. Chem. Soc.* **92**, 5170 (1970).
[9] H. M. Fales, H. A. Lloyd and G. W. A. Milne, *J. Am. Chem. Soc.* **92**, 1590 (1970).
[10] F. H. Field, *J. Am. Chem. Soc.* **91**, 2827 (1969); **91**, 6334 (1969).
[11] D. P. Weeks and F. H. Field, *J. Am. Chem. Soc.* **92**, 1600 (1970).
[12] G. P. Arsenault, J. J. Dolhun and K. Biemann, *Chem. Comm.* 1542 (1970).
[13] S. Wexler and R. P. Clow, *J. Am. Chem. Soc.* **90**, 3940 (1968).
[14] F. H. Field, P. Hamlet and W. F. Libby, *J. Am. Chem. Soc.* **91**, 2839 (1969).
[15] J. D. Baldeschwieler, *Science* **159**, 263 (1968).
[16] G. C. Goode, R. M. O'Malley, A. J. Ferrer-Correia and K. R. Jennings, *Chem. in Britain* **7**, 12 (1971).
[17] J. L. Beauchamp, L. R. Anders and J. D. Baldeschwieler, *J. Am. Chem. Soc.* **89**, 4569 (1967).
[18] J. I. Brauman and L. K. Blair, *J. Am. Chem. Soc.* **90**, 6561 (1968).
[19] J. Diekman, J. K. MacLeod, C. Djerassi and J. D. Baldeschwieler, *J. Am. Chem. Soc.* **91**, 2069 (1969); G. Eadon, J. Diekman and C. Djerassi, *J. Am. Chem. Soc.* **91**, 3986 (1969).
[20] K. R. Jennings, *Int. J. Mass Spec. Ion Phys.* **1**, 227 (1968).
[21] F. W. McLafferty, R. A. Kornfeld, S.-C. Tsai, I. Howe and P. F. Bente, *submitted for publication.*
[22] F. W. McLafferty and H. D. R. Schüddemage, *J. Am. Chem. Soc.* **91**, 1866 (1969).
[23] I. Howe and F. W. McLafferty, *J. Am. Chem. Soc.* **93**, 99 (1971).
[24] For a detailed review, see J. H. Futrell and T. O. Tiernan in *Fundamental Problems in Radiation Chemistry*, P. Ausloos, ed., Wiley-Interscience, 1968.
[25] See, for example, P. F. Knewstubb, *Mass Spectrometry and Ion–Molecule Reactions*, Cambridge University Press, 1969, chapter 1.
[26] F. P. Abramson and J. H. Futrell, *J. Chem. Phys.* **45**, 1925 (1966).
[27] C. Lifshitz and B. G. Reuben, *J. Chem. Phys.* **50**, 951 (1969).
[28] See, for example, H. Yamaoka, P. Dong and J. Durup, *J. Chem. Phys.* **51**, 3465 (1969).

9. Field ionization

9-1. Instrumentation

Field ionization occurs when an atom or molecule is acted upon by an electric field of the order of 10^7–10^8 V/cm.[1] The field-ionization source consists of an anode, known as the emitter, which may be a sharp blade, sharp tip or a thin wire; and a cathode, which is a fine slit or hole (see Fig. 9-1).

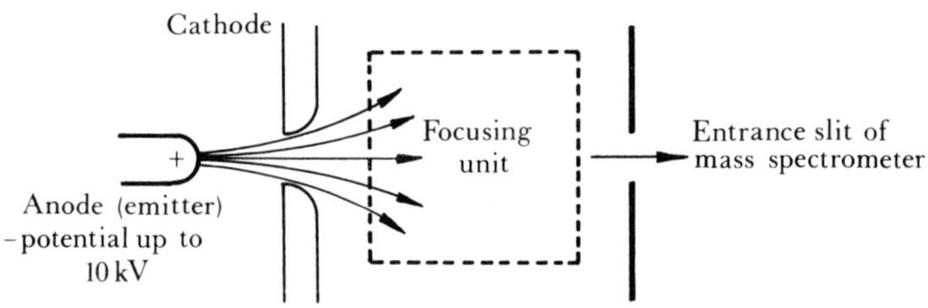

Figure 9-1

The removal of an electron from a neutral molecule by field ionization is not a discontinuous process like ionization by electron impact. In the EI process, ionization only occurs above a sharply defined electron energy at the ionization potential; but in the FI process there is always a finite probability of ionization, which increases continuously with field strength. At sufficiently high field strengths ($\sim 10^8$ V/cm) most molecules reaching the emitter undergo ionization and field ion currents as high as 10^{-6} A may be attained using well-activated wire emitters.

Field ionization is explained in terms of a quantum-mechanical tunnelling effect. In the absence of an electric field the energy of the ejected electron would have to be increased by an amount *I* (the ionization potential) for ionization to occur. In the presence of an electric field the potential wall of the molecule is distorted such that an electron can 'tunnel' (in terms of the laws of quantum mechanics) from the molecule into unoccupied orbitals in the metal emitter. The external electric field supplies the necessary energy to enable the electron to reach unoccupied states of the metal.

The ions produced in this way are immediately accelerated by the field, but then retarded to the translational energies (e.g., 8 keV) which allow separation of ions of differing *m/e*-values exactly as for electron impact (in single or double focusing instruments). In spite of the high ion currents attainable in FI, the intensity at the detector is smaller by a factor of about 100 compared with ionization by EI. This is because in FI the ions leave the emitter over a wide angle compared with the very narrow angle of ions leaving the EI source. After a few hundred nm, the FI ions have acquired a kinetic energy of several keV and divergent ion beams of this energy are difficult to focus into a narrow mass-spectrometer slit.

9-2. Appearance of spectra

Two important features of field ionization are that (a) the molecular ions are, on average, produced with little excess energy compared with electron impact and (b) the residence time of ions in the source is extremely small (10^{-14} to 10^{-12} s). The first factor is responsible for the high probability of finding a molecular ion in FI mass spectra, but since the time needed for

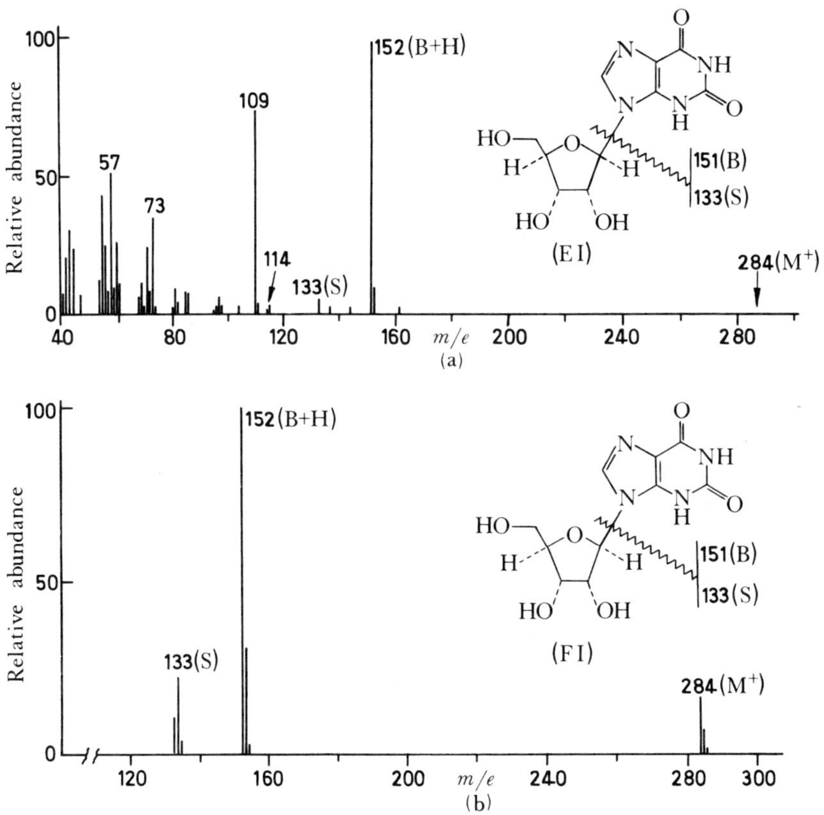

Figure 9-2

ions to reach the collector is essentially the same as that for electron impact, the second factor is barely significant in determining the molecular ion abundance.

The field-ionization spectrum of the nucleoside xanthosine is shown in Fig. 9-2(b), and compared with the corresponding electron impact spectrum (see Fig. 9-2(a)).[2]

Unmodified nucleosides do not yield abundant molecular ions in EI mass spectra and the molecular ion of xanthosine is undetectable. The FI spectrum overcomes this problem: the molecular ion is readily detectable and the assignment of masses to the base (B) and sugar (S) portions of the molecule is a simple matter. In fact the simplicity of FI spectra of nucleosides in general is most gratifying and the molecular ion (or sometimes $M^+ + H$), $(B + H)$, $(B + 2H)$ and S peaks are usually readily identifiable. Since most of the common base components have unique masses, FI spectroscopy offers a reliable means of identifying the base: but stereochemical features are inaccessible by this method. $(M + H)^+$ ions are common in FI mass spectra (see Fig. 9-2(b)) since ion–molecule reactions may occur in the dense layer close to the emitter surface where ionization takes place. However these pressure-dependent $(M + H)^+$ ions are easily distinguishable from M^+ ions.

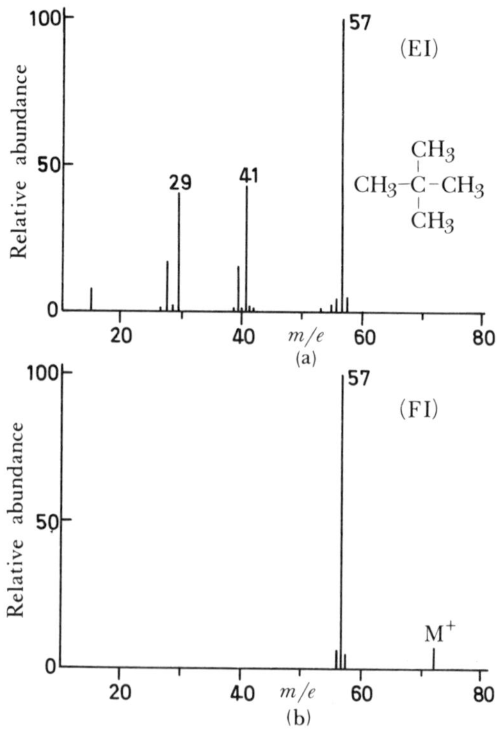

Figure 9-3

It is not uncommon (see Fig. 9-2) for a molecular ion to be undetectable by electron impact yet observable after field ionization, despite the fact that in both processes the times of flight to the ion collector are of the same order of magnitude. Another example is found in the spectra of neopentane (see Fig. 9-3), which yields no molecular ion in the EI spectrum, but eight per cent of the ion current is carried by the molecular ion in the FI spectrum.[1] It therefore appears that the lowest electronic state is not repulsive, but must contain a small potential trough. Presumably vertical ionization (see section 2-1) in the EI process occurs either to a higher electronic state or to the continuum above the dissociation limit of the ground electronic state (e.g., Fig. 9-4).

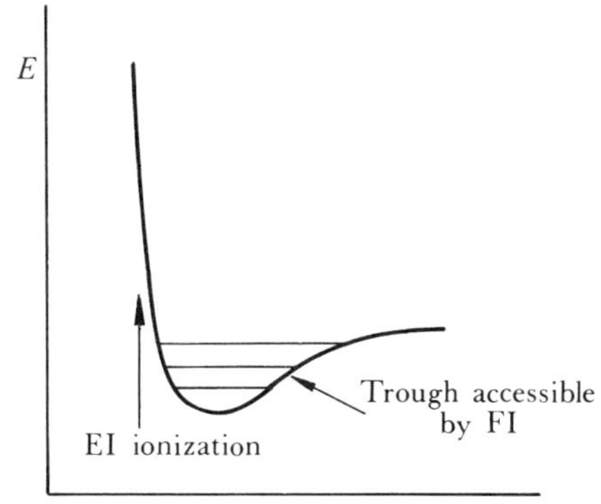

Figure 9-4

Variation in the method of introduction of the sample into the FI source widens the applicability of the technique. Different spectra can be observed if the sample is deposited in solution on the emitter and adsorbed, rather than admitted in the gaseous phase. This field desorption technique overcomes, at least in part, the problem of thermal instability during the sample evaporation from a direct insertion probe prior to the ionization process.[3] A striking illustration of the advantage of this method is shown in Fig. 9-5. The field desorption (FD) spectrum (Fig. 9-5(c)) of D-glucose shows virtually no elimination of water molecules from the $M^+ + 1$ ion, as opposed to the FI spectrum (Fig. 9-5(b)). The EI spectrum exhibits no molecular ion.

Field desorption mass spectrometry is therefore a useful technique for the study of thermally unstable substances of low volatility. Its potential usefulness in structural investigation of natural products is self-evident, especially in the case of mixtures.

9-3. Kinetic studies

Because of the high potential gradient present in a field-ionization source, ion lifetimes in the source have typically a maximum of about 3×10^{-12} s;

whereas ions spend up to 3×10^{-6} s in an electron impact source. The differences between EI and FI spectra may be readily understood in terms of these lifetimes and of the Quasi-Equilibrium Theory (QET).

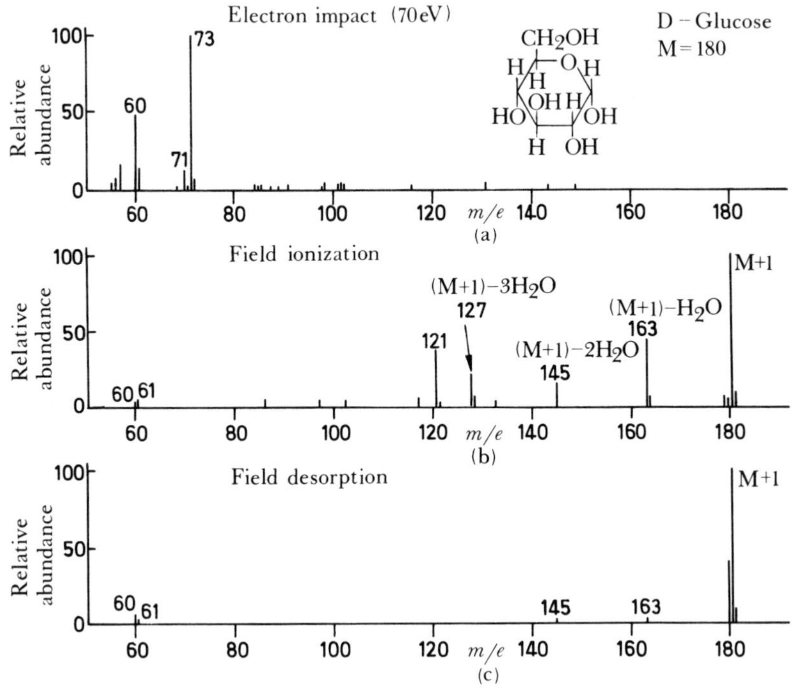

Figure 9-5

For molecular ions having a discrete rate constant k for decomposition and spending a time t_1 in the ion source, the fraction of these ions which decompose to daughter ions A^+ in the source is given by:

$$[A^+] = 1 - e^{-kt_1}$$

If a residence time of 3×10^{-12} s is assumed, then the yields of A^+ for different values of k are given by

k (s^{-1})	10^{13}	10^{12}	10^{11}	10^{10}
$[A^+]$	1·00	0·95	0·26	0·03

It is evident from these figures that a reaction having a rate constant less than 10^{10} s^{-1} will not occur to any significant extent in the field-ionization source and the corresponding normal fragment ion will not be detected at the collector. There is of course a whole range of rate constants of the molecular ion depending on the internal energy, according to the simplified equation:

$$k = \nu\left(\frac{E - E_0}{E}\right)^{s-1}. \tag{9-1}$$

The consequence is that rearrangement reactions (having low frequency factors) have little chance of being observed as daughter ions in field-ionization spectra, particularly as the molecular ions do not have appreciable excess energy to elevate the rate constant to the necessary value according to eq. (9-1). However, reactions having high frequency factors (i.e., direct cleavages, $\nu = 10^{13}$–$10^{14}\,s^{-1}$) are likely to be observed on the FI mass spectrometer time-scale.

These arguments are merely an extension of those developed in chapter 4 in connection with electron-impact mass spectra and are consistent with the general rule that rearrangements will become increasingly important relative to direct cleavage reactions as ion lifetime increases. The consequence of these arguments is that a daughter ion which is relatively intense in both the FI and EI spectrum of a compound cannot arise via a gas-phase rearrangement.[4]

Although in most cases rearrangement reactions are not observed in the FI source, there are instances of 'fast rearrangements'. For example, a low-intensity $M^+ - C_2H_4$ rearrangement peak appears in the FI spectrum of butyrophenone, $C_6H_5COC_3H_7$, although the variation in its relative abundance with temperature and pressure suggests some thermal rearrangement on the emitter.[5] Whatever their origin, low-abundance rearrangement ions (and even multi-step decomposition reactions) may sometimes occur in FI spectra.

It would also be anticipated that hydrogen- and/or carbon-scrambling reactions would not be prevalent before decomposition in a field-ionization source. In support of this, there is no evidence for hydrogen or carbon scrambling before the formation of alkyl ions from 1 and 2 in a field-ionization source.[6]

$CH_3CD_2(CH_2)_5CH_3$	$CH_3(CH_2)_3{}^{13}CH_2(CH_2)_3CH_3$
1	2

In connection with the selection of different mass-spectrometric techniques to sample different ion lifetimes, we should briefly mention the position of *field ionization metastables* in the general spectrum of lifetimes. For the reaction $m_1^+ \rightarrow m_2^+$, 'fast' metastables ($k = 10^{12}$–$10^{8}\,s^{-1}$) formed by decomposition during acceleration will appear, in a single-focusing mass spectrometer, as a broad continuum between m_2 and m_2^2/m_1 and 'slow' metastables ($k = 10^{8}$–$10^{6}\,s^{-1}$) will appear at m_2^2/m_1, as for metastables in the EI spectrum.[4]

A number of different mass-spectrometric techniques have been discussed in previous chapters and the various lifetimes appropriate to these different techniques are illustrated as a general scheme in Fig. 9-6.

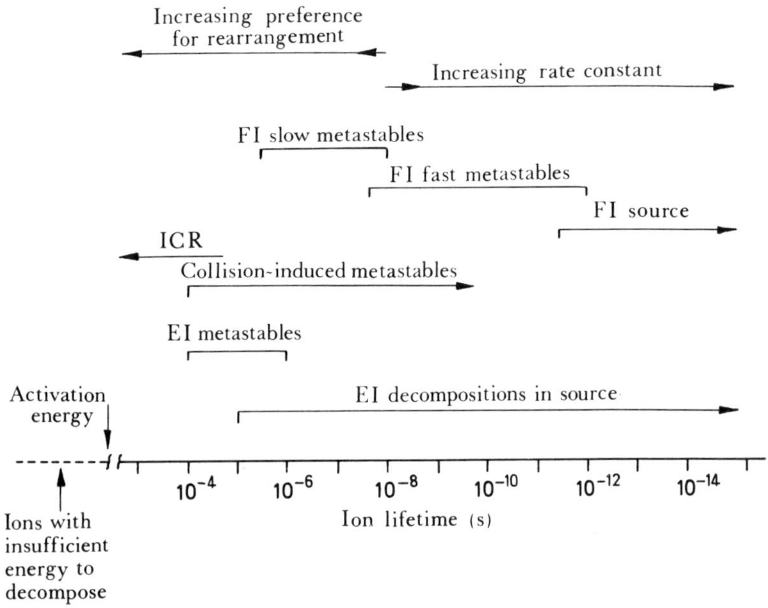

Figure 9-6

In summary, the most important use of field-ionization mass spectrometry is in identification of the parent ion of a sample. Field-ionization spectra are simpler than those obtained by electron impact and usually contain no fragments generated by rearrangement or multiple decompositions. Desorption of the sample from solution onto the emitter may alleviate the problem of thermal degradation before ionization. However, the instability and insensitivity of the field-ionization method is a drawback, and considerable resolution is lost compared with ionization by electron impact.

References

[1] H. D. Beckey, *Angew. Chem. Internat. Edn.* **8**, 623 (1969).
[2] P. Brown, G. R. Pettit and R. K. Robins, *Org. Mass Spec.* **2**, 521 (1969).
[3] H. D. Beckey, *Int. J. Mass Spec. Ion Phys.* **2**, 500 (1969).
[4] H. D. Beckey, *Int. J. Mass Spec. Ion Phys.* **1**, 93 (1968).
[5] E. M. Chait and F. G. Kitson, *Org. Mass Spec.* **3**, 533 (1970).
[6] I. V. Goldenfeld and I. Z. Korostyshevsky, *Int. J. Mass Spec. Ion Phys.* **3**, 406 (1969).

10. Ion structures

10-1. Introduction

The only methods available to the early organic chemists for structure proof were chemical degradation and synthesis, coupled with measurement of confirmatory physical constants, such as melting point, boiling point or refractive index. Nowadays, these methods are complemented by physical techniques, such as X-ray analysis and infrared, ultraviolet and nuclear magnetic resonance spectroscopy.

Mass spectrometry is another important technique for structure elucidation of neutral organic molecules—yielding information on the molecular weight and on the structural entities comprising the molecule. In order to relate fragmentation pattern to structure, chemists have speculated freely upon the structure of the ions in the mass spectrometer; and in view of this, there has been an obvious need for methods to be devised to determine the structures and fragmentation mechanisms of organic ions in the gas phase.

Although it is conceivable that the ir and uv absorption of an ion beam in a mass spectrometer could be scanned, the concentrations of ions are so low that the approach has so far been precluded in practice. The physical methods of structure elucidation mentioned above are therefore excluded and we resort to methods which are in some ways similar to those used in chemical degradation studies; i.e., the degradation products of fast unimolecular (and sometimes bimolecular) reactions are sampled by the mass spectrometer and the energetics and kinetics of these reactions are investigated. Methods that have been employed include:

(a) Heats of formation (section 5-2),
(b) Ion–molecule reactions (chapter 8),
(c) Metastable ion characteristics (section 4-4),
(d) Simple decomposition pathways,
(e) Isotopic labelling (chapter 7),
(f) Substituent effects (section 4-3).

Some, or all, of these methods incorporate potential pitfalls and it is important at the outset to evaluate the kinds of information obtainable by each method, the reliability of the method[1] and whether we could expect the approaches to give the same, or different, answers.

Consider the sequence of events accompanying the ionization and subsequent decomposition of a molecule ABCD. The only circumstance under which we can say with certainty that the structure of the molecular ion $ABCD^{+\cdot}$ is the same as that of the neutral molecule is before the ion has had time to vibrate after its production via electron removal from ABCD in a Franck–Condon process. This is so because the ionization process is fast ($\sim 10^{-16}$ s) compared with the time required for molecular vibration ($>10^{-14}$ s). Exceptions to this could occur if auto-ionization takes place; that is if electron impact produced a highly excited *neutral* which may rearrange before electron expulsion and molecular-ion production:

$$\begin{array}{ccccc} ABCD + e & \longrightarrow & ABCD^{*} + e & \longrightarrow & ABCD^{+\cdot} + 2e \\ & & \Big\downarrow \text{OR} & & \\ & & ACBD^{*} + e & \longrightarrow & ACBD^{+\cdot} + 2e. \end{array}$$

Proceeding on the assumption that such processes are relatively unimportant, we may now consider the relative possibilities of rearrangement and/or decomposition available to the molecular ion, depending on its energy and lifetime. It will have become clear from the preceding chapters (particularly chapter 4) that the unimolecular reaction undergone preferentially by the unrearranged molecular ion is not necessarily the lowest-energy process. The controlling factor is solely a kinetic one; and the fastest reaction for molecular ions of given energy will be the one which has the greatest rate constant at that energy.

After $ABCD^{+\cdot}$ has had time to undergo a sufficient number of vibrations it may, or may not, rearrange to $ACBD^{+\cdot}$. It may rearrange only if its internal energy is greater than or equal to the energy of activation E_r for intramolecular rearrangement. Whether it *will* rearrange is dependent upon the relative magnitudes of the rate constants for decomposition (k_d) and for rearrangement (k_r).

$$\begin{array}{ccc} ABCD^{+\cdot} & \underset{k_b}{\overset{k_r}{\rightleftharpoons}} & ACBD^{+\cdot} \\ \Big\downarrow k_d & & \end{array}$$

The factors influencing rate constants are applied in Table 10-1 to illustrate the various possibilities of rearrangement.

Intermediate cases to those described in Table 10-1 can of course be envisaged. The important point arising is that whether the molecular ion is rearranged or not may depend on the part of the energy distribution that is sampled in the structure investigation.

It is important to realize that in those cases where the molecular ion will rearrange, the relative proportions of $ABCD^{+\cdot}$ and $ACBD^{+\cdot}$ will depend on the ratio k_b/k_r, which will again depend on the segment of the energy distribution which is sampled. Furthermore, we have not considered whether other decompositions will occur from $ABCD^{+\cdot}$ or $ACBD^{+\cdot}$ or both.

Table 10-1
Molecular ion rearrangement at high and low energies

Molecular parameters	M^+ structure	
	Low E	High E
1. $E_r > E_d$; $v_d > v_r$	No rearrangement	No rearrangement
2. $E_r > E_d$; $v_r > v_d$	No rearrangement	Will rearrange
3. $E_d > E_r$; $v_d > v_r$	Will rearrange†	No rearrangement
4. $E_d > E_r$; $v_r > v_d$	Will rearrange†	Will rearrange

† No rearrangement will occur if the energy of $M^{+\cdot}$ is less than E_r.

The above treatment illustrates that (a) the extent of molecular ion rearrangement is a function of internal energy, and (b) evidence for one structure of a molecular ion does not necessarily preclude the existence of other structures in equilibrium with it. These arguments apply equally to fragment ions.

10-2. Methods for determination of ion structures

A. Heats of formation. Ideally, identical heats of formation indicate identical structures at the instant of formation of the ions, but we have seen (section 5-2) that there are potentially serious errors in heat-of-formation measurements. Nevertheless, positive structural conclusions may sometimes be made if it is recalled that appearance potential measurements yield maximum values for heats of formation, whereas heat-of-formation values obtained from ionization potentials of radicals are reliable. Table 10-2 shows heat-of-formation measurements by both methods[2] for $C_4H_9^+$ ions from various sources.

Table 10-2
Heats of formation of $C_4H_9^+$

Reactant	Other product	AP or IP (eV)	ΔH_f (kcal/mole)†
n-$C_4H_9\cdot$		8·01	201 (841)
iso-$C_4H_9\cdot$		8·01	199 (832)
sec-$C_4H_9\cdot$		7·41	183 (765)
$tert$-$C_4H_9\cdot$		6·93	167 (698)
n-C_5H_{12}	CH_3	11·05	186 (778)
iso-C_5H_{12}	CH_3	11·12	185 (774)
neo-C_5H_{12}	CH_3	10·55	170 (711)
n-C_6H_{14}	C_2H_5	10·97	185 (774)
n-C_7H_{16}	C_3H_7	10·95	186 (778)

† Figures in parentheses are in units of kJ/mole.

Even allowing for an error of a few kcals, the data in Table 10-2 are informative. For example, the $C_4H_9^+$ ion from *n*- and *iso*-pentane cannot be formed in the n-$C_4H_9^+$ structure because the heats of formation (186 and 185 kcal/mole respectively) are considerably *less* than the true heat of formation of n-$C_4H_9^+$ (201 kcal/mole). More likely the $M^+ - CH_3$ ions from *n*- and *iso*-pentane are formed in the *sec*-$C_4H_9^+$ structure. By similar arguments, rearrangement has taken place before formation of $C_4H_9^+$ ions at the threshold from *n*-hexane and *n*-heptane. In contrast, the $C_4H_9^+$ ion from *neo*-pentane is clearly formed in the *tert*-butyl structure. It appears from appearance-potential measurements on alkyl ions from a substantial number of *n*-paraffins that the alkyl ions are usually not formed as primary carbonium ions.[3]

The above results for aliphatic hydrocarbons may be contrasted with heat-of-formation measurements for $C_7H_7^+$ produced by ionization of benzyl and cycloheptatrienyl radicals and also via fragmentation reactions (see Table 10-3). It is almost certain that the $C_7H_7^+$ ions are formed in either the benzyl or tropylium structure, and the high heats of formation for $C_7H_7^+$ formation via fragmentation are therefore associated with a kinetic shift and/or reverse activation energy (sections 5-2 and 5-3).

Table 10-3
Heats of formation of $C_7H_7^+$ *ions*[3]

Reactant	Other products	AP or IP (eV)	ΔH_f (kcal/mole)†
$C_6H_5CH_2\cdot$		7·76	216 (903)
cycloheptatrienyl·		6·24	209 (874)
toluene	H	11·8	232 (970)
cycloheptatriene	H	10·4	231 (966)
ethyl benzene	CH_3	11·2	232 (970)
m-xylene	CH_3	11·6	238 (995)

† Figures in parentheses are in units of kJ/mole.

B. Ion–molecule reactions. It was noted in chapter 8 that the activation energies for ion–molecule reactions are zero or very small, whereas unimolecular ionic decompositions have activation energies ranging from zero to as high as ca. 4 eV. It therefore follows that information about the structure of low-energy ions, which have insufficient energy for unimolecular decay, may be obtained via ion–molecule reactions. Ion cyclotron resonance spectroscopy (section 8-2) is ideal for this purpose, since ion–molecule reactions can be followed for ions having lifetimes as high as 10^{-3} s. Virtually all the ions studied, therefore, will have energies ranging from zero up to the lowest activation energy for unimolecular decomposition. That is not to say that the ions may not have sufficient energy for rearrangement, or scrambling; and further ICR studies should determine such possibilities if suitable ion–molecule reactions can be chosen.

The early ICR investigations have been encouraging, but it is emphasized that structures that have been shown to exist at the low energies appropriate to the ICR cell may be in equilibrium with other structures at higher energies. Thus the $C_3H_6O^{+\cdot}$ ion which exists in the enol form 1 at the low energies appropriate to ICR studies may reversibly rearrange to the ionized acetone structure at higher energies.

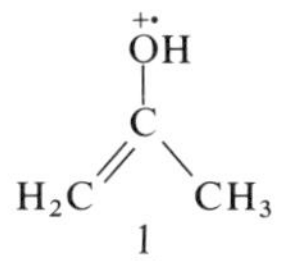

1

C. Metastable ion characteristics. A very useful criterion for comparison of ion structures from different sources at energies just above the lowest activation energy for decomposition is the ratio of competing metastable ion abundances (see section 4-4). The one factor which limits the applicability of this method is the necessity for two or more competing metastable transitions, i.e., two reactions of similar activation energy.

The importance of using metastables simply to exclude identical structures should not be overlooked. For example, the $NO_2C_7H_6^+$ ions (*m/e* 136) generated from the *meta*- and *para*-isomers of compound 2 (e.g., X = Br, CH_3) undergo different metastable transitions.[4, 5] In the case of the *p*-isomers, the *m/e* 136 ion eliminates exclusively NO in a metastable transition (peak at *m/e* 82·6 in Fig. 10-1), whereas the *m/e* 136 ion from the

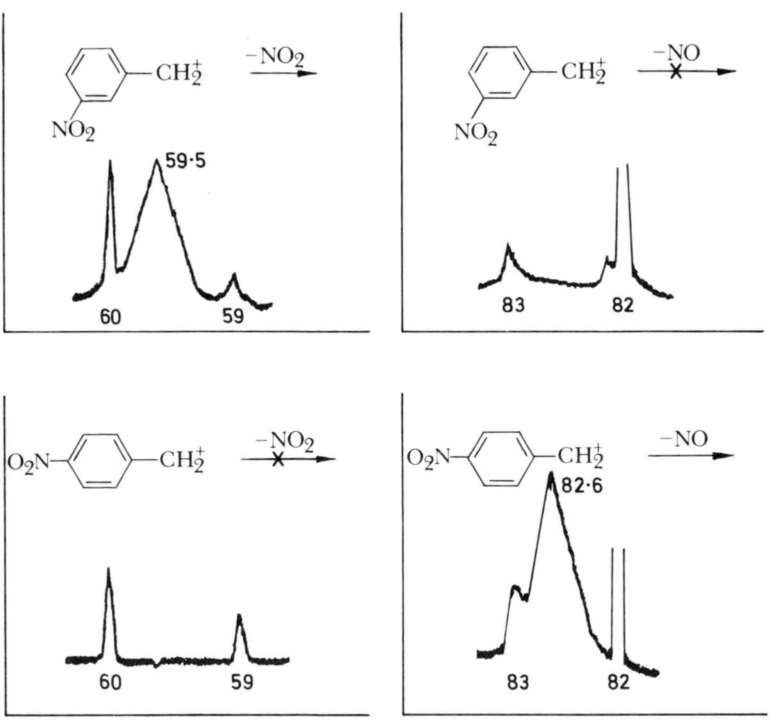

Figure 10-1

m-precursor undergoes exclusively NO_2 loss in a metastable transition (peak at *m/e* 59·5 in Fig. 10-1).

Hence both of the $NO_2C_7H_6^+$ ions cannot have the nitrotropylium structure at energies appropriate to metastable transitions and a likely explanation is that both ions exist in the nitrobenzyl forms 3 and 4. Furthermore, this means that the molecular ions of 2 have not both isomerized to a substituted cycloheptatriene structure at internal energies sufficient to eliminate X *and* to decompose further via a metastable transition.

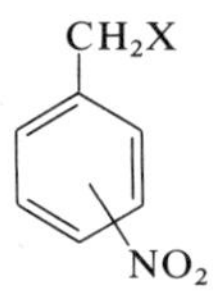

2

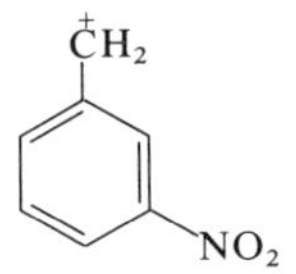

3

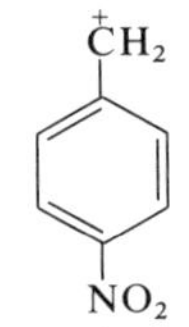

4

The shapes of metastable peaks also yield structural information. For example, the $C_3H_8N^+$ ions (*m/e* 58), formed from a variety of amines, all eliminate ethylene in a metastable transition, but the metastable peaks have either a gaussian shape or are flat-topped, with a kinetic energy release of 0·4 eV.[6] This indicates different structures for the two groups of ions at energies appropriate to metastable transitions, since the reaction must be occurring over two different potential surfaces. Without exception, the release of kinetic energy occurs only from $C_3H_8N^+$ ions formed initially with either of the primary amino structures 5 or 6. All other structures (containing secondary or tertiary amino groups) yielded a gaussian metastable peak for ethylene loss. In addition, all ions formed initially as 5 or 6 have isomerized to a common structure, or mixture of structures, before undergoing metastable transitions, as evidenced by the metastable loss of ammonia and ethylene in the ratio 8:1 in both cases.

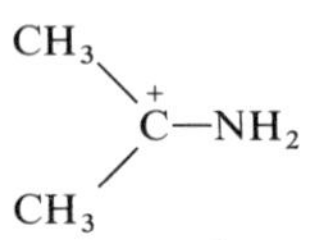

5

$CH_3CH_2\overset{+}{C}HNH_2$

6

D. Simple decomposition pathways. If the decomposition pathways of two isomeric ions in the source are radically different (i.e., not merely moderate differences in relative ion abundances, but obvious differences in the *m/e*-values of fragments), then the structures are different over the part of the energy distribution which gives decompositions in the source. The operation of an *ortho*-effect (section 6-4) which does not occur in *meta*- or *para*-isomers is an example of this kind.

E. Isotopic labelling. As we have seen in chapter 7, isotopic labelling can yield information on ion structures and decomposition mechanisms, particularly in metastable transitions. Deuterium isotope effects are potentially useful in investigation of reaction mechanisms, as in neutral chemistry.

F. Substituent effects. Daughter/parent ion ratios should not be used as a criterion for identity or non-identity of *m*- and *p*-isomers after electron impact (see section 4-3 and also sub-section D above).

In summary, there are many possible pitfalls in the elucidation of ion structures in the gas phase; and suggested ion structures are often only speculative. Competing metastable abundance ratios and ICR spectroscopy are valuable aids to structure elucidation, and isotopic labelling often provides relevant information. It is emphasized that an ion structure at one energy is not necessarily the same at another energy, and different methods may yield different results.

References

[1] See also *Specialist Periodical Reports: Mass Spectrometry*, Vol. 1, The Chemical Society, 1971, chapter 2.
[2] F. P. Lossing and G. P. Semeluk, *Can. J. Chem.* **48**, 955 (1970) and references cited therein.
[3] J. L. Franklin *et al.*, *Ionisation Potentials, Appearance Potentials and Heats of Formation of Gaseous Positive Ions*, NSRDS-NBS 26, U.S. Govt. Printing Office, 1969.
[4] R. Westwood, D. H. Williams and A. N. H. Yeo, *Org. Mass Spec.* **3**, 1485 (1970).
[5] R. H. Shapiro and J. W. Serum, *Org. Mass Spec.* **2**, 533 (1969).
[6] N. Uccella, I. Howe and D. H. Williams, *J. Chem. Soc.* (B), 1933 (1971).

11. Gas chromatography—mass spectrometry

11-1. Introduction

The detection and separation of very small samples of organic compounds from a mixture is readily achievable by gas chromatography, and limited characterization of unknown components is sometimes possible from retention times appropriate to the particular column used. Mass spectrometry is the technique most suited for providing definite structural information from the small quantities of material employed in gas chromatography. The association of the two techniques has therefore provided a means of structurally identifying the components of natural and synthetic organic mixtures.

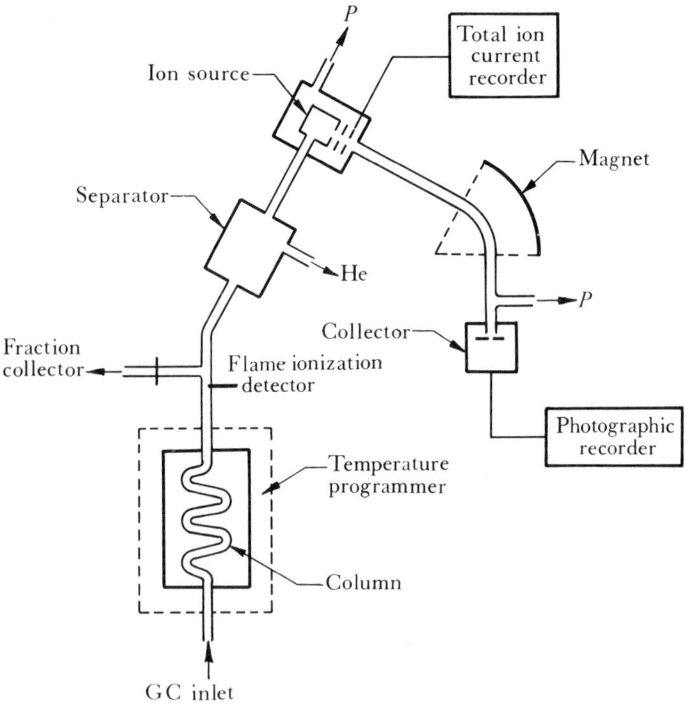

Figure 11-1

Mass spectra of sufficient quality are obtainable for every component that may be separated chromatographically, even though the quantities of components may be less than a μg and available for only a few seconds.

Use of the gc/ms technique has become routine in many laboratories and systems have progressively become more automated. Some systems, particularly in earlier work, involved condensation of the sample emerging from the gas-chromatographic column prior to introduction into the mass spectrometer. The separate collection of samples need not always be a disadvantage, especially if sufficiently large amounts are available. This is because more time is available to run the spectra and the best recording conditions may be ascertained (compared with the directly coupled instruments, where fast scans are taken and variations in recording conditions are not possible). Systems have been used in which the flow of carrier gas is stopped during mass spectrometric scanning of individual fractions.

Direct attachment of gas chromatograph and mass spectrometer (see Fig. 11-1 for schematic diagram) has meant that it is not necessary to trap the components prior to entering the mass spectrometer. In addition, computerized data collection and interpretation cuts down the manual control required during the operation. Several mass spectra of one chromatographic peak can be averaged to reduce any statistical errors and the background subtracted. The electron-multiplier sensitivity may be changed automatically depending on the component concentrations. Without computerization of the gc/ms combination, much of the information contained in a complicated chromatogram might be wasted.†

11-2. Linking the gas chromatograph to the mass spectrometer

It is worthwhile initially to consider the requirements of a successful directly linked gc/ms system:

(a) *The gas chromatograph.* Interchangeable columns should be available which are capable of quick and complete separation of mixtures and which do not leak column material into the mass spectrometer (column bleed). A temperature-programming facility is an advantage.

(b) *The mass spectrometer.* The time of elution of a gas-chromatographic peak normally varies from about 1 s to about 1 min. A mass spectrometer used on-line with a gas chromatograph should be capable of fast scanning, e.g., up to 500 mass units/s. A scanning time about one order of magnitude shorter than the average time of elution of a peak is considered suitable to avoid excessive variation of the amount of effluent compound during the scan (unless an automatic correction for this is employed). The mass spectrometer should carry a total ionization monitor, which acts as a detector for

† See section 12-6 for a more detailed discussion of the application of computers to gc/ms.

the gas chromatograph. The dynamic range of the mass spectrometer for peak recognition should be as high as possible, especially where minor components of mixtures have to be analysed. It is useful to be able to detect metastables.

(c) *GC/MS interface.* The interface should be capable of sustaining a large pressure drop from about 1 atmosphere in the gas chromatograph to about 10^{-6} torr in the ion source of the mass spectrometer. It is desirable to be able to separate the carrier gas from the sample effluent. The transfer of sample effluent across the interface should be rapid and efficient.

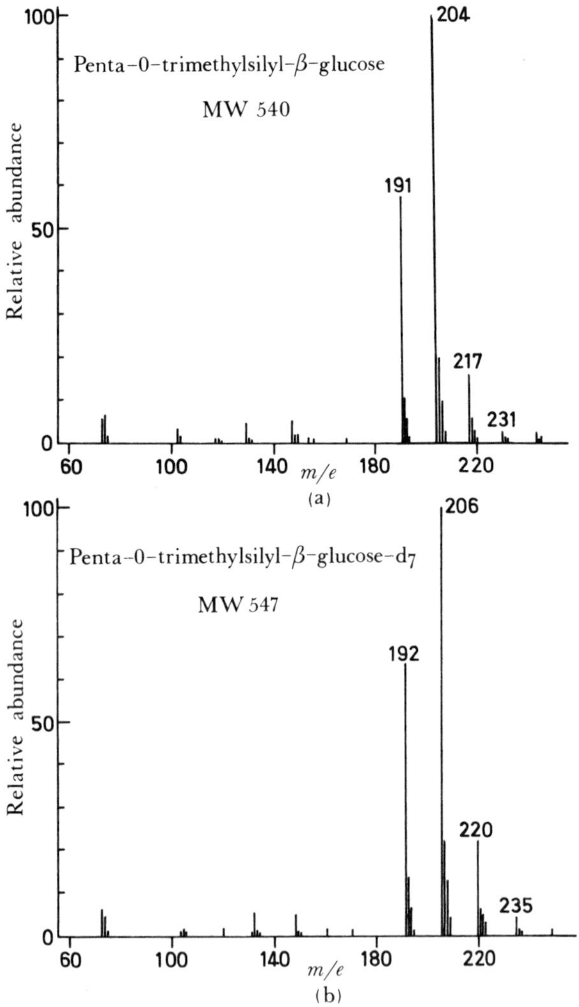

Figure 11-2

(d) *The sample.* The derivatization of samples in gas chromatography is frequently essential in order to confer high volatility. It is desirable, though not essential, that the sample derivatives should give abundant molecular ions.

The main requirements, as outlined above, for the successful operation of a gc/ms on-line system can be met in various ways depending on the needs of the operator. The variety of systems for coupling gc and ms is very extensive indeed and only the most important aspects will be considered below. Review articles[1–3] may be consulted for leading references.

At temperatures approaching 200°C many columns normally used for gas chromatography evolve gases such as CO_2, CO or H_2O and degradation of the column slowly occurs. 'Bleeding' into the mass spectrometer takes place and a background to the mass spectrum is established. If a differential detector were being used, merely a drift in the baseline would occur but with a mass spectrometer attachment the background must be subtracted. This is not entirely satisfactory, since column bleeding usually increases during peak elution. The overall consequence is that temperatures in the gc/ms set-up are usually kept below those in conventional gc. Capillary columns have been used in the gc/ms combination with considerable success.[2, 3]

As mentioned above, a primary requirement for the mass spectrometer of a gc/ms system is a capability for fast scanning. If a number of scans are performed over an eluted peak it is often possible to identify components whose retention times are so similar that they appear as an unresolved, somewhat broadened peak. For example, penta-O-trimethylsilyl-β-glucose and its heptadeuterio form give a single peak in conventional gas chromatography which appears symmetrical. However, combined gc/ms offers a means of separately investigating the elution of the two species to see if the gc column has in fact effected a partial separation.[4] The partial mass spectra (Figs. 11-2(a) and (b)) illustrate that there are three appropriate pairs of ions for investigation: those at m/e 191 and 192, 204 and 206, and 217 and 220.

Multiple partial mass spectra of the required mass region were therefore taken at 15 s intervals during elution of a mixture containing 40 per cent of the TMS-β-glucose-d_7. Plots of uncorrected ion intensities vs. time for the six m/e-values chosen are shown in Fig. 11-3.[4] These curves provide convincing evidence that the compounds were partially resolved on the column and the ratio of areas is approximately equal to the initial ratio of concentrations. It is also clear that the d_7-species is eluted from the column faster than the undeuteriated species. This isotope effect is explainable partly in terms of differences in vapour pressure and of differences in solubility in the column material for the different isotopic species.[5] More information about the nature of bonding to columns remains to be discovered from isotope investigations such as these.[4, 5]

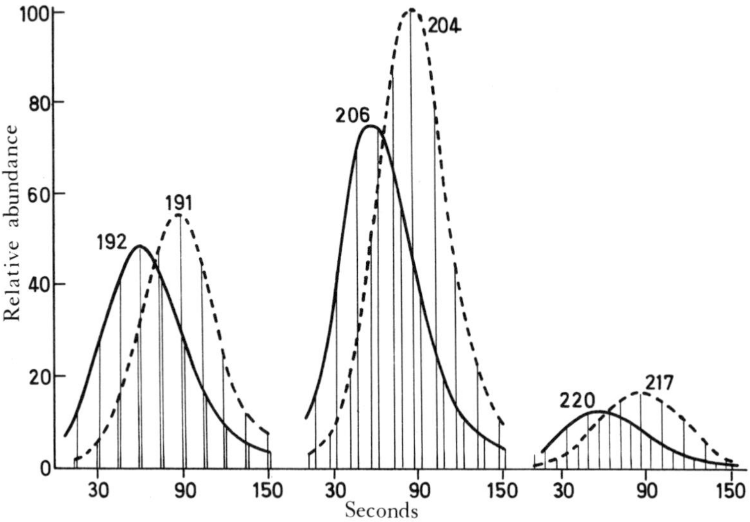

Figure 11-3

In addition to the magnetic-scan mass spectrometers in common use, the time-of-flight (tof) mass spectrometer is suitable for a gc/ms system. The tof instrument is capable of 100 μs scan times, since its principle of operation involves short pulses of ions which are accelerated to uniform kinetic energy and separated according to their different velocities (and hence different times of flight) at the collector and recorded on an oscilloscope. The very fast scan time is not quite the advantage it might seem in the gc/ms field, since

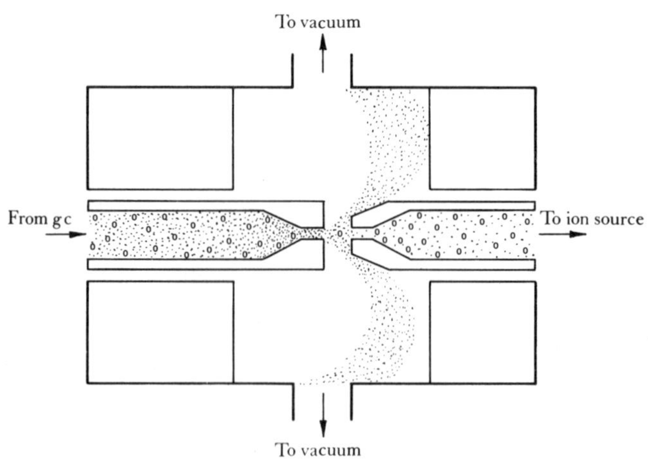

Figure 11-4

multiple scans are normally necessary to eliminate statistical errors in the very short pulses of ions. Nevertheless, the gc/tof combination has been used with success and with sufficiently high resolution. Quadrupole mass spectrometers also possess high scanning speeds and in cases where chromatographic columns of very high efficiency are employed, tof and quadrupole spectrometers are therefore advantageous.

A number of separators for the gc/ms interface have been developed with success according to the requirements outlined above. Here, we will discuss only three of these. The Becker–Ryhage jet separator (Fig. 11-4)[6] effectively eliminates most of the carrier gas (e.g., helium) before the sample enters the ion source, by utilizing the relatively fast diffusion of the low molecular weight carrier gas into a pumped interspace between two aligned orifices. This separator meets the important requirement of having no dead volume and a very small surface for the passing sample. Its efficiency is therefore high and an enrichment factor of 10^2 is attainable. One drawback is the inflexibility for widely different flow rates.

In the Watson–Biemann separator[7] (see Fig. 11-5), the concentration of organic matter in the effluent gas is increased by passage through a porous glass tube. The carrier gas (helium) again diffuses out much faster than large organic molecules. Hence this separator (and also the Becker–Ryhage jet) increases in efficiency with increasing molecular weight of the sample to be examined.

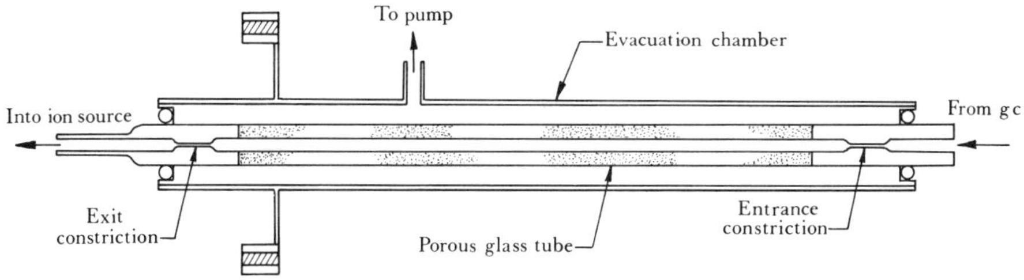

Figure 11-5

The membrane molecular separator[8] employs the preferential diffusion of organic molecules through the membrane and the rejection of most of the helium carrier gas. A schematic diagram of a two-stage membrane separator is shown in Fig. 11-6. Membranes commonly consist of methyl silicone rubber and the partial passage of helium through such a separator, especially above 200°C, requires that two separators in series should be used. With this system, a low helium transmission and high efficiency are achievable and a high pressure drop across the separator can be attained. Modifications of

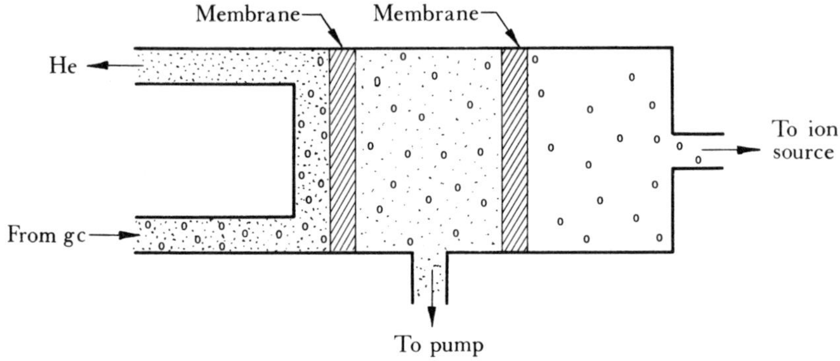

Figure 11-6

the membrane-type separator have been described which employ other types of separator as a second stage.[1]

A drawback in some types of separator is adsorption of eluted compounds on the interface material. Coating the separator with a silanizing reagent alleviates this difficulty. One procedure[9] involves washing the porous interface with water, then sequentially with hot benzene, acetone, one per cent methanolic KOH, methanol and toluene. During subsequent immersion in a five per cent solution of dimethylchlorosilane in toluene, air bubbles are removed from the pores by an aspirator and after 2–3 hours immersion, the interface is drained, rinsed with toluene and methanol and dried at 80°C.

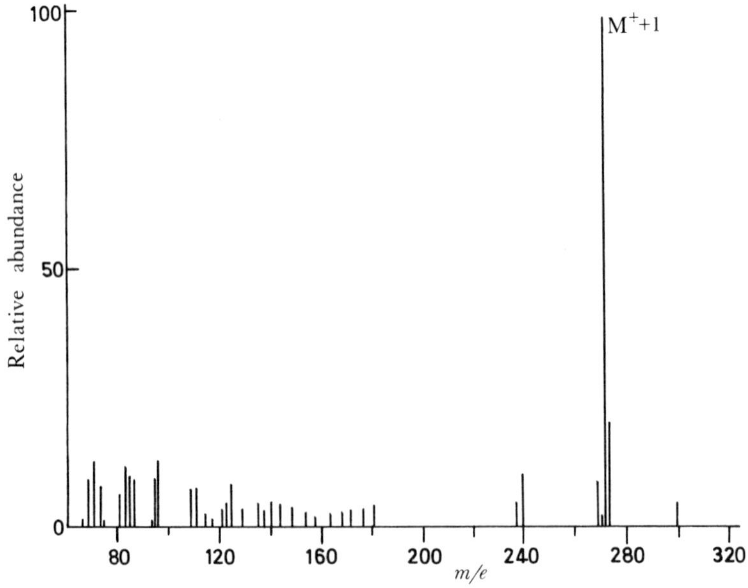

Figure 11-7

The use of a molecular separator may be avoided by construction of a gas chromatography/chemical ionization combination, in which the carrier gas of the gc becomes the ionization reagent in the ci source (see also section 8-1). For example, methane has been used as the carrier/reactant gas in a gc/ci system, operated at a gas rate flow of 5 ml. methane/min.[10] The potential of the technique is illustrated in Fig. 11-7 which shows the ci spectrum of methyl palmitate obtained during a gc/ci run (using a quadrupole mass spectrometer) on a mixture of 1 μg each of the methyl esters of C_{12}–C_{18} even-numbered fatty acids. $M^+ + 1$ is the base peak, with a low-abundance ion at $M^+ + C_2H_5$ (*m/e* 299 in Fig. 11-7), formed via an ion–molecule reaction.

In concluding this section, it should be briefly mentioned that the gas chromatograph may be coupled, via a separator, to a field ionization source.[11] This combination has the advantage that only molecular ions are observed in most cases.

11-3. Derivatives for gc/ms

Many types of compound need to be derivatized before gas chromatographic separation, because of involatility. Many of these derivatives are also suitable for mass-spectrometric identification of the compounds.

A common derivative for alcohols is the trimethylsilyl (TMS) ether. These derivatives are volatile, have low polarity, and give rise to characteristic fragmentations, usually via α-cleavage (see Fig. 11-8). However, TMS derivatives are not always the best choice from a mass-spectrometric viewpoint. For example, free steroids yield spectra that are more structure specific than their corresponding TMS derivatives. If low volatility necessitates the preparation of a derivative, in most cases the preparation of keto compounds from hydroxy steroids is preferable to the preparation of TMS ethers.[12]

Alternative derivatives for alcohols are acetyl esters and methyl ethers. The overriding factor in the choice of derivative is normally a chromatographic one, so that efficient separation can be achieved before mass spectrometric identification. However, all other things being equal, it should be remembered that from the mass spectrometric viewpoint the derivative group should have a small mass increment, and should not undergo facile fragmentations; on these grounds, permethylation (section 6-5D) scores heavily.

O-methyl oximes have proved to be suitable derivatives for aldehydes and ketones,[13] despite the complication of *syn*- and *anti*-isomers. These derivatives yield stable molecular ions and structurally informative fragmentation patterns. For carboxylic acids, it is usual to employ esters, often methyl esters.

11-4. Applications to natural products

The number of applications of gc/ms to natural product studies has increased enormously in recent years and the diversity of these applications has been striking. Components of mixtures ranging from protein hydrolysates to tobacco smoke and fruit volatiles have been investigated by this method. As emphasized at the beginning of this chapter, the gc/ms combination offers a unique means of structurally investigating the components of a complex mixture available in small quantities. Developments of the technique have enabled smaller sample sizes and more complex samples to be characterized.

GC/MS is particularly suited to the separation and identification of natural lipids, and homologous compounds can be separated with ease. Identification of similar compounds having the same number of carbon atoms, particularly olefinic compounds, is more difficult. A method for identification of an unsaturated fatty acid ester of general formula 1 involves oxidation of the olefinic double bond with OsO_4 followed by preparation of the TMS derivative of the *cis*-diol.[14]

$$CH_3(CH_2)_n{-}CH{=}CH{-}(CH_2)_mCO_2CH_3$$

1

The *vic*-diethers are suitable for gas chromatography and yield characteristic ions in the mass spectrum by cleavage between the TMSO groups, as shown in Fig. 11-8. The abundance of these ions is always sufficiently high to allow their identification even when several positional isomers are present and are not separable by gc.

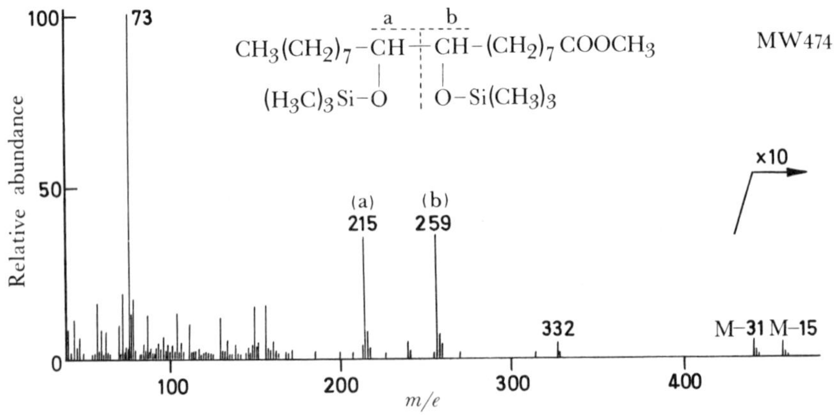

Figure 11-8

An elegant example of the application of gc/ms is provided by the identification of the steroid 5α-pregnane-3α,20α,21-triol (2) in human pregnancy plasma.[15] Only 2 μg of the compound were isolated and no authentic sample was available, so the compound was identified by a few microchemical transformations combined with gc/ms.

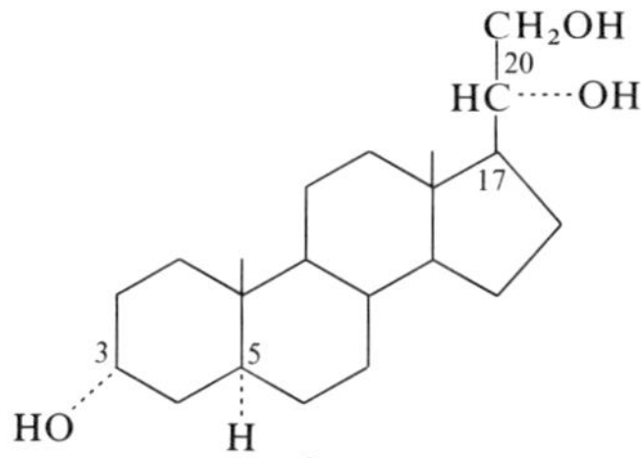

2

GC/MS analysis of the TMS ethers of steroids in the 'monosulphate' plasma fraction identified a component having a molecular weight and mass spectrum corresponding to those of the authentic 3β,20β isomer of 2, but having a different retention time. The unknown compound was therefore tentatively regarded as a pregnane-3,20,21-triol isomer. The configurations at the C-3 and C-20 positions were established by comparing the retention times (SE-30 column) and mass spectra of some simple derivatives of the unknown and the authentic isomer. Table 11-1 lists the retention times of the derivatives, together with the conclusions reached about the unknown.

Table 11-1

Derivative	Relative retention times		Conclusions about unknown
	Unknown	5α,3β,20β isomer	
Parent triol TMS ether	2·06	2·74	
17-CO_2CH_3 3-ketone	0·74	0·74	5α
17-CHO 3-TMS ether	0·56	0·70	3α
Acetonide 3-ketone	1·35	1·23	20α

In the determinations of alkaloidal constituents of plants, it is frequently unnecessary to derivatize the crude extract before analysis by gc/ms. An example of the power of the technique is found in the determination of the alkaloidal constituents of *Aspidosperma neblinae*.[16] Figure 11-9 shows the gas chromatogram of the chloroform eluate from an alumina column. Only 10 μg of alkaloidal material was injected into the gc and some of the components were present in submicrogram quantities. Nevertheless, all the components were readily identified by mass spectrometry with three-second scans over the mass range 14–500. A total of eleven alkaloids were identified from the plant extract. Most of these compounds belonged to the aspidospermine class of alkaloids, which exhibit characteristic fragmentation patterns (section 6-5B).

The important application of mass spectrometry to the sequencing of

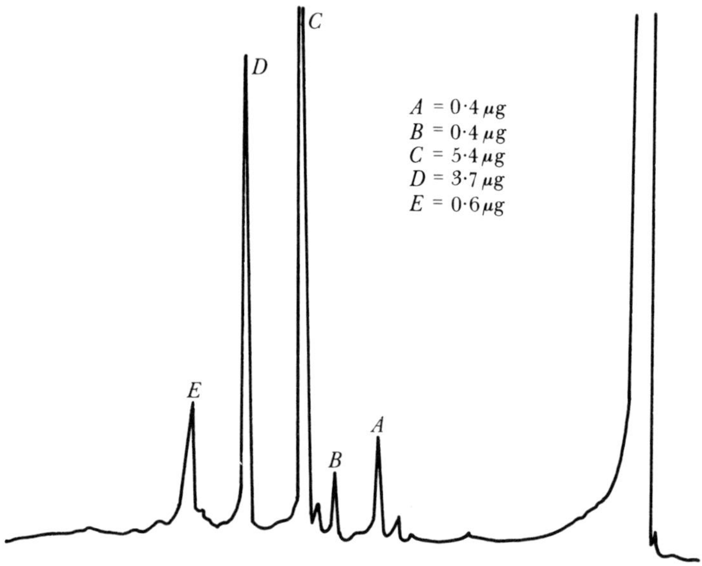

Figure 11-9

peptides has been discussed in section 6-5D. The combined gc/ms system has also been used for identification of the partial hydrolysis products (e.g., di- and tripeptides) of larger peptides. The following combination of gc/ms and liquid-column chromatography has been used for sequence analysis of peptides:[17] at first the peptide is benzylated, thus transforming the sensitive SH groups of cysteine into stable S-benzyl derivatives. After partial hydrolysis, the peptide fragments are easily converted into volatile TFA-peptide methyl esters. The small derivatized peptide fragments are then identified by gc/ms. The higher peptides may be separated by liquid-column chromatography and subsequently identified by mass spectrometry.

Table 11-2 illustrates the elucidation of the amino-acid sequence of the

Table 11-2

cys(Bzl)-tyr-ile-ser-asn-cys(Bzl)-pro-leu-gly
cys(Bzl)-tyr-ile
cys(Bzl)-tyr
tyr-ile*
ile-ser*
ser-asp*
ser-asp-cys(Bzl)
asp-cys(Bzl)-pro-leu
cys(Bzl)-pro*
cys(Bzl)-pro-leu*
cys(Bzl)-pro-leu-gly
pro-leu*
pro-leu-gly*
leu-gly*

nonapeptide shown at the head of the table. Below this are shown the peptide fragments found in the partial hydrolysate by combined gc/ms and liquid chromatography. The peptides marked with an asterisk were determined by gc/ms. Liquid-column chromatography followed by mass spectrometry determined the remainder and also confirmed some of the peptides determined by gc/ms. It should be added that the asparagine present in the original peptide is hydrolysed to aspartic acid during the acid hydrolysis. This procedure is of course more time-consuming than the direct sequence determination of larger peptides after permethylation (section 6-5D). However, at the moment cysteine-containing peptides cannot be satisfactorily handled via the permethylation procedure.

In concluding this section on the gc/ms analysis of natural products it should be mentioned that considerable amounts of hydrocarbons, alcohols, fatty acids, esters of acid resins, sterols and plasticizers are found in filters used in the extraction procedure.[18] These compounds may be present in sufficient quantities to exceed, by a considerable amount, the minimum detection limit for gc/ms. The use of filters without pretreatment should consequently be avoided in the trace analysis of natural products with gc/ms.

11-5. Identification and assay of metabolites

Another important application of the gc/ms combination which has recently received much attention is the identification and/or quantitative analysis of metabolites—either endogenous or from drug administration. Two useful techniques which have been developed for such purposes are (a) mass fragmentography and (b) the integrated ion-current technique.

(a) **Mass fragmentography.** We have already seen (Fig. 11-3) that the use of the mass spectrometer as a detector for selected *m/e*-values from the gc effluent may give evidence of separation of components which might not be forthcoming from other detection methods. More important, it has been noted that the ratio of areas under the intensity vs. time curves corresponds to the ratio of concentrations of the relevant species (approximately 2:3 for the two compounds in Fig. 11-3). In fact the continual and rapid focusing on a few selected mass numbers in turn during elution from the column may be used both for identification purposes and for quantitative analysis. This method is known as mass fragmentography.[19]

The ion intensity of two or more mass numbers can be recorded within a short time interval by use of an accelerating voltage alternator.[19] In a single-focusing instrument, ions of different mass numbers can be alternately brought into focus at the collector by changing the accelerating voltage in the ratio of their masses. It is advantageous if the different mass numbers are close together.

If a component of the gc effluent is present in only very small amounts, it may be preferable to use mass fragmentography rather than complete mass spectral scanning to achieve identification. This is because it may be impossible to extract a meaningful mass spectrum from the relatively high background (due to 'column bleed', etc.) and a gain in sensitivity is obtained by switching between selected abundant ions.

Mass fragmentography has been employed[19] to identify metabolites in human blood of the drug chlorpromazine (3), used in the treatment of mental illness. The mass spectra of chlorpromazine and its side chain derivatives yield abundant ions at *m/e*-values of 246 (see 4), 232 and 234 (a ^{37}Cl isotope peak of *m/e* 232). These three ions were monitored continuously from the gc effluent of the derivatized blood extract (see Fig. 11-10). The abundance of these three ions rises and falls simultaneously for two of the eluted compounds as shown in Fig. 11-10. These elutants are therefore strongly suspected to be chlorpromazine derivatives and the monitoring of suspected molecular

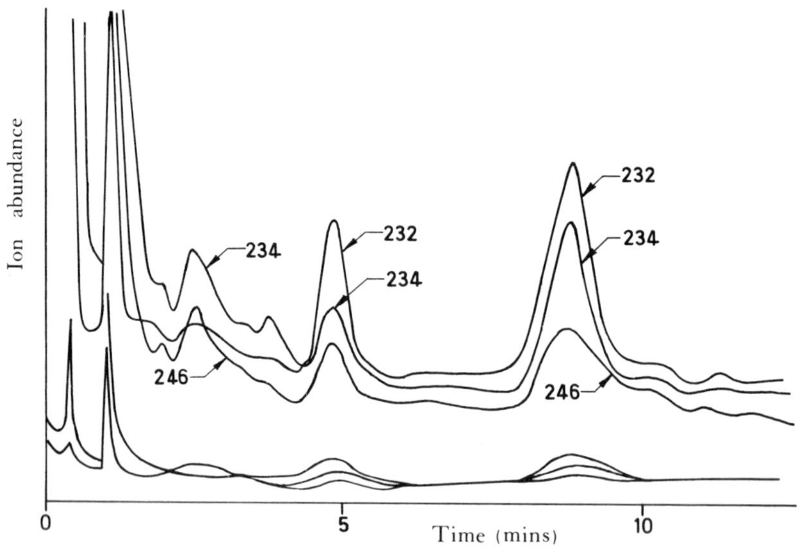

Figure 11-10

ions helped to confirm the structures, along with retention times. The two derivatives were in fact the trifluoroacetates of des- and dides-methylchlorpromazine.

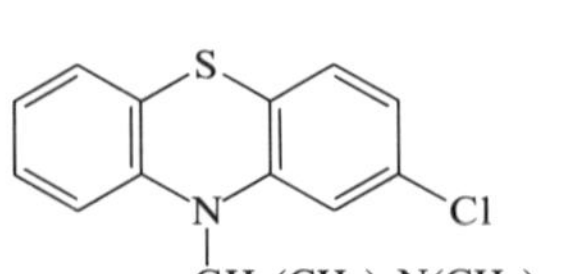
3

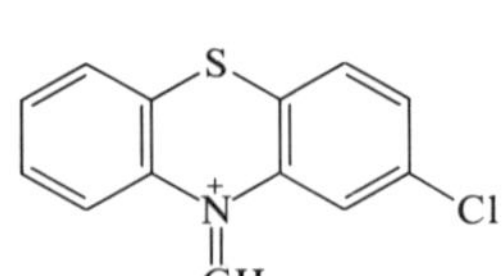
4

Mass fragmentography has also been developed as a method for assay of metabolites. For example, mixtures of biogenic amines have been assayed for quantities in the picomole range (10^{-12} mole).[20] The most abundant peaks in the mass spectra were again chosen for the assay and it was found that a linear calibration curve for peak height vs. amount of injected compound existed over a considerable range. Deuteriated samples of the material to be assayed have been used as internal standards for quantitation,[21] since isotopically-labelled species will have very similar physical properties to the unlabelled species, but ions of different mass number are monitored.

(b) The integrated ion-current technique. A mass spectrometric method has been developed for quantitative detection of metabolites which avoids the use of a gas chromatograph.[22] This 'integrated ion-current technique' simply involves the continual recording of the ion current at a given *m/e*-value (selected as a prominent ion obtained from the compound to be assayed) as the crude biological extract is evaporated from the direct insertion probe. Figure 11-11(a) shows the shape of the envelope enclosing the recorded ion currents at an *m/e*-value of 137 (the molecular ion) obtained during the evaporation at 220° of 1 ng of a pure sample of the biogenic amine, *p*-tyramine (5). If the operating conditions are identical in subsequent experiments, then the area under the curve is proportional to the concentration of *p*-tyramine present on the probe. A working range of 10^{-12} to 10^{-5} g has been reported.[22]

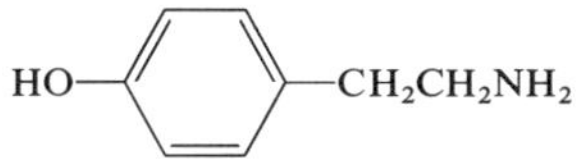

5

When the IIC technique is applied to a crude biological extract the envelope usually ceases to be symmetrical. For example, Fig. 11-11(b) shows the ion current recorded at *m/e* 137 at 2-second intervals from a rat brain extract.[23]

Obviously a potential pitfall occurs if the ion under study arises from two or more components of the mixture. This problem may be alleviated by measuring the IIC curve for two or more ions in the spectrum. For example, measurement of the IIC curve for *m/e* 108 in the assay of *p*-tyramine provides a useful check for the result from *m/e* 137.

Differences in rates of evaporation from the probe may also be employed to identify isomers by the IIC technique.[23] For example, a mixture of *o*-, *m*- and *p*-tyramine yields the IIC curve shown in Fig. 11-11(c). It is clear that the three isomers have been partially separated by this method. After addition of an authentic sample of *p*-tyramine to a rat brain extract no fine structure appeared in the IIC curve. It therefore appears that there are no significant amounts of the *o*- and *m*-isomers present in rat brain.

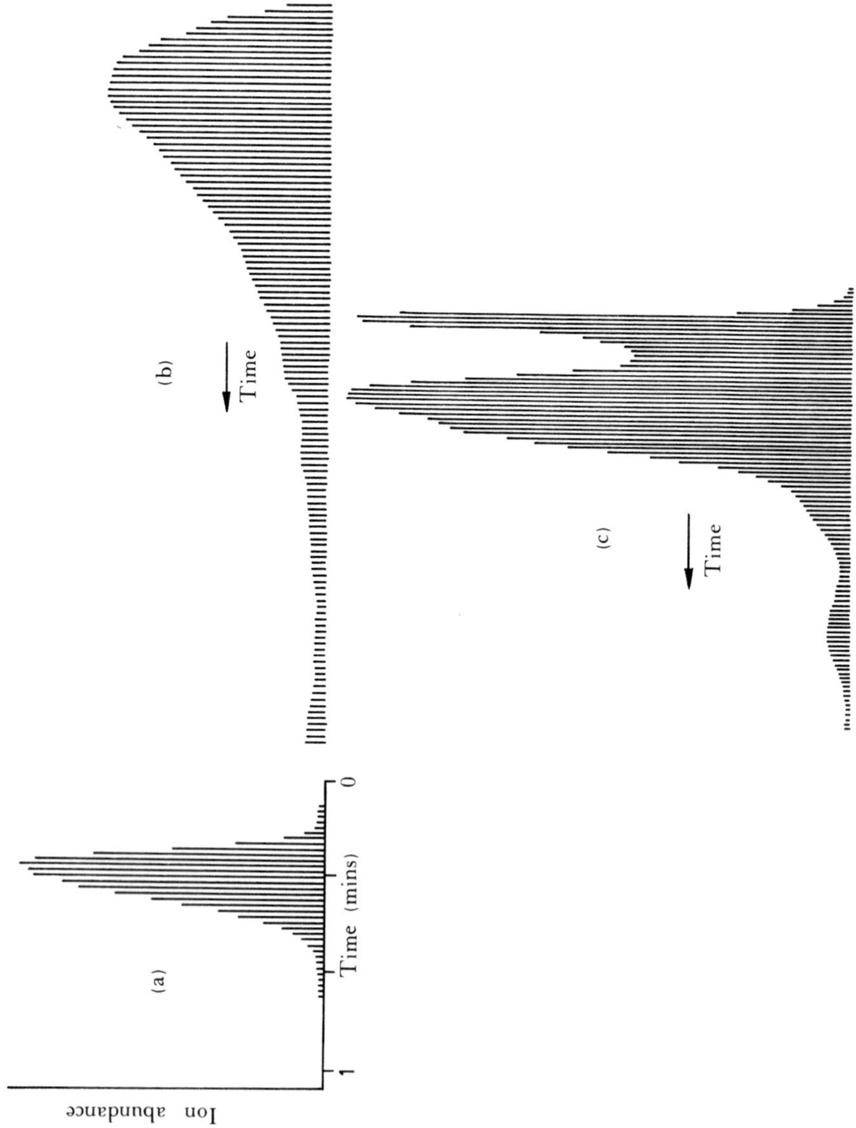

Figure 11-11

The integrated ion-current technique appears to be a useful and simple method for identification and assay of components in a crude biological extract down to the 10^{-12} g level. The technique requires constant ion source conditions and reliable calibration. Deuteriated derivatives of authentic samples should provide useful internal standards for calibration.

References

[1] *Specialist Periodical Reports: Mass Spectrometry*, Vol. 1, The Chemical Society, 1971, chapter 7; see also G. A. Junk, *Int. J. Mass Spec. Ion Phys.* **8**, 1 (1972).
[2] S. Ställberg-Stenhagen and E. Stenhagen in *Topics in Organic Mass Spectrometry*, A. L. Burlingame, ed., Wiley-Interscience, New York, 1970, p. 167.
[3] D. Henneberg and G. Schomburg, *Adv. Mass Spec.* Vol. 5, The Inst. of Petroleum, London, 1971.
[4] C. C. Sweeley, W. H. Elliott, I. Fries and R. Ryhage, *Anal. Chem.* **38**, 1549 (1966).
[5] N. C. Saha and C. C. Sweeley, *Anal. Chem.* **40**, 1628 (1968).
[6] E. Stenhagen, *Z. Anal. Chem.* **205**, 109 (1964).
[7] J. T. Watson and K. Biemann, *Anal. Chem.* **37**, 844 (1965).
[8] See, for example, D. R. Black, R. A. Flath and R. Teranishi, *J. Chromatog. Sci.* **7**, 284 (1969).
[9] P. M. Krueger and J. A. McCloskey, *Anal. Chem.* **41**, 1930 (1969).
[10] G. P. Arsenault, J. J. Dolhun and K. Biemann, *Chem. Comm.* 1542 (1970).
[11] J. N. Damico and R. P. Barron, *Anal. Chem.* **43**, 17 (1971).
[12] K. Kaiser, H. Obermann, G. Remberg, M. Spiteller-Friedman and G. Spiteller, *Monatsh. Chem.* **101**, 240 (1970).
[13] H. M. Fales and T. Luukkainen, *Anal. Chem.* **37**, 955 (1965).
[14] P. Capella and C. M. Zorzut, *Anal. Chem.* **40**, 1458 (1968).
[15] J. Sjövall and K. Sjövall, *Steroids* **12**, 359 (1968).
[16] D. W. Thomas, H. K. Schnoes and K. Biemann, *Experientia* **25**, 678 (1969).
[17] E. Bayer and W. A. Koenig, *J. Chromatog. Sci.* **7**, 95 (1969).
[18] G. V. Unruh, G. Remberg and G. Spiteller, *Chem. Ber.* **104**, 2071 (1971).
[19] C.-G. Hammar, B. Holmstedt and R. Ryhage, *Anal. Biochem.* **25**, 532 (1968).
[20] F. Karoum, F. Cattabeni, E. Costa, C. R. J. Ruthven and M. Sandler, *Anal. Biochem.*, **47**, 550 (1972).
[21] L. Bertilsson, A. J. Atkinson, J. R. Althaus, A. Härfast, J.-E. Lindgren and B. Holmstedt, *Anal. Chem.*, **44**, 1434 (1972).
[22] A. A. Boulton and J. R. Majer, *J. Chromatog.* **48**, 322 (1970).
[23] A. A. Boulton and J. R. Majer, *Can. J. Biochem.* **49**, 993 (1971).

12. Use of the computer in mass spectrometry

12-1. Introduction

As we have seen in the foregoing chapters, mass spectrometry has become a powerful tool for assisting with the structure elucidation of organic molecules. The products of organic ionic reactions, occurring in a few microseconds or less, are recorded in anything from a few seconds to a few minutes and the resulting mass spectrum is ready for analysis by a human interpreter. Analysis of the mass spectrum may of course take a considerable time, depending on the skill and experience of the interpreter, and if the chemist wishes to know the exact mass of the molecular ion (or other ions) the high-resolution measurements will be further time-consuming.

It is therefore clear that computerized data acquisition and interpretation can be a valuable time-saving device in mass spectrometry. A fully automatic, on-line mass-spectrometer/computer system can produce, smooth, and record data much faster than the manually operated mass spectrometer system and ideally can interpret the spectrum if required in a matter of seconds, using stored experience gained by skilled human operators. We have seen in chapter 11 how useful the gc/ms combination is in investigating mixtures of small samples. Computerized acquisition and interpretation of the gc/ms data enables mixture constituents weighing less than a microgram to be rapidly investigated. In an age where the problem of environmental pollution is becoming more serious it is useful to be able to analyse any pollutants quickly and efficiently and to store any data acquired. The mass-spectrometer/computer system would again be appropriate to this problem in many circumstances.

The importance of computers in mass spectrometry therefore requires no further emphasis, but it is also important that the system should not be used merely for acquisition of vast amounts of data. Data in itself is of little value if it cannot be efficiently interpreted. The reliability of such complex equipment can also be a problem.

The choice of the type of mass-spectrometer/computer system adopted is governed not only by the basic research requirements of a laboratory, but also by economic factors. This is perhaps partly the reason for development of a variety of operational systems, automated to different extents.

It is not the purpose of this chapter to give details of hardware, software and mathematical operations used in mass-spectrometer/computer systems. Rather, this chapter attempts to outline the basic principles of operation of the systems used and, in view of the large number of systems reported, the discussion must necessarily be abbreviated. Definitions of some of the terms applicable to this chapter, which may be unfamiliar to the organic chemist, are given below:

On-line: used to describe the direct connection of peripheral equipment (e.g., the mass spectrometer) to the computer.
Real time: in real-time operation, the computer processes data as they arise and keeps in time with the arrival of new data.
Analogue, Digital: an analogue variable is continuously changing, whereas a digital variable is quantized.
Hardware: the actual computer and equipment.
Software: programming schemes used by the computer.
Core: a device in the computer for magnetic storage of data.
Disc: a random-access device used for magnetic recording and storage of programs and data.
Capacity: (of store core or disc) number of words storable, usually quoted in units of 1024 (K).

12-2. Data acquisition and handling

There is a variety of operational mass-spectrometer/computer systems suited to a variety of laboratory requirements and circumstances.[1] Nevertheless, the basic principles of the computer-linked system are relatively simple. The output from a magnetic-scanning, electrical-recording mass spectrometer is a voltage which varies continuously with time (see Fig. 12-1). The continuous electrical output (i.e., analogue signal) from the detector of the mass spectrometer is sampled at precise intervals and then converted into digital (i.e., pulsed) form by an analogue-to-digital (A/D) converter. These digitized signals, consisting of an amplitude and a time (or count number), are then passed to the computer for processing. The A/D conversion process is illustrated in Fig. 12-1.

A variety of sampling rates may be chosen depending on the scan rate and the resolution required. Sampling (digitization) rates of up to 200 kHz (200 000 samples per second) have been employed, although rates are typically lower than this. In one system reported,[2] a sampling rate of 100 kHz was adequate for the real-time digitization of spectra at a resolution of 10 000 and a scan speed of 10 s per decade in mass ($m \rightarrow 0{\cdot}1\ m$), with 15–20 digital samples per peak.

The next step is the thresholding of the data, or rejection of digital samples having an amplitude below a certain threshold limit, so that only the digitized data for peaks remain (see Fig. 12-1). Thresholding is commonly performed by real-time programs in the computer software, but may also be performed by a specialized interface (which includes hardware logic for thresholding) between the A/D converter and the computer. The latter system has the advantage that the volume of data presented to the computer is considerably reduced and more real-time calculations may be performed between successive peaks.

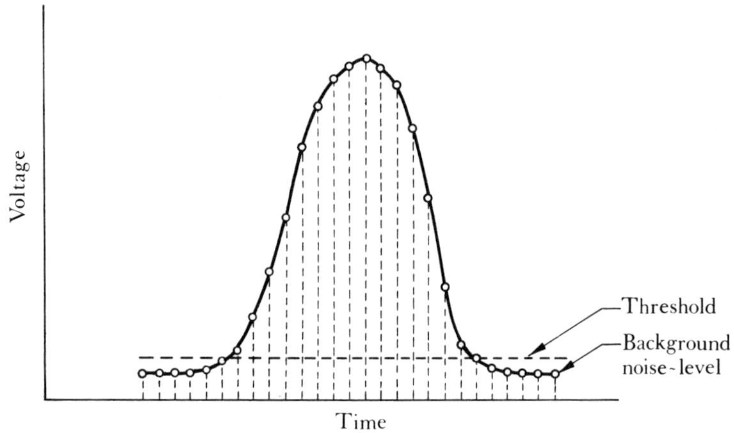

Figure 12-1

After thresholding, the data is now ready to be reduced by the computer to just two quantities for each peak, namely the centroid time and the peak area. This process is normally performed in non-real time by the computer software. Various formulae are available for calculating the centroid of the peak (see below), which is theoretically of Gaussian shape. However, peaks having about 1 per cent of the base peak intensity may only contain as few as 10 ions at maximum scan speeds. This results in irregular peak shapes due to statistical fluctuations and the peak centroid is difficult to estimate.

Accurate masses of all peaks are then calculated by relating the scan time of an unknown peak to those of a marker compound, whose spectrum and exact masses of peaks are known. The mass/time relationship is determined from the marker (reference) compound and the exact masses of the unknown ions are determined by interpolation. The reference generally used is perfluorokerosene (PKF), which produces a series of peaks of accurately known mass at intervals of 12 and 19 mass numbers apart up to about mass 700. The advantage of using perfluorinated compounds lies in the mass deficiency of fluorine (see Table 6-2) which enables easy separation of reference ions from unknown ions of the same nominal mass. Another reference compound, heptacosafluorotributylamine ($N(C_4F_9)_3$) gives a spectrum extending to just above mass 600. For mass determinations up to about m/e 1200 it appears that the perfluorotriazines are potentially useful. For example, perfluorotriheptyltriazine (1) yields a molecular ion at m/e 1185 and intense fragment ions. However, the reference peaks in triazine spectra are more widely spaced than 25 mass units in parts of the spectrum and this can be a drawback in automated data collection.

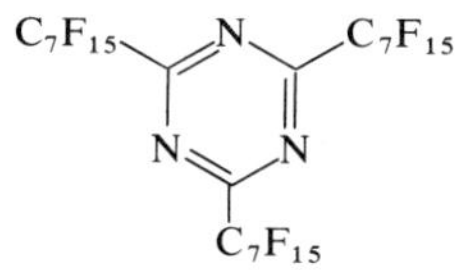

1

The on-line mass-spectrometer/computer system described briefly above may be represented schematically as shown in Fig. 12-2. The mass spectrometer, A/D converter and computer are directly coupled together and can 'communicate' with one another as shown by the arrows. The sequence of steps may be summarized as: (a) output of signal from mass spectrometer, (b) A/D conversion of signal, (c) thresholding, (d) peak centre and area calculation, (e) reference-peak identification and (f) exact mass calculation. Some of the variations on this basic theme will be described below for specific systems.

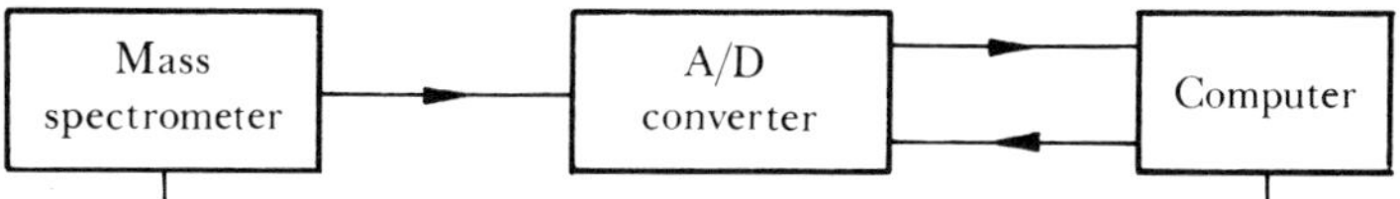

Figure 12-2

A system has been described in which a medium-sized double-focusing mass spectrometer is coupled on line to a small (4K memory words, 32K disc) computer.[3] The system can produce complete elemental composition data within two minutes of completion of the mass-spectral scan. A block diagram of the system is shown in Fig. 12-3.

The system is capable of operating under a variety of conditions, but typically at a resolution of 10 000 and at scan speeds between 40 and 150 s/decade. Analogue/digital conversion rates of up to 20 kHz were possible but 5 and 10 kHz rates were normally used.

Before a mass-spectral scan, the computer establishes the parameters (e.g., scan speed, threshold level) which are to be employed in the data acquisition and processing, by asking the operator a series of questions. The computer then sets the required A/D rate and threshold level, resets the external time counter to zero and waits for the 'start' command.

Thresholding is performed by the interface between the A/D converter and

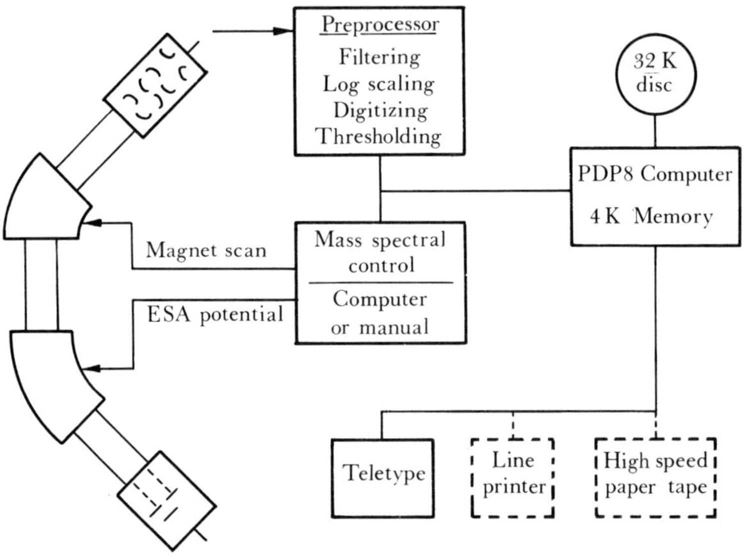

Figure 12-3

the computer. The important function of the interface is to provide the computer with only the peak data and to reject the baseline. A peak is considered to have begun whenever a digital intensity value becomes greater than threshold. The end of the peak is considered to have arrived when the intensities of four successive data points fall below threshold (see Fig. 12-1 for an example of digitized data points from a strong peak).

The best way of computing the centroid of a peak is a problem of statistics and only an approximate solution is possible. As with all statistical problems, the error in the calculated peak mass is reduced as the number of scans increases (the error is approximately inversely proportional to the square root of the number of scans).[3] The data points are first smoothed, using a weighted moving average, and the centroid of each peak is calculated using the equation

$$Y\,(\text{centroid}) = \frac{\int_a^b f(X)X\,\mathrm{d}X}{\int_a^b f(X)\,\mathrm{d}X},$$

where $f(X)$ is the amplitude at position X in the smoothed peak envelope (first data point of peak, $X = 1$). The centroid as calculated above must then be related to a certain elapsed time count. The integral in the denominator gives the peak area and this is used as the abundance value. It is not intended to give details of reference-peak identification and exact mass calculation here, except to say that a time–mass relationship is established by automatically identifying the reference peaks of PFK using an extrapolation technique. The time values of the sample peak are then converted to mass values by using four reference peaks with a Lagrange polynomial equation.

The final operation is to convert the ion masses into all possible elemental compositions within the prescribed error tolerance. This process will be considered in section 12-3, dealing with data interpretation.

The main practical advantages of the system described above over some others in common use lies in the automatic operation of the various phases. Although the computer has a small core memory (4K), the use of punched paper tapes for feeding the various programs into core memory is avoided by use of disc storage. Another useful operation is the automatic identification of PFK peaks.

Systems such as that described above are designed for fast production of a high-resolution mass spectrum. Mass-spectrometer/computer systems have also been developed for low-resolution work. Such systems are particularly useful where comparison of unknown spectra with stored spectra in 'computer libraries' is required (see section 12-5).

Since peaks in low-resolution mass spectra need only be identified to the nearest mass unit, a less precise method is required for establishing the mass scale than for the system discussed above.[3] A method has been described in which the mass scale is fixed by use of an external reference compound.[4] The electron-multiplier output from a single-focusing mass spectrometer is digitized at a rate of 3 kHz while the spectrometer scans from mass 20 to mass 500 in 3 s (see Fig. 12-4 for an illustration of the digitized data). The digital values could either be processed on line by the computer or written onto magnetic tape. After smoothing of the data, the peak centre is determined by evaluating the centre of the peak width at half-height. The resulting array of time vs. intensity values is then converted into mass vs. intensity by comparison of the arrival times of the peaks with those of the known mass peaks of a standard (a high-molecular weight paraffin). The spectrum of the standard is predetermined and the times of arrival of the known mass peaks stored in the computer.

Mass-spectrometer/computer systems may also be established using photographic recording plates. The system incorporates an interface, between the mass spectrometer and the computer, known as a microdensitometer–comparator, which reads the information from the photographic plates and supplies it to the computer. The configuration of the system is indicated schematically in Fig. 12-5.

The basic principles of operation of a photoplate recording computer system are similar to those described above for the electrical recording system. The notable difference is that the digitized data read from the photoplate by the comparator consist of transmission values (dependent on exposure of the photographic emulsion) as a function of distance, whereas the output from an electrical recording instrument is digitized to produce intensities as a function of time. A photoplate system has been described[5]

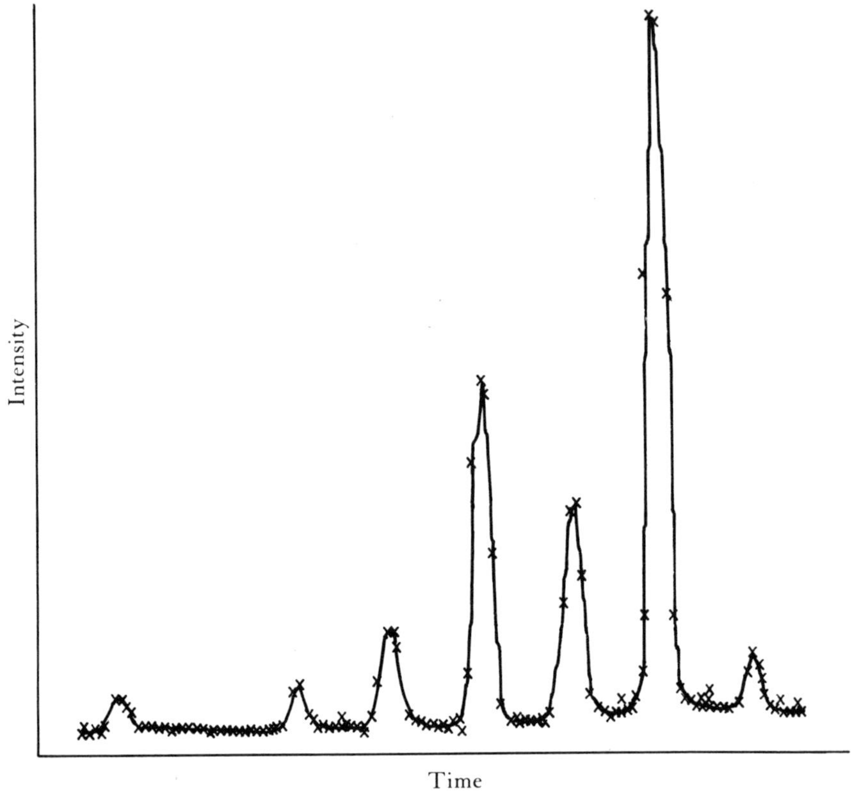

Figure 12-4

in which the comparator reads transmission values every 0·5 μm along the plate and values which are above a given threshold are stored by the computer. The transmission/distance profiles of the peaks are then transformed into intensity/mass functions by determining peak centres and intensities and comparing the centres of unknown peaks with those of the standard (e.g., PFK) mixed with the unknown. It should be noted that the areas under the transmission–distance curve are not proportional to peak intensities, but the functional relationship between the two is known.[1, 5] In this particular investigation, the standard deviation of mass determinations was 0·65 m.m.u. in the range *m/e* 70–250.

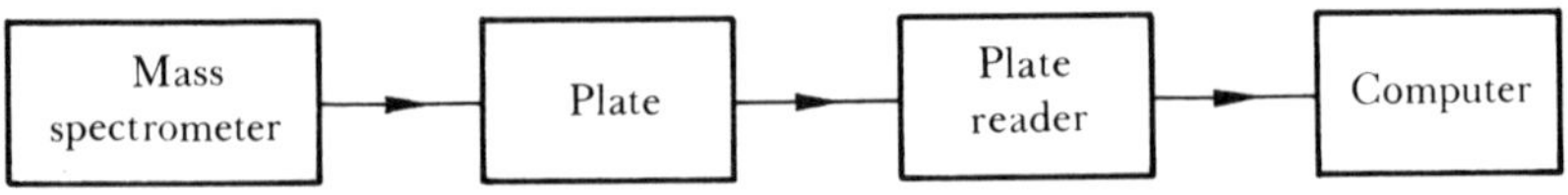

Figure 12-5

12-3. Data processing: high-resolution spectra

The above section has considered the acquisition of raw mass-spectral data and its handling by the computer up to the production of a mass spectrum (ion intensity vs. *m*/*e*). In this section, we will consider the different methods that have been developed for computerized processing of high-resolution spectra with the objective of making structural inferences. The usual approach to the automatic interpretation is to teach the computer to follow the same steps as the mass spectrometrist.

The first step is often the determination of elemental compositions of all ions in the spectrum, within prescribed limits of mass-measurement error. It is important that the operator should specify the maximum number of atoms of each element present, so that the computer does not test for more possibilities than necessary.

Computer programs are now well established for the determination of elemental composition and we do not propose to elaborate here. There are various methods of presentation of elemental composition data. Biemann has suggested an element mapping procedure which consists of a table of *m*/*e*-values, plotted vertically, with a separate vertical column for each set of ions having a different heteroatom content.[6] Figure 12-6 shows the element map of androst-4-ene-3,17-dione. The intensities of all ions are recorded graphically (on a logarithmic scale) by lines of horizontal dashes alongside the elemental composition. The molecular formula ($C_{19}H_{26}O_2$) is provided by the bottom right-hand entry.

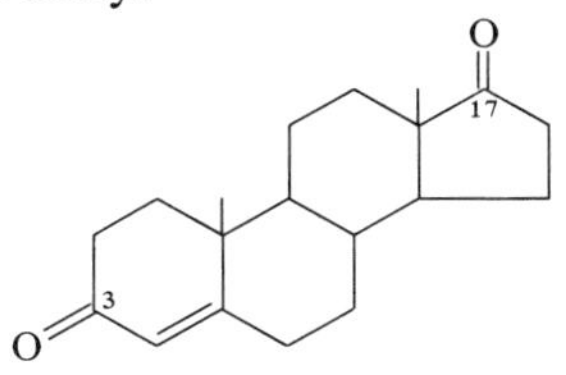

In some systems the computer printout merely lists all the elemental compositions calculated in order of *m*/*e*-values, together with intensity values and other relevant information.

It is, of course, possible to process the elemental composition data further before printout, given that the capacity of the computer system allows it. The first logical step is identification of the molecular ion. It is possible to set a number of criteria which uniquely determine the molecular ion[7] and programs are written accordingly:

(a) The molecular ion may not contain any heavy isotopes.

(b) The number of hydrogens must be even if the number of nitrogens is even or zero, and must be odd if the number of nitrogens is odd.

(c) Ions of lower mass must correspond to loss of reasonable fragments from the molecular ion, and the elemental composition of any lower mass ion must not include more atoms of any one element than the molecular ion.

m/e	CH	CHO	CHO_2
93	----- C7 H9		
95	----- C7 H11		
97		------ C6 H9	
104	---- C8 H8		
105	------ C8 H9		
106	---- C8 H10		
107	------ C8 H11	---- C7 H7	
108	----- C8 H12		
109	-- C8 H13	------ C7 H9	
110		--- C7 H10	
111		---- C7 H11	
117	---- C9 H9		
119	----- C9 H11		
120	---- C9 H12		
121	----- C9 H13	---- C8 H9	
122	----- C9 H14	---- C8 H10	
123		----- C8 H11	
124		-------- C8 H12	
125		---- C8 H13	
128	---- C10 H8		
129	---- C10 H9		
131	--- C10 H11		
133	---- C10 H13	--- C9 H9	
135	----- C10 H15	---- C9 H11	
136		------ C9 H12	
143	--- C11 H11		
144	---- C11 H12		
146	---- C11 H14		
147	---- C11 H15	---- C10 H11	
148	------ C11 H16	---- C10 H12	
149	-- C11 H17	----- C10 H13	
150		----- C10 H14	
159	---- C12 H15	-- C11 H11	
162		----- C11 H14	
163		---- C11 H15	
171	--- C13 H15		
173	---- C13 H17	--- C12 H13	
175		--- C12 H15	
229		---- C15 H21	
271			-- C18 H23
286			-------- C19 H26

Figure 12-6

Such decisions may be made quickly and easily with a computer and further structure interpretation is possible by following the logical steps that might be taken by the spectroscopist in analysing a spectrum,[8] as shown in Table 12-1.

In order to perform the above steps effectively, the computer must be programmed with the behaviour of different compound classes. For example, it must be known which neutral fragments lost from the molecular ion are

Table 12-1
Steps in computer evaluation of elemental composition data

1.	Identification of the molecular ion
2.	Neutral fragments lost from the molecular ion
3.	Identification of characteristic ion series
4.	Selection of the most probable compound classes
5.	Comparison of fragmentation with the characteristic fragmentation pathways of the selected compound classes
6.	Selection of the most probable structures

indicative of which class. Yet we have seen in chapter 6 that this is often not a simple problem, so-called 'characteristic' fragmentations being quenched when alternative lower-energy fragmentation pathways are possible. To date, the best approach appears to rely fairly heavily on operator interpretation in conjunction with computerized high-resolution measurements, as exemplified for antheridiol in section 6-5C. Even in such cases, it should be emphasized that on many occasions the necessary elemental compositions could be determined by the more traditional manual methods (e.g., peak matching on an oscilloscope), thus keeping the equipment at a somewhat cheaper and simpler level.

12-4. Data processing: low-resolution spectra

A *learning machine* has been applied to the problem of determining molecular formulae from low-resolution mass spectra.[9] A learning machine is a computer equipped with programs which enable it to store information that can be used to recognize particular situations. The learning machine begins ignorant of any relationship between the data and the desired answers and develops its own relationships without theoretical basis, by receiving information from known compounds. For example, if a learning machine has to decide whether or not a particular compound contains oxygen, a series of mass spectra for known oxygen-containing compounds is fed into the computer. From the peak intensities a linear discriminant function is calculated which is positive when oxygen is present, and negative when oxygen is absent. If the discriminant function gives the incorrect answer when a known compound is tried, the function is amended until the correct answer is achieved. In this way, the learning machine becomes 'trained' and develops different discriminant functions from known spectra that can recognize a variety of situations. Molecular formulae can be built up in this way.

A program called DENDRAL has been used in the interpretation of low-resolution spectra by the use of 'artificial intelligence'.[10] The basic function of the program itself is to construct a list of all possible, chemically feasible, structural isomers, given the molecular formula (only C, H, N and O are allowed). The number of isomers is considerably reduced by incorporating a BADLIST into the program, consisting of a list of structural features which are either known to be absent or are chemically unfeasible. An example of a BADLIST selected for aliphatic acyclic structures containing C, H, O and N is shown in Table 12-2. This particular BADLIST would be inappropriate for aromatic and heterocyclic compounds since some of the entries listed could exist as subunits of stable structures.

Table 12-2
BADLIST for aliphatic acyclic structures

C=C—O—H	H—O—C—O—H	N—O—N
C=C—N—H	H—O—C—N—H	N—O—O
C≡C—O—H	H—N—C—N—H	O—N—O
C≡C—N—H	|	O—O—O
N≡N—N	H	H—O—C—O—
N=N—O	H—C—N=O	‖
N=N—H	H—O—C=N	O
	O—O	H—O—C—N—
	N—N—N	‖
	N—N—O	O

If it is desired to limit the computer output to isomers which contain a given functionality, these may be incorporated on a GOODLIST. In this event, the program ignores all other structural possibilities. For example, there are over 8000 isomers of $C_8H_{16}O_2$ and a typical BADLIST reduces these to about 4000. However, if the GOODLIST restriction that all isomers should be carboxylic acids (containing CO_2H) is applied, the number of possible isomers is reduced to 39, with a corresponding saving on computer time.

The application of the DENDRAL program to low-resolution mass spectrometry is best illustrated by an example. The chosen mass spectrum and the logical steps in the program are illustrated in Fig. 12-7.

The program is fed a low-resolution mass spectrum of an unknown compound, including any metastable transitions, together with the elemental composition of the molecular ion (in this case $C_9H_{18}O$). Since the number of possible isomers is shown to be 1936 it is necessary to reduce this number by using a subroutine (the PRELIMINARY INFERENCE MAKER) to decide which functional groups are present. The subroutine is tested for ketones, aldehydes, ethers and alcohols in this case. For example, in the case of ketones, the spectrum is examined for fragmentations of the form

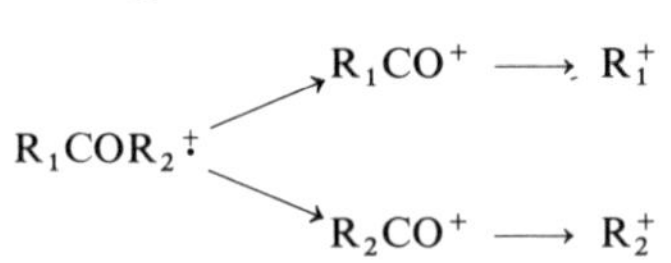

The test for a ketone is positive, thus placing 'ketone' on the GOODLIST, whereas negative responses for aldehydes, ethers and alcohols place them on

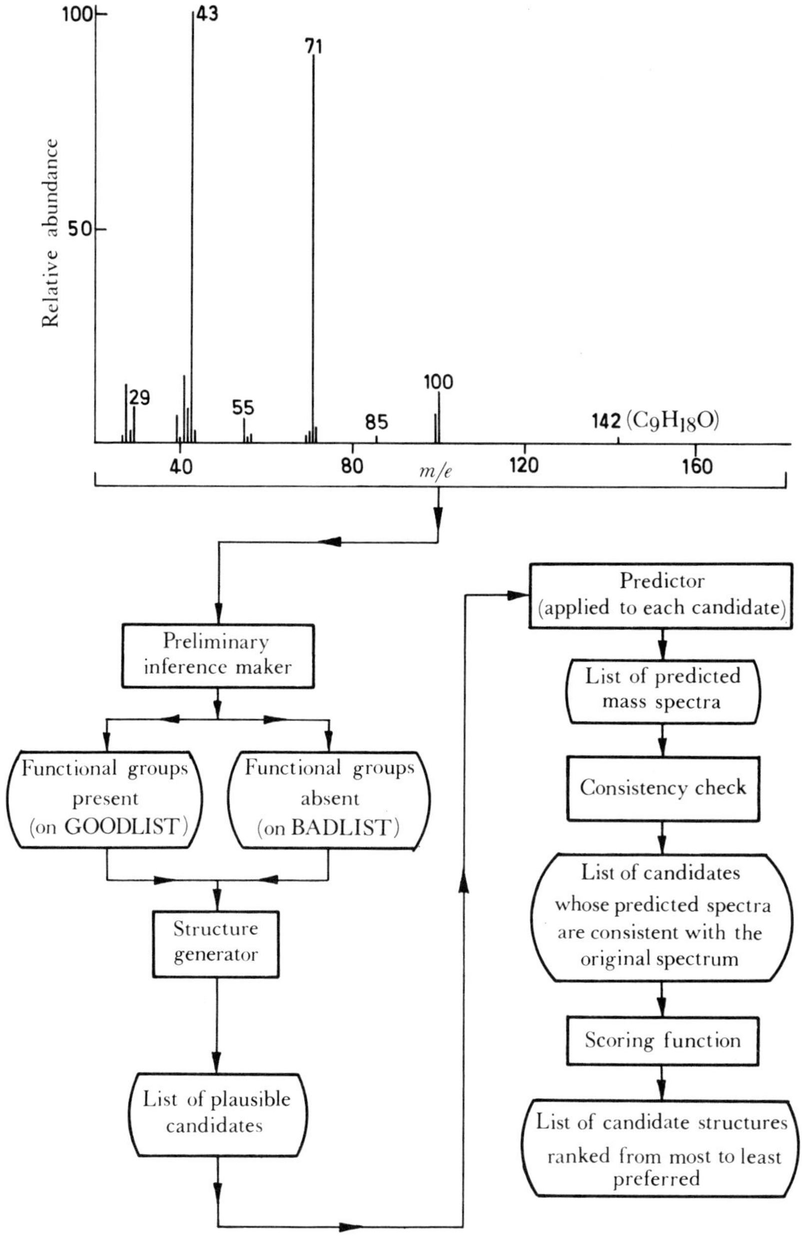

100
50
Relative abundance
29
43
55
71
85
100
142 ($C_9H_{18}O$)
40
80
120
160
m/e
Preliminary inference maker
Functional groups present (on GOODLIST)
Functional groups absent (on BADLIST)
Structure generator
List of plausible candidates
Predictor (applied to each candidate)
List of predicted mass spectra
Consistency check
List of candidates whose predicted spectra are consistent with the original spectrum
Scoring function
List of candidate structures ranked from most to least preferred

Figure 12-7

the BADLIST. DENDRAL generated 82 ketone isomers subject to the above constraints. Further truncation is still desirable and so the PRELIMINARY INFERENCE MAKER searches for other characteristic ions in the mass spectrum after it has identified ketones. Thus inferences are made from the absence of McLafferty rearrangement ions at *m/e* 58, 72 or 86 and any structures which would give these ions are placed on the BADLIST. Acting on this information, the structure generator now generates 8 ketones which are then examined in detail by the rest of the program to eliminate some and rank the rest (see Fig. 12-7).

Each of these structures is then examined by the PREDICTOR which constructs a hypothetical mass spectrum for each structure and by comparison with the spectrum of the unknown accepts or rejects the structure. The list of candidates is now down to 2 ketones and the SCORING FUNCTION ranks these in order by evaluating their success in the tests performed by the CONSISTENCY CHECK. The SCORING FUNCTION was unable to distinguish between the two ketones 2 and 3. In fact the unknown compound is 2.

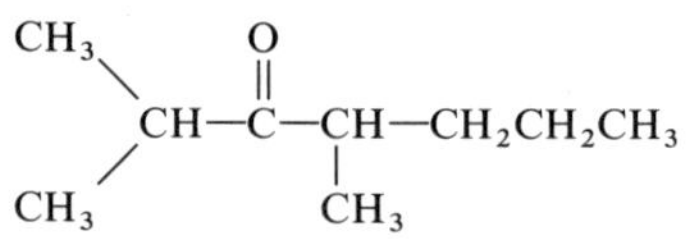

2

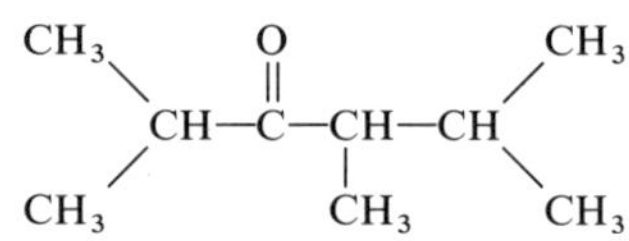

3

In the application of DENDRAL to the mass spectra of aliphatic ethers and amines, it is frequently found that several candidates are generated after the CONSISTENCY CHECK. A further ranking scheme involving nmr spectra is introduced into the program and this usually leads to the identification of a single correct structure. This 'heuristic DENDRAL' program requires a large computer store (40K for each of the three main parts).

12-5. Comparison of mass spectra

The digital computer is an important asset for the identification of an unknown compound by comparing its mass spectrum with a large number of reference mass spectra (e.g., in mass spectral libraries). Rapid progress has been made in the automatic, computerized storage of mass spectra on paper or magnetic tapes and in the discussion below we will only consider the ways in which a mass spectrum may be processed to yield some direct numerical comparison with a series of known spectra.

In any method for comparison of mass spectra it is essential that the spectra are all normalized by similar methods.[11] The method frequently used in tabulations of making the largest peak in each spectrum equal to 100 is not good for two reasons: (a) it places all its weight on the one peak which is likely not to be fully recorded at fast scanning speeds and any error in the base peak affects the whole normalized spectrum, and (b) the largest peak in the spectrum is sometimes not very significant from an identification point of view.

Various normalization procedures have been used. The simplest of these is to make the sum of all the k observed peak heights P_n equal to unity:

$$\sum_{n=1}^{k} P_n = 1. \tag{12-1}$$

Other more acceptable methods normalize a function of each value of P_n to unity:

$$\sum_{n=1}^{k} \sqrt{P_n} = 1 \tag{12-2}$$

$$\sum_{n=1}^{k} P_n^2 = 1. \tag{12-3}$$

Two spectra are compared by evaluating the discrepancy factor D:

$$D = \sum_{n=1}^{k} |P_n(\text{ref}) - P_n(\text{unknown})|,$$

where n is now taken over all observed m/e-values in both spectra. It is evident that $D = 0$ for a perfect correlation and $D = 2$ for complete dissimilarity. For a variety of experimental reasons, no two mass spectra of a pure compound are likely to be identical and therefore it is anticipated that the D-factor for identity will be small, but not zero. For example, a five per cent random error in peak heights gives a D-factor for identity of 0·05. An example of the reliability of the D-factor is shown in Table 12-3, which lists the D-factor obtained by comparing the mass spectrum of methyl butanoate with a series of library spectra using eqs. (12-1)–(12-3).

Table 12-3

Some variations in the D-factor with normalization procedure

Compound	*D*-factor		
	(1)	(2)	(3)
n-nonane	0·84	0·90	0·99
3-hexanone	0·78	0·78	0·85
ethylbenzene	1·39	1·50	1·78
t-butylamine	1·32	1·02	1·43
1-methyl-3-ethylbenzene	1·38	1·46	1·75
n-octane	0·76	0·84	0·92
pentanoic acid	1·18	0·73	1·12
cyclohexene	1·19	1·08	1·31
methyl butanoate	0·20	0·10	0·20

It is found that random errors of up to 20 per cent in the peak heights do not seriously compromise identification and it appears that normalization by eq. (12-2) is marginally the best when the scatter of errors is large. When systematic errors occur (e.g., measurement of spectra on different mass spectrometers, with different source residence times, operating temperatures, etc.), it is found advantageous to normalize and compare peaks within small mass ranges, e.g., 7–20, 21–34, 35–48, etc.

The methods described above give almost certain identification of structures, but are wasteful of computer time when all ions in the spectrum are considered. A search method using the 6 strongest peaks in the spectrum has been employed and a discrepancy factor may be calculated using only the 6 strongest peaks of the unknown and the reference. This method considerably reduces the computer time spent in searching a large library and may be used for rapid identification of components in a gas-chromatograph effluent.[11]

12-6. Gas chromatography/mass spectrometry/computer systems

We have seen in chapter 11 the necessity for rapid scanning of the effluent in a gc/ms system because the time of elution of a component may be short, and total ion intensity may change during a mass-spectral scan. In order that the vast amount of data from a gc/ms sample is not wasted, it is important that a means should be available for storage and processing of this data. Consequently, computerized data acquisition has been applied to this problem.

A data-acquisition system suitable for low-resolution, fast-scanning mass spectrometers has been described.[12] The speed with which data are taken, and the large spectral storage capacity of the on-line computer, make it particularly suited for recording mass spectra of gas-chromatographic effluents. The basic feature of this on-line system is the continuous scanning of mass spectra at four-second intervals. Spectra are recorded during the entire duration of the gas chromatogram, regardless of the emergence of effluent peaks. This feature eliminates the need for constant decision by the operator whether or not to record a given spectrum, and also eliminates the risk that an important spectrum might be missed.

The electron-multiplier output is digitized at a rate of 3 kHz and the result transferred to core store. Real-time processing of this data is carried out to determine all mass-spectral peak heights and centres, which are transferred to magnetic discs along with an index number indicating the scan. The raw data points for each mass spectrum are summed to give the *total ionization* of each scan, which is stored in core memory along with the scan index number. A plot of total ion intensity vs. scan index number reveals the gas chromatogram. The example shown in Fig. 12-8 illustrates this plot for scan indices 95 to 160 of a 320 spectra plot (20 mins). The operator may now use this plot to select the spectra of interest from the disc. The computer will present an individual spectrum in plotted or tabular form (resolution 500). It is possible to identify a minor component occurring as a shoulder on a gas-chromatographic peak. For example, by examination of the arrowed spectra in Fig. 12-8 two methyl esters of long-chain fatty acids were identified.

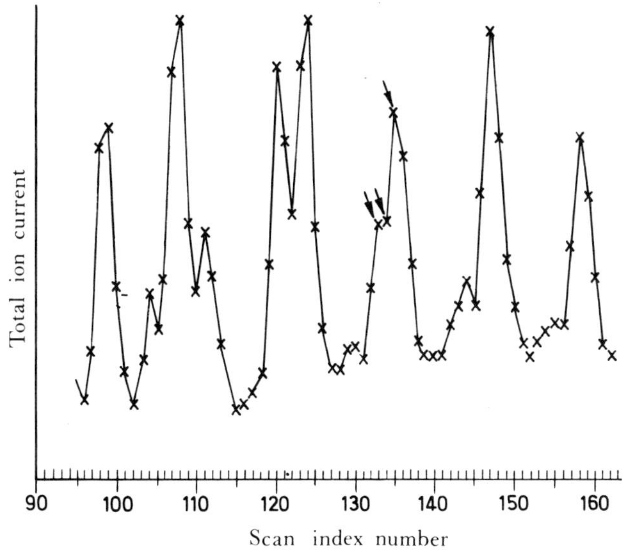

Figure 12-8

The time-averaging of multiple mass-spectral scans of a gc effluent, where very small quantities of material are available, is a useful facility. A computer may be programmed to time-average mass spectral peaks, subtract background peaks, and normalize and print out the spectrum in plotted form.[13] For example, the pesticide 2,2-bis-(*p*-chlorophenyl)-1,1-dichloroethylene (DDE, 4) gives molecular ions of high relative abundance in its mass

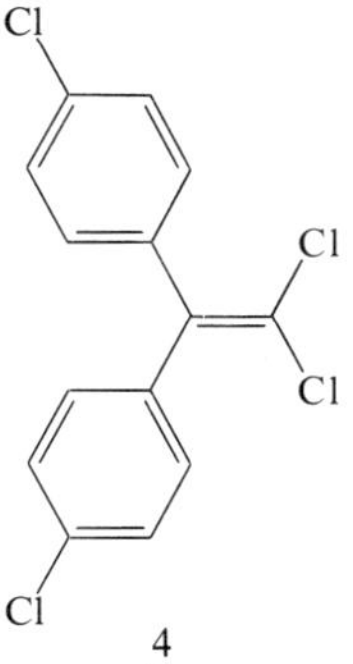

4

spectrum at *m/e* 316, 318, 320 and 322, having the characteristic isotopic distribution for four chlorine atoms. DDE is therefore readily identifiable by scanning just a portion of the spectrum (which accordingly cuts down the scan times). Using multiple scan averaging it is possible to identify this

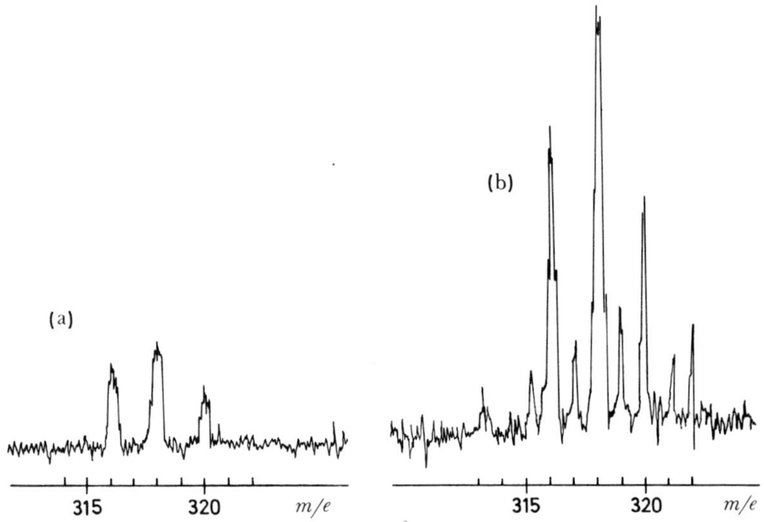

Figure 12-9

molecular ion from amounts as little as 5 ng injected into the gc/ms. Figure 12-9(a) shows the molecular ion region obtained with one scan on a 5 ng injection, while Fig. 12-9(b) was obtained with the same sample size from averaging 12 scans.[13]

References

[1] *Specialist Periodical Reports: Mass Spectrometry*, Vol. 1, The Chemical Society, London, 1971, chapter 6.

[2] H. C. Bowen, E. Clayton, D. J. Shields and H. M. Stanier, *Adv. Mass Spec.* **4**, 257 (1968).

[3] R. J. Klimowski, R. Venkataraghavan, F. W. McLafferty and E. B. Delany, *Org. Mass Spec.* **4**, 17 (1970); see also R. Venkataraghavan, R. J. Klimowski and F. W. McLafferty, *Accounts Chem. Res.* **3**, 158 (1970).

[4] R. A. Hites and K. Biemann, *Adv. Mass Spec.* **4**, 37 (1968); see also *Anal. Chem.* **39**, 965 (1967).

[5] D. M. Desiderio and T. E. Mead, *Anal. Chem.* **40**, 2090 (1968).

[6] K. Biemann, P. Bommer and D. M. Desiderio, *Tetrahedron Letters* 1725 (1964).

[7] K. Biemann and W. McMurray, *Tetrahedron Letters* 647 (1965).

[8] R. Venkataraghavan, F. W. McLafferty and G. E. Van Lear, *Org. Mass Spec.* **2**, 1 (1969).

[9] P. C. Jurs, B. R. Kowalski and T. L. Isenhour, *Anal. Chem.* **41**, 21 (1969).
[10] J. Lederberg, G. L. Sutherland, B. G. Buchanan, E. A. Feigenbaum, A. V. Robertson, A. M. Duffield and C. Djerassi, *J. Am. Chem. Soc.* **91**, 2973 (1969); see also **91**, 2977 (1969); **91**, 7440 (1969) and **92**, 6831 (1970).
[11] L. R. Crawford and J. D. Morrison, *Anal. Chem.* **40**, 1464 (1968).
[12] R. A. Hites and K. Biemann, *Anal. Chem.* **40**, 1217 (1968); see also **42**, 895 (1970).
[13] F. J. Biros, *Anal. Chem.* **42**, 537 (1970).

Index

THIS BOOK HAS BEEN SET IN MONOPHOTO TIMES NEW ROMAN
AND PRINTED AND BOUND IN GREAT BRITAIN BY
WILLIAM CLOWES & SONS, LIMITED, LONDON, BECCLES AND COLCHESTER